Alternative Water Supply Systems

Alternative Water Supply Systems

Edited by

Fayyaz Ali Memon and Sarah Ward

IWA PUBLISHING

Published by

IWA Publishing
Alliance House
12 Caxton Street
London SW1H 0QS, UK
Telephone: +44 (0)20 7654 5500
Fax: +44 (0)20 7654 5555
Email: publications@iwap.co.uk
Web: www.iwapublishing.com

First published 2015

British Library Cataloguing in Publication Data
A CIP catalogue record for this book is available from the British Library

ISBN 9781780405506 (Hardback)
ISBN 9781780405513 (eBook)

Contents

Contributors

Chapter 1

Dr Rodney Anthony Stewart
Associate Professor
Centre for Infrastructure Engineering and Management,
Griffith University, Gold Coast Campus, Australia
Email: *r.stewart@griffith.edu.au*

Dr Oz Sahin
Research Fellow
Centre for Infrastructure Engineering and Management,
Griffith University, Gold Coast Campus, Australia
Email: *o.sahin@griffith.edu.au*

Mr Raymond Siems
Research Assistant
Centre for Infrastructure Engineering and Management,
Griffith University, Gold Coast Campus, Australia
Email: *raymond.siems@griffithuni.edu.au*

Mr Mohammad Reza Talebpour
PhD Candidate
Centre for Infrastructure Engineering and Management,
Griffith University, Gold Coast Campus, Australia
Email: *r.talebpour@griffith.edu.au*

Dr Damien Giurco
Associate Professor
Institute for Sustainable Futures, University of Technology,
Sydney, Australia
Email: *damien.giurco@uts.edu.au*

Chapter 2
Mr Mohammad Reza Talebpour
PhD Candidate
Centre for Infrastructure Engineering and Management,
Griffith University, Gold Coast Campus, Australia
Email: *r.talebpour@griffith.edu.au*

Dr Oz Sahin
Research Fellow
Centre for Infrastructure Engineering and Management,
Griffith University, Gold Coast Campus, Australia
Email: *o.sahin@griffith.edu.au*

Mr Raymond Siems
Research Assistant
Centre for Infrastructure Engineering and Management,
Griffith University, Gold Coast Campus, Australia
Email: *raymond.siems@griffithuni.edu.au*

Dr Rodney Anthony Stewart
Associate Professor
Centre for Infrastructure Engineering and Management,
Griffith University, Gold Coast Campus, Australia
Email: *r.stewart@griffith.edu.au*

Mr Michael Hopewell
Urban water planner
Gold Coast Water, Gold Coast City Council, Gold Coast City, Australia
Email: *mhopewell@goldcoast.qld.gov.au*

Chapter 3
Dr Alan Fewkes
Principal Lecturer
School of Architecture, Design and the Built Environment,
Nottingham Trent University, Nottingham, UK
Email: *alan.fewkes@ntu.ac.uk*

Chapter 4
Mr Richard Kellagher
Technical Director
HR Wallingford Ltd, UK
Email: *r.kellagher@hrwallingford.com*

Mr Juan Gutierrez-Andres
Principal Engineer
HR Wallingford Ltd, UK
Email: *j.gutierrez-andres@hrwallingford.com*

Chapter 5

Cath Hassell,
Director
ech2o Consultants Ltd, London, UK
Email: *cath.hassell@ech2o.co.uk*

Dr Judith Thornton,
Research Development Officer,
Institute of Biological, Environmental and Rural Sciences,
Aberystwyth University, UK
Email: *jutl3@aber.ac.uk*

Chapter 6

Dr Ilan Adler
Teaching Fellow
Department of Civil, Environmental and Geomatic Engineering
University College London, London, UK
Email: *ilan.adler.09@ucl.ac.uk*

Dr Luiza C. Campos
Senior Lecturer
Department of Civil, Environmental and Geomatic Engineering
University College London, London, UK
Email: *l.campos@ucl.ac.uk*

Dr Sarah Bell
Senior Lecturer
Department of Civil, Environmental and Geomatic Engineering
University College London, London, UK
Email: *s.bell@ucl.ac.uk*

Chapter 7

Mr Jonathan Thayil-Blanchard
Graduate Research Assistant
Department of Civil and Environmental Engineering
University of South Florida, Florida, USA
Email: *jthayilblanchard@gmail.com*

Dr James R. Mihelcic
Professor
Department of Civil and Environmental Engineering
University of South Florida, Tampa, Florida, USA
Email: *jm41@usf.edu*

Chapter 8
Dr Sarah Ward
Senior Research Fellow
College of Engineering Mathematics and Physical Sciences
University of Exeter, UK
Email: *sarah.ward@exeter.ac.uk*

Dr Fernando Dornelles
Institute of Hydraulic Reseach
Federal University of Rio Grande do Sul, Brazil
Email: *fds_eng@yahoo.com.br*

Dr Marcio H. Giacomoni
University of Texas at San Antonio, USA
Email: *marcio.giacomoni@utsa.edu*

Dr Fayyaz Ali memon
Associate Professor
College of Engineering Mathematics and Physical Sciences
University of Exeter, UK
Email: *f.a.memon@exeter.ac.uk*

Chapter 9
Pacia Diaz
PhD Candidate
Department of Civil and Environmental Engineering
University of South Florida, Tampa, Florida, USA
Email: *phernan2@mail.usf.edu*

Jennifer Isenbeck
Facilities Manager, Sodexo
Email: *Jennifer.isenbeck@sodexo.com*

Dr Daniel H. Yeh
Associate Professor
Department of Civil and Environmental Engineering
University of South Florida, Tampa, Florida, USA
Email: *dhyeh@usf.edu*

Chapter 10
Dr Eran Friedler
Associate Professor
Faculty of Civil and Environmental Engineering
Technion – Israel Institute of Technology, Haifa, Israel
Email: *eranf@tx.technion.ac.il*

Dr Amit Gross
Associate Professor
Blaustein Institute for Desert Research
Ben-Gurion University of the Negev, Sede Boqer Campus, Israel
Email: *amgross@bgu.ac.il*

Chapter 11

Miss Melissa Toifl
Environmental Scientist
CSIRO Land and Water, Melbourne, Australia
Email: *mtoifl@csiro.au*

Dr Gaëlle Bulteau
Research and Development Engineer
Centre Scientifique et Technique du Bâtiment
Nantes, France
Email: *gaelle.bulteau@cstb.fr*

Dr Clare Diaper
Visiting Fellow
Water Science Institute
Cranfield University, Bedford, UK
Email: *clarediaper@hotmail.com*

Dr Roger O'Halloran
Principal Research Scientist
CSIRO Land and Water, Melbourne, Australia
Email: *rogeroh@optusnet.com.au*

Chapter 12

Dr Marc Pidou
Academic Fellow in Resource Recovery
Water Science Institute
Cranfield University, Bedford, UK
Email: *m.pidou@cranfield.ac.uk*

Chapter 13

Dr Fayyaz Ali Memon
Associate Professor
College of Engineering Mathematics and Physical Sciences
University of Exeter, UK
Email: *f.a.memon@exeter.ac.uk*

Dr Abdi Mohamud Fidar
Chief Engineer
KT & I PLC, Addis Ababa, Ethiopia
Email: *afidar2@gmail.com*

Dr Sarah Ward
Senior Research Fellow
College of Engineering Mathematics and Physical Sciences
University of Exeter, UK
Email: *sarah.ward@exeter.ac.uk*

Dr David Butler
Professor in Water Engineering
College of Engineering Mathematics and Physical Sciences
University of Exeter, UK
Email: *d.butler@exeter.ac.uk*

Dr Kamal Alsharif
Associate Professor of Environmental Science and Water Policy
School of Geosciences, College of Arts and Sciences
University of South Florida, USA
Email: *kalshari@usf.edu*

Chapter 14

Dr Amit Chanan
Chief Operating Officer
State Water Corporation
Sydney, Australia
Email: *amit.chanan@statewater.com.au*

Saravanamuth Vigneswaran
Professor and Director
Centre for Technology in Water & Wastewater,
University of Technology Sydney, Australia
Email: *Saravanamuth.Vigneswaran@uts.edu.au*

Jaya Kandasamy
Associate Professor
School of Civil and Environmental Engineering,
University of Technology Sydney, Australia
Email: *Jaya.Kandasamy@uts.edu.au*

Stuart Khan
Associate Professor
School of Civil and Environmental Engineering,

University of New South Wales, Sydney, Australia
Email: *s.khan@unsw.edu.au*

Chapter 15

Siân Hills
Water Reuse Consultant
SHSWS, London & Wales, UK
Email: *sian.hills@hotmail.co.uk*

Christopher James
Plant Manager
Old Ford Water Recycling Plant,
Dace Road, London, UK
Email: *christopher.james@thameswater.co.uk*

Chapter 16

Dr Peter Schouten
Visiting Fellow
Land and Water – CSIRO
Dutton Park, Queensland, Australia
Email: *peter.schoutenau@gmail.com*

Dr Ashok Sharma
Principal Research Scientist
Land and Water – CSIRO
Highett, Victoria, Australia
Email: *ashok.sharma@csiro.au*

Stewart Burn
Professor and Program Leader
Land and Water – CSIRO
Highett, Victoria, Australia
Email: *stewart.burn@csiro.au*

Mr Nigel Goodman
Research Scientist
Land and Water – CSIRO
Highett, Victoria, Australia
Email: *nigel.goodman@csiro.au*

Ms Shiv Umapathi
Water Resources Engineer
SA Water Centre for Water Management and Reuse,
School of Natural and Built Environments,
University of South Australia, Australia
Email: *shivanita.umapathi@mymail.unisa.edu.au*

Chapter 17
Dr Valentina Lazarova
Suez Environnement
38 rue du president Wilson,
78230 Le Pecq, France
Email: *valentina.lazarova@suez-env.com*

Chapter 18
Dr Bernard Keraita
International Researcher
Copenhagen School of Global Health,
University of Copenhagen, Copenhagen, Denmark
Email: *bernard.keraita@sund.ku.dk*

Mr Javier Mateo-Sagasta Dávila
Senior Researcher
International Water Management Institute (IWMI)
Colombo, Sri Lanka
Email: *J.Mateo-Sagasta@cgiar.org*

Dr Pay Drechsel
Principal Researcher
International Water Management Institute (IWMI)
Colombo, Sri Lanka
Email: *p.drechsel@cgiar.org*

Dr Mirko Winkler
Scientist
Department of Epidemiology and Public Health,
Swiss Tropical and Public Health Institute, Basel, Switzerland
Email: *mirko.winkler@unibas.ch*

Ms Kate Medlicott
Technical Officer, Water, Sanitation, Hygiene and Health
WHO-Department of Public Health and Environment
WHO Headquarters, Geneva, Switzerland
Email: *medlicottk@who.int*

Chapter 19
Caryssa Joustra
PhD Candidate
Department of Civil and Environmental Engineering,
University of South Florida, Tampa, Florida, USA
Email: *cjoustra@mail.usf.edu*

Dr Daniel H. Yeh
Associate Professor
Department of Civil and Environmental Engineering,
University of South Florida, Tampa, Florida, USA
Email: *dhyeh@usf.edu*

Chapter 20

Paula Kehoe
Director of Water Resources
San Francisco Public Utilities Commission,
San Francisco, California, USA
Email: *Pkehoe@sfwater.org*

Sarah Rhodes
Project Manager
San Francisco Public Utilities Commission
San Francisco, California, USA
Email: *Srhodes@sfwater.org*

John Scarpulla
Water Resources Planner
San Francisco Public Utilities Commission,
San Francisco, California, USA
Email: *Jscarpulla@sfwater.org*

Chapter 21

Dr Sarah Bell
Senior Lecturer
Department of Civil, Environmental and Geomatic Engineering
University College London, London, UK
Email: *s.bell@ucl.ac.uk*

Preface

Scientists, engineers and policymakers are still searching for a consensus on the extent of climate change, its possible causes and the severity of potential implications. An upward trend in the frequency and intensity of extreme weather events (such as droughts and floods) has already challenged the capacity and resilience of the conventional centralised water management infrastructure.

Leaving aside climate change uncertainties, anticipated global population growth alone will have significant implications for most of the sectors heavily dependent on freshwater availability. It is estimated that by 2050, urban areas are likely to see three billion additional inhabitants and the demand for agricultural production and energy could double. Water is needed for generating energy and producing food and by 2030, water demand is projected to increase by 30%.

Meeting the ever increasing demand for wholesome freshwater through conventional centralised systems for both potable and non-potable applications has already become an unrealistic aspiration. This is due to competing demands on limited financial resources and limited flexibility of existing water infrastructure for expansion and adaptation. Demand management or water efficiency measures alone are not sufficient and the conventional water supply still requires augmentation using alternative sources.

The emphasis on alternative approaches to supply 'fit for purpose' water is emerging and alternative water supply (AWS) systems are becoming a visible practice in many water stressed regions. AWS will continue to remain an active research area. A paradigm shift is already taking place and low grade (in terms of quality) water is now increasingly seen as a resource rather than liability.

In time, the wider uptake of AWS systems appears to be inevitable and requires an evidence-based understanding of their interactions with existing infrastructure, end users and the environment. Consequently, the need arises to assess health implications, quantify risk, develop mitigation strategies, undertake holistic cost benefit analyses and provide improved structured decision support to meet the specific needs of different stakeholders.

This book mainly builds on a number of case studies on AWS systems in operation in different parts of the world, both in developed and low-income countries. Both the pilot and full scale systems implemented at domestic and community level are discussed. Thematically, the book content can be divided into four distinct sections.

Section I consists of 9 chapters with the majority addressing aspects related to rainwater harvesting (RWH) systems. These aspects include: their effectiveness in meeting non-potable demand and attenuating storm water flows; system capacity design approaches; energy implications and issues relating to community-based RWH systems for potable applications.

A considerable volume of condensate can be harvested from air-conditioning systems in large commercial buildings located in hot climatic regions. The collected condensate can partly meet non-potable water demand. Condensate recovery and reuse is an emerging research area and an introductory discussion and examples of sites where it has been implemented are covered in the last chapter of *Section I.*

Although of all the types of AWS systems, RWH appears as the most popular option (due to several factors, including its relatively better quality and minimal treatment requirements), the year-long reliability of supply cannot be guaranteed and this is where greywater recycling systems perform better. Greywater is broadly defined as wastewater generated from showers, baths and hand wash basin and normally excludes wastewater streams from toilets and kitchen sinks. The supply of greywater is fairly continuous and stable. However, a level of treatment is required to render greywater fit for intended applications. Greywater recycling, treatment technologies, risk identification, risk mitigation strategies and energy implications are discussed in *Section II.*

Section III provides an overview of treated and untreated wastewater reuse systems, their energy footprint and environmental implications. Also described, in this section, are some of the approaches to minimise associated health risks both in the urban context of the developed world and for communities in low-income countries. Techniques such as sewer mining, treatment and local reuse are also discussed in this section.

Finally, *Section IV* discusses the need for integrated decision support to facilitate the inclusion and operation of AWS in buildings. Furthermore, it presents some of the institutional and legal challenges and approaches for the implementation of AWS programmes and provides reflections on the drivers and barriers within a socio-technical context.

The book attempts to provide an unbiased perspective and shares the current research and practice in the domain of AWS. The book includes contributions from a team of near 50 professionals coming from nearly 20 different countries and contexts. Inherently, you will find a range of styles, formats and lenses through which to consider the most important challenge of addressing water insecurity through AWS. The views and opinions expressed in the book are solely of the authors and do not necessarily represent any formal position of their respective organisations or named institutions. Finally, writing a chapter for a book like this is no mean feat and we would like to thank all contributors for their support and dedication.

Fayyaz Ali Memon
Sarah Ward

Section I

Rainwater Harvesting and Condensate Recovery Systems

Chapter 1

Performance and economics of internally plumbed rainwater tanks: An Australian perspective

Rodney Anthony Stewart, Oz Sahin, Raymond Siems, Mohammad Reza Talebpour and Damien Giurco

1.1 INTRODUCTION

Water security is becoming a global issue of concern. In developed nations like Australia, high population growth and strong economic development are increasing demand, while supply is under threat from environmental degradation and climate change. Centralised reservoir and distribution networks have long served major metropolitan centres with potable water supply. However, the capture capacity of traditional supply sources is approaching a limit in many areas, leading to a host of new supply options coming into consideration (WWAP, 2012). Correspondingly, water security is considered as one of the six key risks in Australia under a changing climate. The Intergovernmental Panel on Climate Change asserts that climate change will lead to a reduction in water supply for irrigation, cities, industry and riverine environments in those areas where stream flow is expected to decline (for example in the Murray-Darling Basin, Australia) and annual mean flow may drop 10 to 25% by 2050 and 16 to 48% by 2100 (Hennessy *et al.* 2007).

Rainwater tank systems, collecting and distributing water at a decentralised level, are one potential solution to assist in bridging supply-demand gaps. The basic principle of these decentralised systems is the capture of precipitation collected from the available roof area, which flows by gravity into a storage tank, where it can serve demand for water end-uses. Historically, internally plumbed rainwater tanks (IPRWTs), serving water end-uses inside the house, have only been prevalent in rural areas in the absence of centralised supply infrastructure. In the last 10 to 20 years, amid new concerns over water security, a variety of water businesses, governments and other stakeholders have been advocating the use of IPRWTs in urban areas. However, almost universally these systems have been recommended and implemented without a proper understanding of their underlying viability and performance. In an urban setting, there are a multitude of alternative water supply options and any chosen supply system must be both competitive and sustainable.

This chapter details an investigation into the economics and performance of IPRWTs conducted in Australia's South-east Queensland (SEQ) region and examines these findings in an international context. The study utilises a combination of modelling and empirical data to generate a range of unit life cycle costs (LCC) under different scenarios and conducts a sensitivity analysis on pertinent variables.

1.2 BACKGROUND

The practice of rainwater harvesting (RWH) can be traced back at least 4000 years BC (Gould and Nissen-Peterson, 1999; Mays *et al.* 2007), with systems employing cisterns fed with rainwater attached to single households in ancient civilisations such as Jordan, Rome, Greece and Asia. In more modern times, they have primarily been used in the rural domain where the construction of centralised infrastructure was not feasible. In the new age of sustainability, RWH has enjoyed something of a renaissance; systems have again penetrated into cities where the bulk of the world's population resides. In excess of 100,000,000 people worldwide are estimated to be using a RWH system of some form (Heggen, 2000).

RWH systems can be separated into a number of subcategories based on how they are configured. They may be communal, whereby a number of residences are connected to a tank that is fed from a large roof area, or installed on an individual basis to stand-alone households. Many systems only supply outdoor uses such as garden irrigation and pools, while the popular trend recently has been to internally plumb systems to supply a range of in-home end-uses to maximise savings (via substitution) from centralised sources. The advent of modern appliances requires that the water supply to the house be pressurised. Therefore, the vast majority of IPRWTs contain a pump that can extract water from tanks and deliver it under pressure to the house. These pumps may operate at different levels based on a flow rate or be single speed. More complex pressure vessel setups may also be employed. Switch systems that allow end-uses to be supplied by either the tank or central mains supply are commonplace, so that when a tank is empty essential supply is maintained. This chapter focuses on typical IPRWTs installed on single detached residential households configured with single speed pump and switch systems supplying water for toilets, clothes washers and external use.

There are many purported benefits of RWH and the herein focused upon contemporary IPRWT systems; the predominant benefit being a reduction in urban water demand. For residents, this can offer reduced water bills and decreased reliance on mains supplies. For communities and governments, this can delay the need for centralised infrastructure upgrades and reduce peak stormwater volumes (Coombes *et al.* 2003). By decreasing the amount of water required from central supplies,

RWH can also assist in raising groundwater levels; an urgent task in many urban locations. The major negatives associated with RWH arise from a lack of reliable supply and potentially poor water quality; both of which can be circumvented with the right system setup in the presence of a backup or mains supply.

1.2.1 IPRWT systems in Australia

IPRWT systems have been utilised for generations in rural Australia (EHAA, 1999; Marsden Jacob Associates, 2007). Deployment in urban areas was widely discouraged for many years with a number of local governments banning rainwater tanks in the 1960s, citing water quality as a prohibitive hazard (White, 2009). A severe drought that ran from 2000 until 2009 affected large portions of south-eastern and south-western Australia (CSIRO, 2011), leading to critical depletion of freshwater reservoirs. This triggered the introduction of legislation and Government-backed incentives to install IPRWTs in urban households. They were championed as 'green' and 'sustainable' solutions to the water security crisis, with limited research available to verify such notions at the time.

As of 2007, about 20% of Australian households had some form of RWH system (ABS, 2007). Retamal *et al.* (2009) provide a comprehensive description of a range of IPRWT configurations in Australia and their advantages and disadvantages. The majority of residential dwellings being constructed use fixed speed pumps with potable switch systems or tank top-up systems. The more elaborate and efficient designs, incorporating pressure vessels and variable speed pumps, are rarely considered by house builders as they are predominantly concerned with satisfying mandated building code requirements at least capital cost (in locations where IPRWTs are mandated). The IPRWTs examined in this study were mandated by the Queensland Government to be installed in new houses built or those substantially renovated.

1.2.2 RWH and IPRWTs around the globe

RWH in one form or another is practiced very widely around the globe. Two purpose driven groups can be considered: those that are using rainwater as a supplement to already existing water supply systems and those using rainwater as basic supply (König & Sperfeld, 2006). IPRWTs similar to those examined in Australia's SEQ require a certain socioeconomic level to be present in homes. Some of the nations with widespread IPRWTs are listed below. It should be noted that in Australian literature the distinction between RWH and IPRWTs is explicit, while in much of the international literature this is not the case.

- *Germany*: Regarded as a leader in IPRWT technology, some 35% of new buildings are installed with a RWH system (EA, 2010). Germany has groundwater over abstraction problems in many regions and RWH systems have been promoted through legislation and incentives as a means to reduce this issue (Herrmann & Schmida, 2000). 1.5 million systems are estimated to be supplying toilet flushing, clothes washers (washing machines) and garden irrigation (Galbraith, 2012).

- *United Kingdom*: RWH was a traditional water source before central mains supply became widespread. Modern RWH systems have only been introduced recently. Adoption is supported and encouraged by the *Code for Sustainable Homes* under which all new houses must have a rating of 3, with IPRWT installation a means of raising this score. The UK Rainwater Harvesting Association (2006) reports that approximately 4000 RWH systems are installed in the UK each year with approximately 100,000 already in existence. These systems are commonly internally plumbed to supply toilet flushing as well as garden irrigation (EA, 2010).
- *Malaysia*: Introduced after the 1998 drought, rainwater use is encouraged for domestic purposes under Water Services Industry legislation (Shaari *et al.* 2009).
- *Sri Lanka*: RWH was initially popular rurally and is now also promoted in cities through the country's Urban Development Authority (2007).
- *China*: Gansu province began research and implementation, with 17 provinces now adopting RWH. Over 5.6 million tanks supply potable water to 15 million people (UNEP, 2001).
- *Bermuda*: Mandated by law for all buildings, rainwater is the primary source of domestic water (Rowe, 2011).

Table 1.1 Cost elements and effectiveness considerations for IPRWTs.

Cost element	Effectiveness element
Rainwater tank	Roof catchment area
Tank installation and fitting	Tank size
Water pump	The use of rainwater for outdoor and indoor use
Operating cost	Annual rainfall
Maintenance and pump replacement	Impact of climate variability
Tank requirements (first flush, gutter guard)	Rainfall pattern

Source: adapted from Tam *et al.* (2010)

Trends around the world appear similar, with urban penetration increasing with advocacy from governments. König and Sperfeld (2006) noted that amortisation (pay back) of IPRWTs increases with the cost of mains water, therefore those nations with the highest cost of mains water are typically the highest adopters of IPRWTs technology. In terms of the cost and effectiveness of IPRWTs, regardless of location or configuration, there are a number of factors that determine the cost and effectiveness of IPRWTs, which are summarised in Table 1.1. Any location will have its own make-up of these variables. However, the relationships that govern many of these variables will be very similar between locations. A well-documented investigation conducted in one area can provide insight into the performance and economics of IPRWTs on a wider scale. This is undertaken in the following sections in an Australian context.

1.3 AUSTRALIAN CASE STUDY

The study presented in this chapter identified that Australian water businesses have been implementing a range of alternative water supply schemes, in an attempt to conserve centralised supplies of potable water. However, they undertook such schemes with only best guess potable savings figures and alternative source demand values to serve as justification. Seeking a more rigorous assessment process, the present study followed an evidence-based approach whereby the water consumption of IPRWTs was monitored through end-use studies and costs were evaluated using actual cost and performance data. The end goal of this assessment process was to arrive at an accurate total resource perspective unit cost ($/m^3) for IPRWTs in order to better inform decision-making regarding their use. The IPRWT performance and economic analysis was completed alongside evaluations of three other alternative supply schemes, including desalination and recycled water. Readers are referred to Stewart (2011) if they seek information on the latter two schemes.

1.3.1 Context of investigation

In 2007, the Queensland state government introduced new legislation, namely the Queensland Development Code Mandatory Part 4.2 (QDC). This stipulated that all detached residential households needed to achieve potable water savings (DIP, 2009). Under this legislation, water savings targets are mandated for new detached houses in Queensland, ranging from 16 to 70 m^3 per household per year (m^3/hh/y), depending on the local government area. The widely accepted solution to reduce potable water use was through the installation of a 5 m^3 polymer rain tank plumbed to the toilet, laundry and external taps of detached, single residential households. A minimum of 100 m^2 of roof area must divert rainwater into the tank. Internal fixtures supplied from a rain tank are required to have a backup supply of potable water using a trickle top-up or automatic switching system. Gardiner (2009) notes that, of more than 300,000 tanks in SEQ, about 30,000 were installed under the QDC. Inspections revealed that in most cases house builders chose the least cost IPRWTs with a single speed pump and switch system. Three successive wet years in SEQ saw reservoirs return to capacity and pressure on water supply decrease. Consequently, the Queensland State Government removed the requirement for new houses to have IPRWT from late 2012 due to a number of reasons. These included the need to recoup the construction cost of bulk water infrastructure (such as desalination) constructed during the drought, reduced water consumption due to behaviour change and housing affordability.

1.3.2 Data gathering and end-use study experimental procedure

Data gathering was conducted to inform modelling and the LCC analysis. Eighty-seven (n = 87) Gold Coast City (GCC) detached households (a single dwelling on a single lot) without IPRWTs were sampled during two cross-sectional periods during 2010. This case serves as the business-as-usual water supply scheme for the purposes of this study and is used for baseline potable water savings comparisons. The sample provides a reasonable representation of household types with a strong mix of family types, income categories and household occupancies.

High-resolution smart metering equipment was employed to enable the collection of water consumption data and subsequent end-use analysis. The relationship between smart metering equipment, household stock inventory surveys and flow trace analysis is shown in Figure 1.1. Essentially, a mixed-method approach was used to obtain and analyse water-use data. Two aligned main processes were adopted: (1) physical measurement of water use via smart meters with subsequent remote transfer of high-resolution data; and (2) documentation of water-use behaviours and compilation of water appliance stock via individual household audits and self-reported water-use diaries.

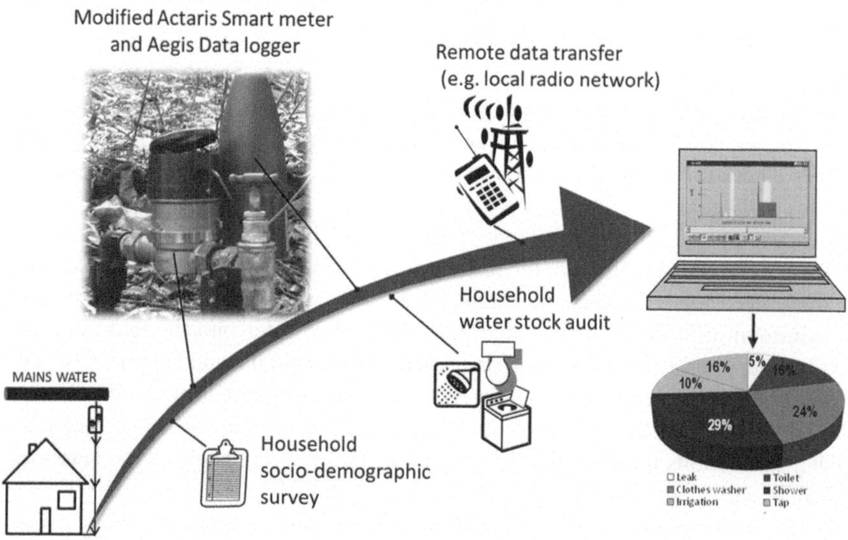

Figure 1.1 Schematic process for acquisition, transfer and analysis of flow data (Beal & Stewart, 2014).

1.3.2.1 Instrumentation

Standard local government residential water meters were replaced with high resolution water meters. These meters measured flow to a resolution of 72 pulses per litre, or one pulse every 0.014 litres. The smart meters were connected to data loggers programmed to record pulse counts at 5-second intervals. Each logger was wired to a meter, labelled and activated prior to installation to reduce reliance on plumbing contractors to prepare and activate the equipment; all equipment was installed by approved plumbing contractors.

1.3.2.2 Data transfer and storage

As the loggers were wireless, data was transferred remotely to a server at Griffith University through a General Packet Radio Service (GPRS) network (such as a 2G or 3G phone network) via email. Removable SIM cards were inserted in each logger and tested prior to installation. The data was transferred weekly, creating approximately 120,000 data records, sent to email addresses before being downloaded and processed. Raw data files in the *ASCII* format were modified to *.txt* files for flow trace analysis.

1.3.2.3 End-use analysis process

End-use data in the *.txt* file format were analysed using Trace Wizard version 4.1 (Aquacraft, 1997). Water diaries and stock appliance audits were used to help identify flow trace patterns for each household. A template was created for each household and data for a sampled 2-week period were analysed. Trace Wizard was employed in conjunction with water audits and diaries to analyse and disaggregate consumption into a number of end-uses, including toilets, irrigation, showers, clothes washers and taps. A Microsoft Excel spread sheet was utilised as a final output for more detailed statistical trend analysis and chart production.

1.3.2.4 End-use results summary

There was a notable difference in irrigation between the two seasonal periods monitored. Winter 2010 irrigation end-use was 9.4 litres per person per day (lpd), representing only 7% of total consumption. This was less than half of the 21.9 lpd recorded in summer 2010, supporting historical bulk reading data that irrigation in GCC is greater during summer. The average sampled total per capita residential consumption value of 156.5 lpd was very close to the Queensland Water Commission (2009) reported SEQ monthly per capita residential consumption average for the 2010 period (140–160 lpd). This indicated that the end-use results were representative and useful for comparisons. A summary of the summer and winter 2010 end-use breakdown for the single detached, potable-only reticulated scheme end-use values is presented in Table 1.3. Readers are referred to Stewart

(2011) and Beal *et al.* (2011) for a full description of the end-use data used in this current study. This sample of potable-only homes situated on the Gold Coast is used for comparison with the potable plus IPRWT supplied households discussed below.

1.3.2.5 Rain tank pump energy pilot study

A pilot study of 5 GCC houses with an IPRWT system was also conducted by the research team to determine the energy intensity (kW/m^3) of the pumping system at an end use level, which would be used for the LCC calculations. The pilot study indicated that the pump energy intensity ranged from 1.04 kW/m^3 for irrigation events, to 1.67 kW/m^3 for half flush toilet events (Talebpour *et al.* 2011). For the purpose of the economic modelling discussed later, an overall IPRWT energy intensity value of 1.5 kW/m^3 was taken to be representative for typical Gold Coast City IPRWT configurations.

1.3.3 IPRWT modelling

Two software packages were used to model the performance of IPRWTs installed to QDC specifications: (1) Rainwater TANK and (2) RainTank. A brief description of the method of analysis applied for each of these approaches is provided below.

1.3.3.1 Rainwater TANK model

The Rainwater TANK model is an Excel-based spreadsheet linked to a FORTRAN executable application (Vieritz *et al.* 2007). Rainwater TANK simulates the capture of rain by an urban roof. The primary aim of the model is to assess how the rainwater tank can meet the water demand of the urban allotment. The tank water volume for the current day is determined from a mass balance as expressed in Equation 1.1.

$$\text{TWtank} = \text{Yest\_TW} + \text{TopUpW} + \text{TankInflow} - \text{IWUtank} - \text{EWUtank} \quad (1.1)$$

where:

TWtank = water volume (m^3)
Yest_TW = yesterday's tank water volume (m^3)
TopUpW = top-up or trucked water volume (m^3) for the current day
TankInflow = flow of rainwater into the tank from the roof for the current day (m^3)
IWUtank = internal Water Use for tank water (m^3) for the current day
EWUtank = external Water Use for tank water (m^3) for the current day

The key assumptions and mathematical formula for the model are described in Vieritz *et al.* (2007). In summary, the initial water level in the tank is set to a user-defined top-up point. Within each daily time step, the order of calculations depends on the 'Run' setting chosen. The rain tank is assumed to be any regular shape, whereby the volume is calculated by multiplying the tank's basal area and its height. Household water end-uses have a fixed amount of water used per day (nominated by the user; here informed by the end-use data gathered from the 87 GCC houses). The primary assumption with respect to internal water use is that the demand must be always fulfilled. This internal water use is assumed to be constant for each day of the run. When the tank runs out of water, the model will automatically meet the internal demand using potable water, thereby providing an estimate of the supply shortfall (Vieritz *et al.* 2007).

1.3.3.2 RainTank model

The second rainwater tank modelling software utilised was RainTank (Jenkins, 2009), which is designed to simulate the collection and use of water from a rain tank connected to the roof of a house. The model uses daily rainfall and consumption information for the house, based on the location of the house and tank site. The model uses a continuous simulation of rainfall and runoff from the house roof to the rain tank and a daily water consumption model for water stored in the rain tank. The conceptual arrangement of the RainTank model includes the following elements (Jenkins, 2007):

- *Roof area*: the total area of the house roof that drains into the tank;
- *Tank volume*: the total volume of the rain tank, including the air space that is available for stormwater detention;
- *Rainwater storage*: the part of the tank that is available for storage of rainwater collected from the roof, which is equal to the tank volume minus the air space available for stormwater detention;
- *Air space for stormwater detention*: the top section of the tank that is available for stormwater detention is defined as a percentage of the tank volume. As it takes some time for the water within this air space to drain out of the tank, this water is used first to supply the daily consumption before the remaining volume is withdrawn from the tank;
- *Initial loss*: rain that falls at the start of a rain event is often absorbed into the pores of the roofing material or is trapped on the roof by surface tension effects, evaporating before any runoff can occur. The model assumes a constant initial loss for each rain day throughout the simulation period;
- *Drainage system efficiency*: during intense rain events runoff often overflows the drainage system elements before it can reach the rain tank. Although a function of intensity, the model assumes a constant drainage system efficiency;
- *First flush loss*: the initial runoff from a roof surface often contains a higher concentration of contaminants than the remaining part of the storm runoff. Many RWH systems allow for the inclusion of a first flush device, which

discards an initial volume of rainwater. No runoff enters the rain tank when the daily roof runoff is less than or equal to the value defined by the first flush loss.

1.3.3.3 Modelling input parameters

The purpose of using two modelling software programs was to compare results and confirm, or otherwise, rain tank yield, with all scenarios being run under each model. The key input parameters can be found in Table 1.2. There were some minor variations between models. A detailed discussion of the RainTank and Rainwater TANK analysis methods and scenario input parameters can be found in Stewart (2011).

Table 1.2 Input parameters for the Rainwater TANK and RainTank models, respectively.

Parameter	Value	Parameter	Value
Climatic region	Southport (Gold Coast)	Roof area	100 m^2
Model years	1980–2008; 1996 dry; 1983 wet	Tank volume	5000 l
Switch system	Automatic with override at 15%	Initial volume	0 l
Residents per household	2.8	First flush volume	15 l
Per capita consumption	156.5 lpd	Tank Height	2 m
End-uses	Two external taps, toilet, cold water laundry		

1.3.3.4 IPRWT end-use breakdown

Table 1.3 presents a summary for the sample of potable-only houses and those also having IPRWT. The actual consumption and associated proportion of water consumption for the two water supply sources across the end use categories for these two types of detached residential households is also provided in this table. For the potable-only houses the total demand was 162.3 lpd. For the IPRWT houses, the total potable and rain water use was calculated to be 115.90 lpd (68.6%) and 53.0 lpd (31.4%) respectively, leading to a total per capita water use of 168.9 lpd. Total demand for rain tank supplied end-uses was 76.6 lpd, with 53.0 lpd supplied by the rain tank and another 23.6 lpd having to be sourced through potable mains due to depleted rain tank supplies (i.e., tank has switched to potable water supply). This implies that the utilisation ratio for the rain tank is approximately 70% (i.e., 30% of demand from IPRWT end uses needs to be covered by potable water) for the 'average' conditions modelled.

Table 1.3 Summary of rainwater and potable water end-uses.

Supply source end-use category	Potable only homes		Potable with IPRWT homes	
	lpd	%	lpd	%
Potable – non-IPRWT end-uses:				
Shower	50.0	30.8	50.0	29.6
Tap	33.8	20.8	33.8	20
Dishwasher	2.3	1.4	2.3	1.3
Bathtub	3.5	2.2	3.5	2.1
Leak (potable line)	2.7	1.7	2.7	1.6
Total (Potable A)	*92.3*	*56.9*	*92.3*	*54.6*
Potable – IPRWT and mains plumbed:				
Clotheswasher (potable line)	32.4	20.0	12.8	7.6
Toilet (potable line)	21.9	13.5	4.9	2.9
Irrigation (potable line)	15.7	9.6	5.9	3.5
Total (Potable B)	*70.0*	*43.1*	*23.6*	*14.0*
Total Potable (A + B)	*162.3*	*100*	*115.9*	*68.6*
IPRWT supply:				
Clotheswasher (cold)	na	na	22.3	13.2
Irrigation (IPRWT taps)	na	na	13.9	8.2
Toilet (IPRWT sourced)	na	na	15.7	9.3
Leak (IPRWT sources)	na	na	1.1	0.7
Total IPRWT	*na*	*na*	*53.0*	*31.4*
Total (all supplies)	162.3	100	168.9	100

1.3.4 Life cycle cost analysis

The per capita end-use water balance laid the foundations for an evidence-based assessment of the potable water savings from installing IPRWT systems, as well as their overall demand. The water savings over the life cycle (LC) can be aligned with the Net Present Value (NPV) LCC of the scheme, including all capital and operating costs. Greenhouse gas emissions from energy generation will incur further costs and it is likely that both water customers and water utilities will pay for these costs in higher energy prices (Fane *et al.* 2011). This NPV LCC analysis resulted in a unit cost (\$/m^3) for IPRWT systems, based on their ability to derive such potable water savings. Note that all costs presented are in Australian (AUD) dollars (1 AUD = 1.00 USD as at April 2013, xe.com (2013)).

The NPV LCC analysis includes a very limited financial assessment on the wider environmental and societal benefits of IPRWTs. These costs and benefits are discussed and arguments provided alongside the formulated unit costs for the various schemes. The scope of the analysis does not consider the funding package (government revenue, bank debt, bonds) applied and the interest costs associated with each scheme. The NPV LCC assessment considers capital costs and recurrent expenses to be funded through government or business revenues.

1.3.4.1 IPRWT capital cost estimates

The cost of the additional works required to meet QDC MP 4.2 is included in the building contract cost of a new dwelling and is ultimately borne by the homeowner. The average capital works cost of IPRWT installations in new dwellings, including the cost of the tank, delivery, installation and plumbing, plus incidentals such as a concrete slab, tank stand and potable water switching devices is available in a number of studies (WBM Oceanics, 2005; Coombes, 2007; NWC, 2007; Tam *et al.* 2010). This study extracted capital costs from these studies and used the most representative average or median value for application in this NPV LCC assessment (Table 1.4). The reticulation of IPRWT installations is cost prohibitive for existing houses and there is no requirement or indeed general desire for existing households to implement them.

Table 1.4 Capital cost (AUD) of installing an internally plumbed rainwater tank system.

5 m³ RWT (AUD*)	Pump (AUD)	Plumbing (AUD)	Installation (AUD)	Total (AUD)	Source
1150	355	730	550	2785	Tam *et al.* (2010)
1091	650	727	548	3016	NWC (2007)
1388	770	–	–	–	WBM Oceanics (2005)
–	–	–	–	2765	Coombes (2007)
1150	650	729	549	3078	This study

*1 AUD = 1.00 USD as at April 2013, xe.com (2013).

1.3.4.2 IPRWT operating and maintenance costs

Recent monitoring and the pilot study suggest an average energy intensity value of 1.5 kWh/m³ for the most common pump and switch systems (Retamal *et al.* 2009; Talebpour *et al.* 2011). In this study, a 7.3% inflation rate (Table 1.5) was adopted for electricity, which represents the average for the past five years; there is no evidence of reduced electricity price inflation expectations in the medium term. A GHG cost implication of running the pump and an assigned cost of \$20/t CO_2 was applied in this study. As reported by DERM (2007), an assigned 1.046 kg CO_2-e/kWh was determined as the level of carbon generated from the pump system.

Table 1.5 NPV LCC base case financial model parameters.

Parameter	Value	Parameter	Value
Life cycle period	25 years	Carbon emissions	1.046 kh CO_2/kWh (DERM 2007)
Discount rate (base case)	7%	Carbon price	\$20/t with 4% escalation
Capital costs	See Table 1.5	Pump replacement	15 years (replace once)
Energy intensity	1.5 kWh/kW	Tank replacement	25 years
Electricity tariff	\$0.1713/kWh	Tank reliability factor	0.9
Electricity price inflation	7.3%	Pump replacement labour	3 hours @ \$70/hour
Inflation for pump & tank	3%	Tank replacement labour	4 hours @ \$60/hour
Inflation for labour	4%		

There is still limited evidence on the life span of urban water rain tanks and pump systems as they have not been widely implemented in urban areas until recently. Current documentation from suppliers indicates a 25-year structural life span for polymer rain tanks, which represent the majority of stock. Pumps are often reported as having a life span of approximately 15 years. These life spans are applied for the purposes of the NPV LCC analysis, however, there is anecdotal evidence to suggest poor manufacture is leading to shorter life spans. Tank and pump replacement will also generally require a labour cost contribution, as most homeowners would not be suitably skilled or feel comfortable installing these components. IPRWTs have a number of components that need to be readily checked and maintained, including first flush systems, leaf protection mesh and filters, to name a few. In this study, a AUD\$20 annual miscellaneous maintenance amount was proposed (NWC, 2007; Tam *et al.* 2010), which considers that homeowners would replace filters and so on (thus no labour cost).

Anecdotal evidence suggests that some homeowners may unknowingly or knowingly have a tank or switch system that is not functioning. Given the design of switch systems, the water supply reverts to potable supply when the pump has failed or the power is turned off. Owners will therefore still receive water even if their pump is not functioning and may choose to turn them off completely if the noise upsets them or they do not have sufficient funds to replace the pump. Based on recent discussions with researchers and field technicians, a rain tank reliability reduction factor was applied in the NPV LCC analysis. A reduction factor of 0.9 (i.e., 1-in-10 connections estimated as not providing water savings at any time for the base case scenario) was therefore applied for the base case NPV LCC assessment.

1.3.4.3 NPV LCC base case financial model parameters

The financial parameters utilised for the base case scenario are summarised in Table 1.5. Readers should note that the NPV LCC analysis was considered on a per connection basis to determine a unit cost for potable water savings resulting from the installation of IPRWT on detached houses in this scheme. Additionally, the boundary of the unit cost analysis covers only those costs attributed to the customer installing the IPRWT (costs and potable water savings to customer). There are a number of follow-on benefits of IPRWT that have not been considered herein due to their difficulty to monetise, such as reductions in daily and peak demand in the pipe network due to demand being assumed by the IPRWT. If IPRWTs had high rates of diffusion in urban areas, there are potential pumping and infrastructure deferral savings that accrue to the water utility. However, there is presently insufficient evidence to quantify the monetary link between IPRWT and reductions in water distribution network demand and infrastructure deferral opportunities.

1.3.4.4 Life cycle cost results

The difference between the potable water supplied to a traditional potable-only household and the potable demand met by the IPRWT scheme is considered to be the water saving attributed to the IPRWT in this study. As detailed in Table 1.3 this is 46.4 lpd (162.3–115.9 = 46.4 lpd) or 47.4 m^3/hh/y based on the average household occupancy of 2.8 persons in the city ((365 × 2.8 × 46.4)/1000 = 47.4 m^3/hh/y). The initial IPRWT capital outlays make up the majority share (refer to Table 1.6; 3.36/4.06 = 82.7%) of the total unit cost for this scheme. Initial capital cost expenditures at the building stage are the most critical component, followed by pump and tank replacements at the end of their life.

Table 1.6 NPV LCC base case assessment for IPRWTs on a per connection basis.

Financial item description	Value	Unit
Life cycle potable water savings per connection	1067	m^3/connection
NPV LCC per connection	4326	$/connection
Capital cost component of total unit cost	3.36	$/m^3$
Operating cost component of total unit cost	0.70	$/m^3$
Total unit cost	4.06	$/m^3$

1.3.5 Sensitivity analysis

A sensitivity analysis was used to explore the unit cost implications for a range of scenarios where input parameters were modified within a realistic range

(Table 1.7). The following variations of critical NPV LCC input parameters were considered:

- *Scenario A (SA)*: discount rates set at 4%, 6%, 7% (base case) and 9%;
- *SB*: 1% increase in base case operating cost component annual inflation rates (such as Consumer Price Index (CPI));
- *SC*: 1% decrease in base case operating cost component annual inflation rates;
- *SD*: IPRWT water reliability factor reduced from 0.9 to 0.8 (i.e., no supply at 1 in 5 houses at any time);
- *SE*: reduced life spans for rain tank (25 years reduced to 15) and pump (15 years reduced to 10).

Table 1.7 The influence of variable discount rates on NPV LCC model parameters for IPRWT's unit cost.

Scenario	Parameter modified discount rate (i)	Unit cost (AUD$/m^3)			
		4%	6%	7%	9%
SA	Discount rate change alone	4.62	4.22	4.06	3.80
SB	1% increase in base operating cost inflation rate	4.88	4.40	4.22	3.92
SC	1% decrease in base operating cost inflation rate	4.40	4.05	3.92	3.70
SD	IPRWT water reliability reduced from 0.9 to 0.8	5.20	4.74	4.56	4.28
SE	Reduced life spans of RWT and pump	6.50	5.64	5.30	4.76

The sensitivity analysis indicated a range of unit costs for the IPRWT scheme between AUD$3.70–6.50/m^3 (base case = AUD$4.06/m^3). Table 1.7 illustrates that scenario SE, where the life span of the RWT and pump was reduced, led to the highest unit costs (Table 1.7). Reducing the average base case IPRWT infrastructure life spans from 25 to 15 years for the RWT and 15 to 10 years for the pump is highly probable due to a range of reasons. Firstly, while most manufacturers report long life spans, the industry has a number of low quality manufacturers producing rain tanks with thin wall thicknesses that are prone to breakages. Also, low cost pumps are now available that may not be as reliable as the long-established products. Another issue of concern is that the management of the IPRWT system is presently the responsibility of homeowners, many of whom rent out the household and do not readily inspect the tank or pump operation. Urban home occupants are typically unfamiliar with external pumps and tanks and may not be sufficiently competent to maintain these systems, thereby reducing their reported life span. The second most influential parameter on the unit cost is related to the reliability of actually receiving the water saving or demand from the IPRWT (scenario SD in Table 1.7). It is a real possibility that this scenario might eventuate given the same arguments presented for the life span of the IPRWT.

Additionally, as a household ages and equipment requires replacement, homeowners will need to consider whether to replace their pump and switch system. Given that the IPRWT system is designed so that potable water is automatically supplied when the pump or switching system has become non-operational, there is a lack of incentive for many homeowners to replace broken equipment. A new pump with a switching system is approximately $600 installed, which in monetary terms equates to over four years of utility variable water charges related to the savings made by the IPRWT.

1.4 INTERNATIONAL COMPARISONS

The method and analysis found in the case study presented above may prove informative to those outside of Australia. However, due to most parameters being location sensitive, direct financial comparisons cannot be made between locations. A number of studies worldwide have investigated the LCC of various RWH systems, though few have extended this to an incremental or levelised cost. A major barrier to comparison with other studies relates to dwelling types. In Australia, approximately 80% of residences are detached houses with a surrounding garden or lawn (Pink, 2010), which results in most IPRWTs being on an individual household scale. This is in stark contrast to most European, Asian and Middle Eastern nations. For example, in the European Union just 34.4% of citizens live in detached houses (Eurostat, 2012). Therefore, in these nations it is much more common for IPRWTs to be on a communal scale, collecting rainwater from a single roof area to serve multiple households.

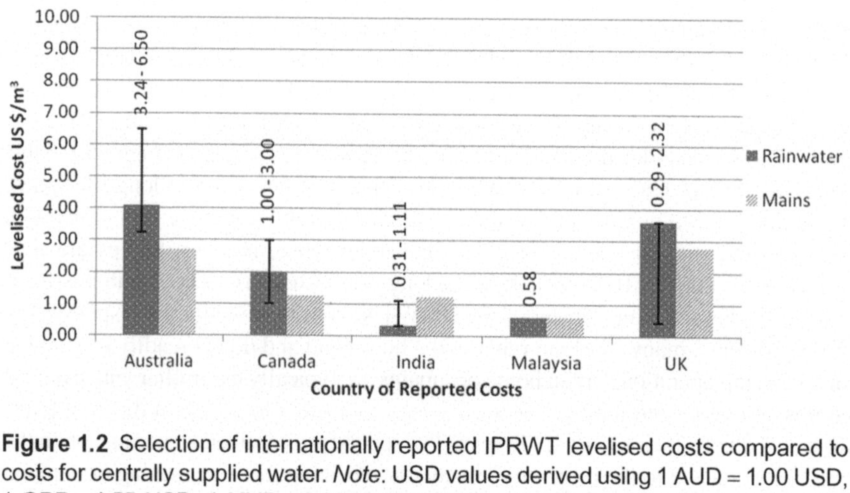

Figure 1.2 Selection of internationally reported IPRWT levelised costs compared to costs for centrally supplied water. *Note*: USD values derived using 1 AUD = 1.00 USD, 1 GBP = 1.55 USD, 1 MYR = 0.337 USD, 1 INR = 0.0185 USD, 1 CAD = 1.00 USD conversion rates (as at April 2013, xe.com (2013)).

Nonetheless, a snapshot of internationally reported of IPRWTs levelised costs is presented in Figure 1.2 and compared to the mains water costs in those locations at the time each study was conducted (Brewer *et al.* 2001; Vishwanath, 2001; Shaaban & Appan, 2003; CRDWS, 2007). Figure 1.2 indicates that there is a wide range of reported values for the unit cost of IPRWTs. With the exception of India, the mean unit cost of rainwater supply is higher than the mains water supply.

1.5 DISCUSSION

The study presented in this chapter determined that an IPRWT could save 47.4 m^3/hh/y of potable water and had a unit cost of AUD$4.06 /$m^3$. Other modelling studies in Queensland have reported yields of 26–144 m^3/hh/y, with an average of 78 m^3/hh/y (Coombes *et al.* 2003; MWH, 2007; NWC, 2007). However, Coombes *et al.* (2003), for instance, assumed that rainwater was used for hot water and rainfall data was taken in pre-Millennium drought conditions. In 2009, the Water Corporation (MJA, 2009) released a factsheet indicating that IPRWTs had a unit cost of $4.00–13.00/$m^3$. Turner *et al.* (2007) indicated a unit cost of AUD$3.96/$m^3$ while Marsden Jacobs's (2007) comprehensive investigation on the cost-effectiveness of IPRWTs indicated a unit cost of AUD$2.29/$m^3$ (50 m^2 roof area) to AUD$5.47/$m^3$ (200 m^2 roof area) for a 5 m^3 tank in Brisbane (plumbed both internally and externally). The base case unit costs determined by the study presented herein are close to those reported in the literature, particularly the value reported by Turner *et al.* (2007). A sensitivity analysis showed that the reliability of supply and the life spans of tanks and pumps pose the major hurdles to the overall cost effectiveness of IPRWTs. Governments could consider additional regulatory and quality assurance frameworks to manage these problems.

Rainfall is obviously the key factor that is non-property-specific in harvested rainwater yield. However, harvested rainwater yield in terms of actual harvested rainwater used by the household is highly dependent on the regular use (emptying) of the tank; a half-full tank will only capture 50% of its total potential during a rainfall event. The water demand management campaign in SEQ has been highly effective in reducing household water consumption and this is extending to prudent use of rainwater, therefore reducing the maximum potential of the harvested rainwater to reduce potable water use. Households with high water consumption are also tending toward higher reductions from potable supply as they are probably using more harvested rainwater; allowing the tank to empty and refill more frequently.

IPRWTs can reduce total daily per capita potable demand by approximately one-third. They also have some flow-on reduction to the peak hour (8–9 am) demand (litres per person per hour of the day) for potable water. Peak demand parameters drive the design of most centralised pump and pipe infrastructure for distributing water. Therefore reductions in peak demand may mean reduced requirements to upgrade or duplicate existing major trunk mains, reservoirs and pump stations.

Such infrastructure deferral benefits from IPRWTs have not been considered in the analysis presented here because they are not yet fully understood and have not been financially quantified. Nonetheless, the infrastructure deferral benefits of decentralised systems such as IPRWTs should also be considered alongside the herein developed unit costs for potable water savings.

House owners with an IPRWT will likely have a lower quarterly water bill due to reduced consumption. This represents a small proportional saving since the majority of water bills in Australia are composed of fixed charges (water service and wastewater charges). Given the lower peak demand contribution from these households discussed above they could potentially be entitled to a reduction in fixed charges.

Currently Australia's major population centres are not beset by drought, which has seen central reservoirs return to high levels and water security fears decrease. In Queensland, the QDC MP 4.2 legislation has now been suspended to allow the state government to raise revenue from water sales and building costs. However, for the GCC consumer the price of water has risen to \$3.29/m^3 since the completion of the case study (GCCC, 2012), which falls inside the lower end of the range of levelised LCC costs calculated. In this way, if IPRWTs are not currently cost effective for consumers in some locations, it is very likely they will become so in the future as water prices rise.

1.6 SUMMARY AND CONCLUSIONS

This chapter has presented the results of a combination of end-use monitoring and modelling, which indicated that IPRWTs fed from a 100 m^2 roof area with a 5 m^3 tank can save 47.4 m^3/hh/y (Table 1.3), when supplying irrigation, laundry and toilet flushing end-uses for a 2.8 person household, in comparison to potable-only households in Australia's SEQ region. Additionally, a life cycle cost analysis has shown that IPRWTs can produce water at AUD\$4.06/m^3 (Table 1.6) over a 25 year life cycle. This is in excess of the AUD\$3.29/m^3 currently charged for potable water through central supply lines in GCC, but with water costs forecast to continue rising well above inflation, IPRWTs may be cost competitive in the near future.

Furthermore, a sensitivity analysis revealed a range of costs from \$3.70/m^3 to \$6.50/m^3 (Table 1.7). This analysis identified that the most critical factors were the lifespan of the pump and tank, followed by the reliability of supply. In order for IPRWTs to be financially effective in any location around the world, these factors must be adequately controlled.

In summary, the study presented highlighted that IPRWTs can be a suitable potable source substitution measure, helping governments and communities to strengthen their water security. However, IPRWT may not be a least cost measure and must be carefully designed and installed (pump, tank and roof quality and sizing) to ensure that they deliver desired outcomes, when compared to other alternatives with a similar unit cost (such as desalination plants, which are usually managed centrally). Most importantly, this study highlights the importance of a detailed assessment of the performance and finance of particular water scheme(s) before embarking on state or citywide mandated policy or incentive schemes.

REFERENCES

Aquacraft (1997). Trace Wizard. Boulder, Colorado.

Australian Bureau of Statistics (2007). Environmental Issues: People's Views and Practices. Canberra.

Beal C., Stewart R., Huang T. and Rey E. (2011). SEQ residential end use study. *Water: Journal of the Australian Water Association*, **38**(1), 80–84.

Beal C. D. and Stewart R. A. (2014). Identifying residential water end-uses underpinning peak day and peak hour demand. *Journal of Water Resources Planning & Management*, DOI: 10.1061/(ASCE)WR.1943-5452.0000357

Brewer D., Brown R. and Stanfield G. (2001) Rainwater and Greywater in Buildings: Project Report and Case Studies. Technical Note TN 7/2001. BSRIA, Berkshire, UK.

Coombes P. (2007). Energy and economic impacts of rainwater tanks on the operation of regional water systems. *Australian Journal of Water Resources*, **11**(2), 177–192.

Coombes P., Kuczera G. and Kalma J. (2003). Economic, water quantity and quality impacts from the use of a rainwater tank in the inner city. *Australian Journal of Water Resources*, **7**(2), 111–120.

CRDWS (2007). Rainwater Harvesting in Greater Victoria, Capital Regional District Water Services, Victoria, British Columbia. https://www.crd.bc.ca/docs/default-source/water-pdf/technologybrief-rainwaterharvestingforgreatervictoria.pdf?sfvrsn=2 (accessed 24 April 2014).

CSIRO (2011). The millennium drought and 2010/11 floods factsheet. South Eastern Australian Climate Initiative. http://www.seaci.org/publications/documents/SEACI-2Reports/SEACI2_Factsheet2of4_WEB_110714.pdf (accessed 24 April 2014).

Department of Environment and Resource Management (2007). ecoBiz Queensland Tables of Conversions and Units. http://www.daintreehour.com/miscellaneous/p01295al.pdf (accessed 24 April 2014).

Department of Infrastructure and Planning (2009). South East Queensland Regional Plan 2009–2031. Department of Infrastructure and Planning, Queensland Government.

EHAA (1999). Rainwater Tanks: Their Selection, Use and Maintenance. Department of Environmental Heritage and Aboriginal Affairs, Government of South Australia.

Environment Agency (2010). Harvesting Rainwater for Domestic Uses: an Information Guide. http://www.chs.ubc.ca/archives/files/Harvesting%20rainwater%20for%20domestic%20uses%20an%20information%20guide.pdf (accessed 24 April 2014).

Eurostat (2012). Europe in Figures: Housing Statistics. http://epp.eurostat.ec.europa.eu/statistics_explained/index.php/Housing_statistics (accessed 24 April 2014).

Fane S., Blackburn N. and Chong J. (2011). Sustainability Assessment in Urban Water Integrated Resource Planning for Urban Water – Resource Papers. Waterlines Report Series No 41 (March 2011), National Water Commission, Canberra. http://archive.nwc.gov.au/__data/assets/pdf_file/0017/10385/41_IRP.pdf (accessed 24 April 2014).

Galbraith K. (2012) Saving up for a Dry Day. The New York Times online. http://www.nytimes.com/2012/11/15/business/energy-environment/15iht-green15.html?_r=2& (accessed 24 April 2014).

Gold Coast City Council (2012). Water and Wastewater Pricing 2012–13. http://www.goldcoast.qld.gov.au/documents/bf/water-and-wastewater-pricing-2012–13.pdf (accessed 22 March 2013).

Gould J. and Nissen-Peterson E. (1999). Rainwater Catchment Systems for Domestic Supply: Design, Construction and Implementation. Intermediate Technology Publications, London.

Heggen R. J. (2000). Rainwater catchment and the challenges of sustainable development. *Water Science and Technology*, **42**(1–2), 141–145.

Hennessy K., Fitzharris B., Bates B. C., Harvey N., Howden S. M., Hughes L., Salinger, J. and Warrick R. (2007). Australia and New Zealand. In: Climate Change 2007: Impacts, Adaptation and Vulnerability. Contribution of Working Group II to the Fourth Assessment Report of the Intergovernmental Panel on Climate Change, M. L. Parry, O. F. Canziani, J. P. Palutikof, P. J. van der Linden and C. E. Hanson (eds), Cambridge University Press, Cambridge, UK, pp. 507–540.

Herrmann T. and Schmida U. (2000). Rainwater utilisation in Germany: efficiency, dimensioning, hydraulic and environmental aspects. *Urban Water*, **1**(4), 307–316.

Jenkins G. (2007). Use of continuous simulation for the selection of an appropriate urban rainwater tank. *Australian Journal of Water Resources*, **11**(2), 231–246.

Jenkins G. (2009). RainTank, version 2.0. Griffith University, Gold Coast, Queensland.

König K. W. and Sperfeld D. (2006). Rainwater Harvesting – A Global Issue Matures. Association for Rainwater Harvesting and Water Utilisation. Darmstadt, Germany.

Marsden Jacob Associates (2007). The Cost-Effectiveness of Rainwater Tanks in Urban Australia. Waterlines Report No. 1 (March 2007). Canberra, Australia: National Water Commission. (accessed 24 April 2014] http://archive.nwc.gov.au/—data/assets/pdf_file/0014/10751/cost-effectiveness-rainwater-tanks-body-waterlines-0407.pdf

Mays L. W., Koutsoyiannis D. and Angelakis A. N. (2007). A brief history of urban water supply in the antiquity. *Water Science and Technology*, **7**(1), 1–12.

MWH (2007). Regional Water Needs and Integrated Urban Water Management Opportunities Report. Report 4 of the SEQRWSSS IUWMA Taskgroup, MWA-01.

Pink B. (2010). 2009–10. Year Book Australia. Canberra: Australian Bureau of Statistics.

Queensland Water Commission (2009). SEQ Desalination Siting Study Supplement Reports Appendix I; WaterSecure Gold Coast Desalination Facility – 45, 135 and 270 megalitres per day expansion Study. [Not publically available]

Retamal M., Glassmire J., Abeysuriya K., Turner A. and White S. (2009). The Water-Energy Nexus: Investigation into the Energy Implications of Household Rainwater Systems. Report of the Institute for Sustainable Futures, University of Technology, Sydney.

Rowe M. P. (2011). Rain water harvesting in Bermuda. *Journal of the American Water Resources Association*, **47**(6), 1219–1227.

Shaaban A. J. B. and Appan A. (2003). Utilising rainwater for non-potable domestic uses and reducing peak urban runoff in Malaysia. *Proceedings of the 11th International Rainwater Catchment Systems Association Conference, August 2003*, Texococo, Mexico.

Shaari N., Che-Ani A. I., Nasir N., Tawil N. M. and Jami M. (2009). Implementation of Rainwater Harvesting in Sandakan: Evolution of Sustainable Architecture in Malaysia. Regional Engineering Postgraduate Conference, October 2009, Kuala Lumpur, Malaysia.

Stewart, R. (2011). Verifying the End Use Potable Water Savings from Contemporary Residential Water Supply Schemes. Canberra: National Water Commission Waterlines Report No. 61 (October 2011, Australian Government. http://archive.nwc.gov.au/_data/assets/pdf_file/0016/18007/61-Verifying_potable_water_savings.pdf (accessed 24 April 2014).

Talebpour M. R. Stewart R. Beal C. Dowling B. Sharma A. and Fane S. (2011). Rainwater tank end usage and energy demand: A pilot study. *Water: Journal of the Australian Water Association*, **38** (1), 97–101.

Tam V., Tam T. and Zeng S. (2010). Cost effectiveness and tradeoff on the use of rainwater tank: an empirical study in Australian residential decision-making. *Resources, Conservation and Recycling*, **54**, 174–186.

Turner A., Willetts J., Fane S., Giurco D., Kazaglis K. and White S. (2007). Review of Water Supply – Demand Options for South East Queensland. Report of the Institute for Sustainable Futures, University of Technology, Sydney, Sydney and Cardno, Brisbane.

UKRHA (2006). Enhanced Capital Allowance Scheme as it Applies to Rainwater Harvesting Systems. The UK Rainwater Harvesting Association, Industry Fact Sheet No. 3.

United Nations Environment Programme (2001). Rainwater Harvesting and Utilisation: An Environmentally Sound Approach for Sustainable Urban Water Management: An Introductory Guide for Decision-Makers. http://www.unep.or.jp/ietc/publications/urban/urbanenv-2/9.asp. (accessed 24 April 2014).

Urban Development Authority (2007). Urban Development Authority (Amendment) Act No. 36. Parliament of the Democratic Socialist Republic of Sri Lanka, Colombo.

Vieritz A., Gardner T. and Baisden J.(2007). Rainwater tank model designed for use by urban planners. Proceedings of the Australian Water Association's Ozwater Convention and Exhibition. 4–8 March 2007, Sydney. (CD). ISBN 978-0-908255-67-2.

Vishwanath S. (2001). Rainwater harvesting in urban areas. Rainwater Club, Bangalore. http://www.rainwaterclub.org/docs/R.W.H.industries.urban.pdf (accessed 24 April 2014).

WBM Oceanics (2005). Pimpama Coomera Waterfuture Masterplan Implementation – Rainwater Tank Optimisation Study. Report by WBM, Urban Water Cycle Solution, Bligh Tanner and Noosphere Ideas.

White, I. W. (2009). Decentralised Environmental Technology Adoption: The Household Experience with Rainwater Harvesting. Unpublished PhD thesis, Griffith University, Australia.

World Water Assessment Programme (2012). The United Nations World Water Development Report 4: Managing Water under Uncertainty and Risk. UNESCO, Paris.

xe.com (2013). Foreign Exchange Currency Conversions: Mid-market rates, http://www.xe.com/ (accessed 24 April 2014).

Chapter 2

Evaluating rain tank pump performance at a micro-component level

Mohammad Reza Talebpour, Oz Sahin, Raymond Siems, Rodney Anthony Stewart and Michael Hopewell

2.1 INTRODUCTION

The global freshwater crisis and its associated risks have been greatly appraised. Many of the worlds developed nations are faced with water supply and quality dilemmas, while more than one billion people in the developing world are without consistent water supply (WWAP, 2012). In Australia, the availability of freshwater is expected to decline due to climate change (CSIRO, 2011), while demand for the water is set to increase under a growing population (Pink, 2010). This supply-demand gap means that new sources of water must be identified, evaluated and developed. It must be considered that water supply systems are not only impacted by climate change, but that they also contribute to it through the consumption of energy (Flower *et al.* 2007). This energy-water-climate nexus dictates that water supply systems that are selected to augment traditional reservoir-based supply must both provide water and consume energy efficiently to achieve sustainability.

Internally plumbed rainwater tanks (IPRWT), supplying water to residential households, are a member of the alternative water supply source spectrum. IPRWT systems typically contain a pump to generate the necessary flow and pressure required for water end-uses in and around the home. Consequently, these pumps are responsible for the operational energy consumption of rainwater tank systems. This consumption generates a cost to the homeowner (through electricity and carbon tariffs) and to the environment through associated greenhouse gas emissions.

A number of Australian and international studies have determined the energy intensity of IPRWT pumps on a theoretical basis, while a few empirical studies have also been completed. Of empirical home monitoring studies, many have been completed to determine the net system energy intensity, but not at an end-use level. This chapter covers a recently completed investigation into rain tank pump performance conducted in South-east Queensland (SEQ), Australia. The study is the first known empirical in-home evaluation of rain tank pumps at an end-use level. This evaluation incorporates water and energy data captured at high resolution from 19 homes over a 6-month period, combined with socio-economic and stock inventory data.

2.2 BACKGROUND

Internationally, rainwater tank systems are experiencing a renaissance as they are perceived to be a low cost source substitution option for many end-uses or micro-components of water demand (toilet, clothes washer, irrigation). In times of poor water security, rainwater tank systems are often mandated or subsidised by the Australian Government for new urban developments or retrofits to existing buildings. Government policies and installation guidelines are often framed with a narrow view of rainwater tank systems water savings, with limited consideration of their design with respect to the energy they consume. In Queensland, Australia, after the regions' combined dam levels fell to under 14% in 2007, the state government introduced the Queensland Government (2008) Development Code Mandatory Part 4.2 (QDC MP 4.2). This mandated that all new detached residential households achieve water savings targets of between 16 and 70 m^3/hh/year, depending on the local region (DIP, 2009). The most common way to satisfy these requirements has been the installation of a 5 m^3 polymer rainwater tank, plumbed internally to supply the toilets, clothes washer cold feed, as well as external taps (Stewart, 2011). QDC MP 4.2 triggered the widespread uptake of IPRWT across most of SEQ and other urban areas of Australia, in the absence of detailed research to advise best practice design measures.

2.2.1 Pump energy intensity and associated costs

Table 2.1 lists the key factors influencing the cost and effectiveness of IPRWT. Of these, energy intensity (the energy consumed by the system pump to deliver water to the intended end-use) influences the systems' operational cost and contributes towards greenhouse gas emissions.

Energy intensity quantification is among the least transparent of these factors because it is not a fixed or one-off cost, instead it is a function of many variables (Retamal *et al.* 2009). These variables include pump systems (pump and related equipment), end-use water demand and pipe head loss due to friction. This chapter primarily considers the interaction between pump systems and end-use water demand and its influence on system efficiency.

Table 2.1 IPRWT cost elements and effectiveness considerations.

Cost element	Effectiveness element
Rainwater tank	Roof catchment area
Tank installation and fitting	Tank size
Water pump	The use of rainwater for outdoor and indoor use
Energy Intensity	Annual rainfall
Maintenance and pump replacement	Impact of climate variability
Tank requirements (first flush, gutter guard)	Rainfall pattern

Source: Adapted from Tam *et al.* (2010)

2.2.2 Common configurations for rainwater tank systems

Rainwater tank systems can be setup in a variety of different configurations, which dictate the end-uses plumbed, pump system installed and how this system performs. Many early systems installed in urban Australia in the last century were only designed to supply water for low pressure outdoor non-potable uses. These relied on gravity head, negating the need for a pump. This non-potable use was partly due to commonly held fears over rainwater quality at the time (White, 2009).

The next configuration to gain popularity was the trickle-top up system. This is where the mains water supply is fed into a rainwater tank when the level of harvested rainwater in a tank falls below a certain volume. These were advocated because they maintained constant supply through rainwater pipes and prevented the backflow of rainwater into mains pipes (Coombes *et al.* 2003). However, these systems are by nature inefficient because they require re-pressurisation of water after it has already been in a supply-ready state. Many of these pump dependent systems were configured to supply water to internal end-uses such as toilet flushing and clothes washing.

Mains switch systems are now the most common IPRWT configuration. This is where plumbing infrastructure is arranged to supply water to end-uses from both mains supply and rainwater supply, with a governing switch at the intersection point. When there is sufficient rainwater supply, the switch allows rainwater only into supply. If the pump cannot supply an adequate flow rate for an end-use, or if the rainwater tank is empty, then the switch allows mains water to flow. Readers are referred to Retamal *et al.* (2009) for detailed explanations and explanatory diagrams.

There are two common pump types; single speed and variable speed. Variable speed pumps are designed to vary output based on the flow-rate requirement, while fixed speed pumps operate at a single output level regardless of the requirements of an end-use event. Single speed pumps are generally cheaper than their variable speed counterparts. Other pump systems that are available to system owners include pressure vessels, venturi pumps, gutter storage and header tanks. However, these are uncommon and not widely used in Australia (Retamal *et al.* 2009).

Table 2.2 Summary of previous IPRWT energy intensity studies conducted in Australia.

Study	Method	Sample size	Component examined	Pump type(s)	Energy intensity (kWh/m³)
Cunio and Sproul (2009)	Modelled	NA	Net system	Single speed	0.10–0.20
Hallman *et al.* (2003)	Modelled	NA	Irrigation Toilet	Single speed	0.24 0.36
Retamal *et al.* (2009)	Modelled	NA	Irrigation Toilet Clothes washer	Single speed	0.4–0.8 1.7–2.7 0.5–0.9
de Haas *et al.* (2011)	Modelled	NA	Net system	Unknown	0.8–1.40
Hall *et al.* (2011)	Modelled	NA	Net system	Unknown	2.3
Hood *et al.* (2010)	Empirical	24	Net system	Single speed	1.40
Umapathi *et al.* (2013)	Empirical	20	Net system	Mixed*	1.52
Ferguson (2012)	Empirical	52	Net system	Mixed*	0.70–3.00
Beal *et al.* (2008)	Empirical	5	Net system	Single speed	2.00–3.90
SEWL (2008)	Empirical	31	Net system	Mixed*	0.59–11.61
Retamal *et al.* (2009)	Empirical	10	Net system	Mixed*	0.9–2.3
Hauber-Davidson and Shortt (2011)	Laboratory	8	Net system	Mixed*	0.4–1.6
Tjandraatmadja *et al.* (2011)	Laboratory	3	Toilet Clothes washer Dishwasher Tap	Single speed	0.6–5.3
Cunio and Sproul (2009)	Laboratory	2	Toilet Clothes washer	Mixed*	0.07–1.70

*These studies considered both single and variable speed pumps.

2.2.3 Previous studies

A significant number of Australian and international studies have been conducted to date, determining the energy intensity of IPRWT and evaluating pump performance. Known Australian studies are summarised in Table 2.2. The outcomes of these studies were mainly based on datasets collected using three methods, namely: empirical, modelled and laboratory. Studies using the modelling methods mainly rely on the manufacturers specifications for analyses, while empirical methods use actual data collected from homes. Laboratory methods use data obtained from a sample home built in laboratory conditions.

There is a significant disparity between the majority of the modelled and empirical values. The lowest modelled energy intensities align closely with manufacturer quotations, which are considered to be unrealistic. This can be due to, as reported by Retamal *et al.* (2009), the models used by manufacturers to determine energy consumption generally underestimating the energy consumed by the pumps in practice. It should also be noted that the energy intensity values utilised in some life cycle studies are significantly lower on average than the empirically reported values (Coombes *et al.* 2003; Marsden Jacob Associates, 2007 and Tam *et al.* 2010). Financial assessment of IPRWT can also be improved by correct determination of pump energy intensity and an understanding of the factors governing the performance.

Reported international studies display the same incongruity between theoretical and empirical values. Chiu *et al.* (2009) (Taiwan), Ghisi and de Oliveira (2007) (Brazil), Ward *et al.* (2012) (United Kingdom) and Campling *et al.* (2008) (Belgium) report theoretically derived values of 0.06, 0.18, 0.54 and 0.60 kWh/m^3 respectively, while Parkes *et al.* (2010) (United Kingdom) reports an empirical value of 3.45 kWh/m^3.

It is clear that many studies, with reliable sample sizes, have only evaluated net system energy intensity (energy intensity of total water consumption). However, there is no known in-home empirical study that has been conducted at an end-use level. Limited modelling has been carried out at this resolution in a lab environment (a 'lab home'). However, the correlation between a model home in a lab environment and conditions in a real home is unknown. The case study presented in the following section attempts to assist in narrowing this gap in knowledge on IPRWT energy intensity values.

2.3 AUSTRALIAN END-USE PUMP PERFORMANCE STUDY

As of 2007, over 20% of Australian households had some form of rainwater tank system (ABS, 2007). In SEQ, Gardner (2009) estimated there were over 300,000 systems with over 30,000 IPRWT systems installed under the QDC MP 4.2 alone. This number would have increased steadily until the termination of the QDC MP 4.2 legislation in late 2012. Given the number of IPRWT in SEQ and across Australia, it was important to investigate the energy implications of these systems.

2.3.1 Research objectives

Developing an understanding of the energy intensity of various pumping configurations across a range of end-use events is essential in order to optimise the design of, and policy for, future IPRWT installations. Based on this overarching goal of this research study, the specific objectives devised were to:

(1) Determine the rate of energy and water usage for the four end-uses supplied by the IPRWT (those being: toilet half flush, toilet full flush, clothes washer, irrigation);
(2) Determine the energy intensity of each water end-use category for each sampled household and the overall study sample;
(3) Compare and discuss energy and water usage as well as energy intensity values for the sampled household and overall study sample.

2.3.2 Methodology

A mixed methods approach was adopted to determine the energy intensity and evaluate the performance of IPRWT pumps through an in-home monitoring study of 19 households spread across Gold Coast City (GCC). Quantitative recording of water and electricity usage was combined with socioeconomic and stock inventory data recorded through participant surveys and interviews.

Prior to the commencement of the full study, a two-week 5-home pilot study was conducted (Talebpour *et al.* 2011). This allowed the verification of the experimental methodology. The methodology was required to reliably disaggregate high resolution water and electricity data into individual events to allow classification under one of four end-uses (toilet full flush, toiler half flush, clothes washer and irrigation events). The pilot study proved successful and therefore the same method was employed for the full study.

2.3.2.1 Sample selection process

Owners and occupants of homes constructed under QDCP MP 4.2 (since 2007) across GCC were engaged to participate in the study. Potential participants were identified through bulk emails, letters and home visits. All potential participants were required to complete an *Intention to Participate* form. Upon completion of the recruitment and consent process, a *Water Audit* was conducted. These processes collected data on:

- Socio-demographics, including the number of occupants, their ages and when the house was occupied;
- Water consumption habits, including typical frequency of use and time of use;
- Stock-inventory, including make and model of clothes washer, toilet(s), irrigation equipment and determination of swimming pool ownership.

These data allowed the suitability of participants to be assessed and gave the opportunity for a wide range of demographics to be selected, such that subsequent data analysis would better reflect the broad range of usage conditions present in GCC homes. Interestingly, despite QDC MP 4.2 legally requiring that at least

100 m^2 of roof area be plumbed to drain into the rainwater tank (QG, 2008), it was identified that a large percentage of properties were noncompliant, with areas ranging from 60–100 m^2.

2.3.2.2 Study sample

In total, 19 households were selected to participate in the study. A brief overview of each household's descriptive information is summarised in Table 2.3. All were owner occupied, with IPRWT containing single speed pumps and automatic switch systems. These composed the overwhelming majority of systems encountered when selecting participants and is expected to widely reflect the population of IPRWT installed in SEQ. The mean household occupancy of the study sample was 3.2 persons.

2.3.2.3 Water and energy data capture

Three modified Actaris CTS-5 high resolution water meters (0.014 L/pulse) and one EDMI Mk7c electricity meter (0.1 Wh/pulse) were installed at each home. The location of meters is shown in Figure 2.1. One smart water meter and wireless data logger were installed at each home's mains water box, to record all mains consumption. The two other smart water meters and one wireless data logger were attached to the rainwater tank system; one smart meter before the tank input switch and one after. This allowed the amount of tank water supplied for an end-use event to be identified. The electricity meter was installed to record the energy consumption of the pump and switch systems.

Two loggers were installed at each house due to the distance between the locations of the mains water meter box and the location of the meters at the tank. The DataCell-R loggers recorded data at 5 second intervals, with daily data transmission occurring through the mobile GPRS network via email to Griffith University's Smart Meter Information Portal (SMIP). This data was then available for download in text format.

2.3.2.4 Data preparation and processing

Before the raw data feeds could be processed, a number of small errors needed to be repaired. These included discontinuities (from logger maintenance down time and clock resets) and multiple logger formats (due to some loggers being replaced with upgraded models). To address these issues, a number of MATLAB (MathWorks, 2012) scripts were written to perform these repairs. Additionally, data filtration was required to separate rainwater tank events from mains-only events, with the former being of primary interest. Additional MATLAB scripts were also written to perform this function.

Following the data preparation stage, a trace analysis was conducted. For this task Trace Wizard (Aquacraft, 1997) was employed to decompose the usage

Table 2.3 Summary of the 19 households selected for participation in this study.

Home ID	Number of occupants			Total land size (m^2)	Pump size (W)	Clothes washer capacity (kg)	Clothes washer type	Toilet flush type	Volume of toilet half flush (l)	Volume of toilet full flush (l)
	Total	Adults	Children							
Home 1	3	2	1	701–1000	770	5 to 7	Front loader	Dual	2.5	5
Home 2	2	2	0	501–700	770	5 to 7	Top loader	Dual	4.5	7
Home 3	6	2	4	1000+	800	5 to 7	Front loader	Dual	4.5	9
Home 4	1	1	0	501–700	770	7+	Top loader	Dual	4	8
Home 5	2	2	0	501–700	770	5 to 7	Top loader	Dual	3	6
Home 6	4	2	2	1000+	800	5 to 7	Top loader	Dual	2.5	4
Home 7	2	2	0	701–1000	Unknown	5 to 7	Front loader	Dual	2.5	7
Home 8	4	3	1	1000+	890	5 to 7	Front loader	Dual	3.5	7

Home 9	2	2	0	501–700	770	5 to 7	Front loader	Dual	3	6
Home 10	5	2	3	701–1000	770	7+	Front loader	Dual	4	7
Home 11	2	2	0	701–1000	770	5 to 7	Front loader	Dual	3.5	6
Home 12	3	2	1	501–700	770	5 to 7	Top loader	Dual	3.5	6
Home 13	4	2	2	701–1000	1100	5 to 7	Top loader	Dual	3	4.5
Home 14	2	2	0	501–700	770	7+	Front loader	Dual	2	5
Home 15	3	2	1	501–700	770	5 to 7	Front loader	Dual	4	8
Home 16	3	2	1	501–700	770	7+	Front loader	Dual	4.5	7
Home 17	4	2	2	501–700	770	5 to 7	Top loader	Dual	3.5	6
Home 18	4	4	0	501–700	770	5 to 7	Top loader	Dual	3.5	7
Home 19	3	2	1	701–1000	770	5 to 7	Front loader	Dual	3	7

information present in the data feed into classified end-use events. The program gives a visualisation of the data feed and allows the creation of templates. These user-created templates contain the characteristics that differentiate one end-use from another (e.g., the duration or flow rate). After a template is created, the program classifies all consumption data in a file based on these characteristics, which is an iterative process that is necessary to attain a high accuracy with the results. The supplementary information that was collected through the stock inventory and socio-demographic surveys plays an important part in this process. Further information on the trace analysis can be found in Willis *et al.* (2009), Beal *et al.* (2011) and Stewart (2011).

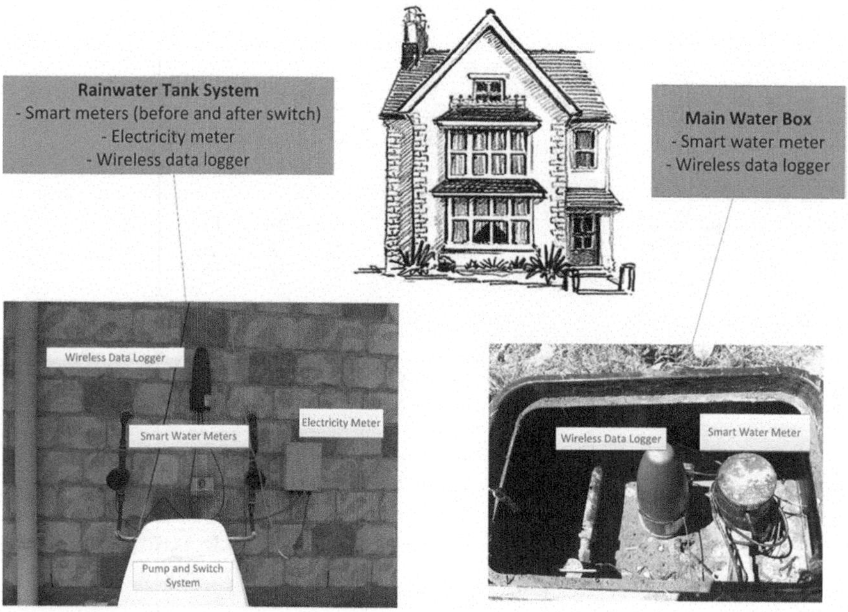

Figure 2.1 Internally plumbed rainwater tank system, meters and logger setup.

The Trace Wizard database files were used to manually extract data on an event by event basis from the original logger files, matching the water usage with its corresponding electricity consumption data. These events were inserted into pre-formatted Excel (Microsoft Corporation, 2010) templates, which allowed detailed statistical analysis. In the master sheet for each home, box and whisker plots of flow rate and electricity consumption were automatically generated to identify outlier events. Parameters were taken from each event, with the data also forwarded to an aggregated population master sheet. Thus, both individual home event populations and total data collected under each end-use were available for analysis.

2.3.3 Results and analysis

2.3.3.1 Rainwater use event sample size

For the data analysis, it was planned to capture 20 events under each end-use from the 19 study homes to give a data population of 1520 events. However, the number of rainwater use events available in any given timeframe depended on household water usage patterns and climatic conditions. Therefore, a total of 1210 events were captured and analysed during the study period.

An abundance of toilet half-flush and full-flush events was present in the data logs for all households, allowing a full component of 20 events to be captured from each home. Participant surveys indicated that many homes rarely irrigated, but it was still thought that 20 events could be captured from each home in a 6 month period. However, during the 4 months of the data collection period, GCC experienced 150% of average rainfall, while during the whole data collection period GCC experienced 128% of average rainfall (ABoM, 2013). This is thought to have significantly reduced the number of irrigation events that took place in many households.

All homes indicated in the participant surveys that they used a clothes washer at least twice per week. Despite this, 20 events could not be found for 6 homes – four of which had no clothes washer events at all. Discussions with residents indicated that this was due to hot washes being the cycle of choice (with IPRWT only supplying the cold source tap). Hand washing was also preferred by some homes.

2.3.3.2 Total sample water end-use results

Table 2.4 displays the mean values for the 4 mandated IPRWT end-uses under QDC MP 4.2. These values have been calculated by aggregating all the events for each end-use and dividing by the total number of events. This, arguably, is the best reflection of how pumps are behaving from an overall sample perspective (rather than taking the mean of the mean of each end-use from each home).

Table 2.4 Mean event characteristics from aggregated event data population.

End-use Category	Tank supplied (l)	SD* for tank supplied water	Pump energy (Wh)	SD* for pump energy	Event duration (s)	Energy intensity (Wh/l)	SD* for energy intensity
Toilet half-flush	3.22	0.88	5.79	1.31	49.8	1.88	0.53
Toilet full-flush	5.84	1.32	9.06	1.46	71.97	1.61	0.36
Clothes washer	70.15	36.51	85.35	39.20	2982.40	1.32	0.46
Irrigation	221.86	124.05	234.37	111.02	1450.34	1.13	0.26

*Standard deviation

The statistical analysis reveals that toilet-half flush events are the most energy intensive. Toilet full-flush events are the next most intensive, followed by clothes washer events, while irrigation events are the least energy intense (or most energy efficient). These figures constitute the first known in-home empirically-derived indicators of pump energy consumption at an end-use level.

In order to understand the determining factors of pump event efficiency (pump energy and flow rate behaviours), a number of indicators were taken from the aggregated event data population, presented in Table 2.5. The immediately obvious trend is the difference between the flow rate (l/s) and the pump energy (Wh/s). While both vary in accordance with energy intensity, the strength of the correlation is dissimilar. The peak and mean electricity (Wh/s) consumption varies less than 10% between the four events types (0.18 Wh/s and 0.19 Wh/s; and 0.15 Wh/s and 0.17 Wh/s, respectively), while the peak and mean flow rates show a similar change in magnitude to the overall energy intensities. The peak rainwater tank supply (l/s) varies by 40% between end-uses (0.14 l/s and 0.21 l/s) and the mean flow rate (l/s) varies by 36% (0.10 l/s and 0.17 l/s). However, overall energy intensity (Wh/l) has a similar 45% range from the most efficient to least efficient end-use (1.80 Wh/l and 1.02 Wh/l).

Table 2.5 Mean end-use event flow rate and energy characteristics.

End-use category	Peak tank supply (l/s)	Mean tank supply* (l/s)	Peak pump energy (wh/s)	Mean pump energy* (Wh/s)	Event duration (s)	Energy intensity (Wh/l)
Toilet half-flush	0.14	0.10	0.18	0.15	49.80	1.88
Toilet full-flush	0.14	0.11	0.18	0.16	71.0	1.61
Clothes washer	0.18	0.13	0.19	0.16	3264.70	1.32
Irrigation	0.21	0.17	0.19	0.17	1453.10	1.13

*Mean is calculated based on non-zero data entries for a given event.

However, it should also be noted that there are homes having the same pump model that have very different energy intensity values. This indicates that there is a range of other factors that are also influencing energy intensity ratings, such as appliance flow rate demands and individual usage habits, which are examined in the subsequent sections.

2.3.3.3 Individual home end-use results

Whilst clear trends emerged in the previous section when considering the captured water end-use events in an aggregated form by end-use, a home-by-home analysis revealed very large variation between systems. The energy intensity from the 380 captured toilet half-flush events ranged from 0.96 Wh/l to 3.65 Wh/l. The results for the average of each home are summarised in Table 2.6.

Table 2.6 Home-by-home results for toilet half-flush events.

Home ID	Average Water consumption (l)	Average electricity consumption (Wh)	Average event duration (s)	Average energy intensity (Wh/l)
Home 1	2.16	4.56	50.00	2.11
Home 2	4.44	6.61	44.25	1.49
Home 3	4.31	6.79	50.00	1.58
Home 4	**4.02**	**4.24**	**30.00**	**1.05**
Home 5	3.21	7.84	59.00	2.44
Home 6	2.67	7.40	44.75	2.77
Home 7	2.27	4.63	41.00	2.04
Home 8	3.35	6.26	35.75	1.86
Home 9	2.23	3.91	32.50	1.75
Home 10	4.17	6.31	45.75	1.51
Home 11	3.18	4.81	36.50	1.51
Home 12	3.49	5.95	37.75	1.70
Home 13	2.80	4.99	32.50	1.78
Home 14	**1.83**	**6.07**	**98.50**	**3.32**
Home 15	4.23	5.59	40.75	1.32
Home 16	4.64	8.51	74.00	1.84
Home 17	2.63	5.97	118.00	2.27
Home 18	3.19	5.75	43.50	1.80
Home 19	2.33	3.80	32.50	1.63
Average	3.22	5.79	49.84	1.88

In general, pumps were able to supply the high volume and high flow rate toilet half-flush events more efficiently than the low volume and low flow rate flushes. To illustrate this point, Home 4 and Home 14 results are highlighted. Home 14 has a water efficient 1.8 l half-flush, which is supplied at a very low flow rate (98.5 s average duration to fill 1.8 l). This low flow rate led to an energy intensity of 3.32 Wh/l on average over 20 events. In contrast, Home 4 has a high volume toilet half flush that is supplied quickly. Hence, the toilet has a high flow rate and can be supplied very efficiently at 1.1 Wh/l. In this particular example, this relationship poses a problem, with water efficient events being relatively energy inefficient due to a low flow rate when using the widely utilised single speed pumps. It should be noted, as illustrated in Table 2.3, that the volumes of half and full-flush toilet (4 and 8 l, respectively) in Home 4 are much higher than the volumes of half and full flush toilets in Home 14 (2 and 5 l respectively).

The total energy intensity of a pump is a function of the pump start-up energy usage, energy used during the pump operation and the water consumption. In this example, as the pump start-up energy requirement was the same due to the same pump model and size being used in both homes, the water consumption and the duration of the pump operation were the defining factors in determining the energy intensity. Therefore, it may be concluded that if for any reason the intensity increases, or decreases, these two factors would directly influence the energy intensity of the pump.

The data collected from each home indicated that toilet flush events are relatively consistent in nature, with the standard deviation of energy intensity for each home less than 0.3 Wh/l. The 380 event sample of full-flush events, summarised in Table 2.7, mirrored the trends found in their half-flush counterparts. The homes with the most and least efficient half-flush also exhibited the most and least efficient full-flush events. Again, flow rate was the primary determinant. Full-flush events were more consistent than half-flush events and the range of the entire sample varied from 0.97 Wh/l to 2.68 Wh/l, while the standard deviation of energy intensity was less than 0.2 Wh/l on average on a home-by-home basis.

Table 2.7 Home-by-home results for toilet full-flush events.

Home ID	Average water consumption (l)	Average electricity consumption (Wh)	Average duration (s)	Average energy intensity (Wh/l)
Home 1	4.49	7.03	73.25	1.56
Home 2	6.57	9.91	65.00	1.51
Home 3	7.77	10.73	76.25	1.38
Home 4	7.48	7.64	50.50	1.02
Home 5	5.33	10.03	61.75	1.88
Home 6	3.85	8.85	54.00	2.30
Home 7	6.86	9.29	69.75	1.35
Home 8	6.69	11.21	59.75	1.68
Home 9	5.02	7.47	55.50	1.49
Home 10	6.69	9.49	67.50	1.42
Home 11	5.43	7.39	52.50	1.36
Home 12	2.98	7.21	55.25	2.42
Home 13	4.55	6.57	40.50	1.44
Home 14	4.75	10.58	123.25	2.22
Home 15	6.90	8.64	60.75	1.25
Home 16	6.90	10.80	91.50	1.56
Home 17	5.24	9.16	150.25	1.75
Home 18	6.48	10.62	75.75	1.64
Home 19	6.88	9.60	65.50	1.40
Average	5.84	9.06	70.97	1.61

With respect to pump type, Home 6 and Home 13 had the same pump model, however, their mean energy intensities for toilet half-flush and full-flush end-uses vary by approximately 40%. This suggests that when examining these popular systems, the appliance water demand characteristics and user habits are very important predictor variables on system energy intensity. For example, a manual adjustment of flow rate by a resident may easily increase (or decrease) the energy intensity of a toilet system with comparison to the same system installed at another location. Similarly, variation of default flow rate settings between manufacturers would affect the energy intensity of the toilet system.

Clothes washer (CW) events were the most time consuming to classify due to the many different usage permutations. Under QDC MP 4.2, clothes washers are installed to take water from both hot water systems (mains supplied) and cold water systems (tank supplied when tank water is available). Hence, depending on the wash cycle selected, an event may be made up of entirely of non-tank supplied hot water or entirely of tank supplied cold water, or a combination of the two. This results in large variability between events, and between homes due to usage habits. CW events were expected to have mean flow rates much higher than toilet events, however, only the peak flow rate was higher on average. Typical clothes washer events were found to be broken up into many short periods of water demand, leading to lower pump efficiencies. A typical CW event segment is shown in Figure 2.2. Talebpour *et al.* (2011) reported that a 118 l cold wash operates at an energy intensity of 1.09 Wh/l. However, the current study has found that the energy intensity of a front loader CW usually is around 20% higher than a top loader clothes washer. The details of events captured from each home are displayed in Table 2.8, with 262 captured events in total.

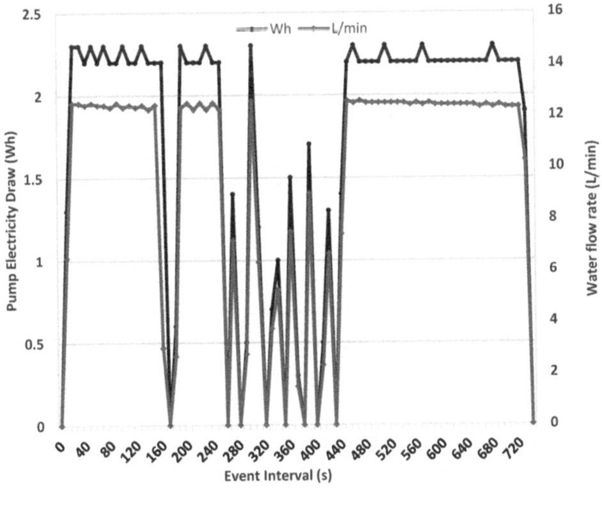

Figure 2.2 Cycle segment of a typical clothes washer (CW) event.

Table 2.8 Home-by-home results for clothes washer water-use events.

Home ID	Average water consumption (l)	Average electricity consumption (Wh)	Average wash duration (s)	Average energy intensity (Wh/l)
Home 2	94.43	94.15	1410.75	1.00
Home 5	155.83	157.00	2189.50	1.01
Home 7	48.19	66.04	4235.50	1.37
Home 8	35.24	57.54	1271.50	1.63
Home 9	43.02	56.30	2823.75	1.31
Home 10	54.41	148.32	1945.25	2.73
Home 11	70.31	91.99	6797.50	1.31
Home 12	51.90	46.85	642.00	0.90
Home 13	126.67	125.74	2074.00	0.99
Home 14	70.93	71.26	1058.25	1.00
Home 15	43.62	49.69	1565.75	1.14
Home 16	91.95	114.74	6039.50	1.25
Home 18	108.97	129.87	2276.00	1.19
Home 19	43.02	45.42	4135.75	1.06
Average	70.15	85.35	2982.40	1.32

Similar to toilet flush events, water efficiency is often at odds with energy efficiency for the CW. There was no discernible correlation between event duration and the number of inflow periods occurring in a whole wash, with the overall end-use events energy efficiency. Flow rate during inflow periods (omitting non-zero data entries) was the major determinant of energy efficiency. The home with the least energy efficient clothes washer events (Home 10) drew small amounts of water over almost the entire wash cycle leading to high mean energy intensity of 2.73 Wh/l. The most energy efficient CW homes (Homes 2, 5, 13, and 14) drew higher volumes of water over short inflow periods, resulting in them being more energy efficient.

Irrigation events are by nature erratic, with a wide range of applications (e.g., car washing, pool filling, garden watering) and end-use appliances (such as hose nozzles, sprinklers, taps). The energy intensity of the 168 captured events ranged from 0.65 Wh/l to 2.5 Wh/l. The wide range of events identified in the sample allowed for the behaviour of pumps under different conditions to be evaluated. The best illustration of the influence of flow rate on pump performance was found by comparing two irrigation events from the same system (Figure 2.3). Event 1 has a large flow rate at 0.174 l/s, which the pump supplies at a median electricity draw of 0.16 Wh/s. Event 2, on the other hand, has a low flow rate of just 0.036 l/s, but the same median electricity draw of 0.16Wh/s, making it nearly 500% less efficient than Event 1. This basic example illustrates the importance of carefully matching irrigation systems with the size of a rainwater tank pump.

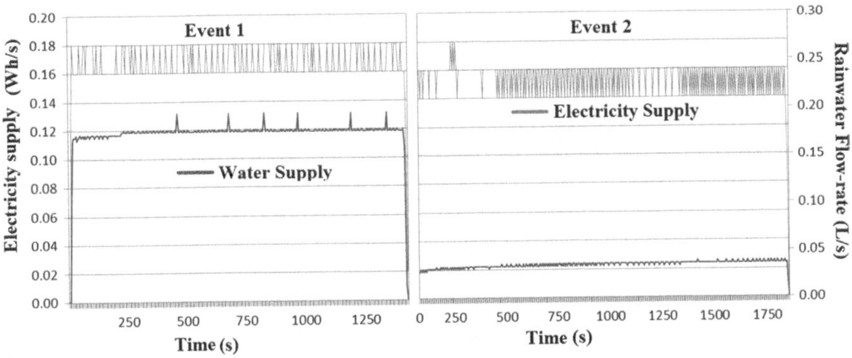

Figure 2.3 Pump performance comparison of two irrigation water-use events taken from the same IPRWT system.

2.4 ALTERNATIVE SUPPLY SPECTRUM COMPARISONS

To put these energy intensity findings into context, Figure 2.4 compares the energy intensity of the four QDC MP 4.2 rainwater end-uses to the energy intensity of pumping and treating centralised water in Brisbane and GCC, using figures from Kenway (2008). Also included in the comparison are the mean costs of reverse osmosis (RO) desalination plants in Australia reported by the CSIRO (Hoang *et al.* 2009).

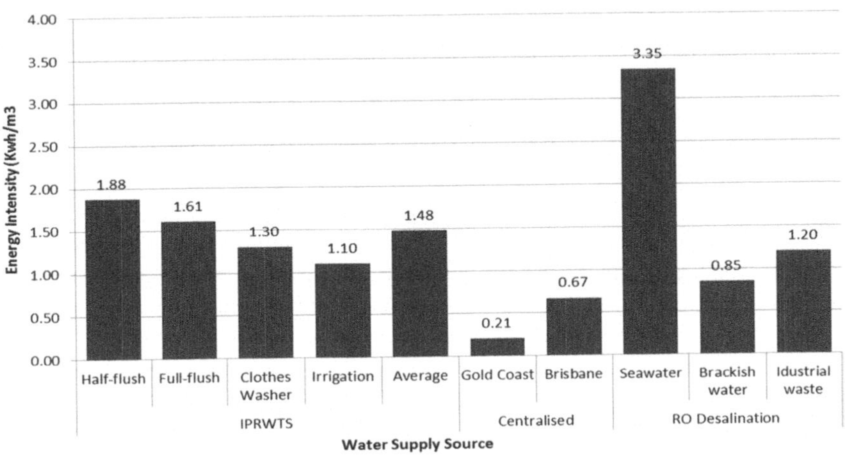

Figure 2.4 Energy intensities of Australian water supply.

It can be seen in Figure 2.4 that the energy cost of supplying water from IPRWT is approximately 2–6 times greater than supplying water from surface reservoirs through centralised networks in SEQ. It should be noted, however, that the Gold Coast value of 0.21 Wh/l is particularly small due to the large pressure head attained from the elevation of the Hinze dam (major supply source) relative to GCC. The Brisbane value is a more representative value for Australia as a whole (Kenway, 2008).

The single speed pump rainwater energy intensity compares well when compared to RO seawater desalination, being greater than twice as efficient. Brackish and industrial effluent RO desalination are 85% and 30% less energy intense than the average rainwater value. It is worth noting however, that in many coastal areas these are limited or non-existent input sources (Hoang *et al.* 2009).

2.5 DISCUSSION AND CONCLUSIONS

The study described in this chapter employed a mixed method approach in determining the energy intensities of pumps supplying water end-uses fed by internally plumbed rainwater tank systems (IPRWT) in South East Queensland, Australia. 19 homes across Gold Coast City were monitored for a period of 6 months, with supplementary data (appliance stock inventories and socio-demographic surveys) collected to enhance data analysis and processing.

From the analysis of the aggregated samples of each water end-use on a home-by-home basis, it has been identified that high-flow rate events have the lowest energy intensity due to the pump system working closer to its optimal range. Consequently, homeowners and occupants could be advised to better match appliances (clothes washers and toilets) and irrigation systems with a suitable pump size and type. High flow rate irrigation events should also be encouraged when used in conjunction with single speed pumps. Methods such as trickle irrigation systems coupled with 770–1100 W single speed pumps would consume energy very inefficiently based on the data captured. Slow leaks are also likely to attract high energy intensities, so occupants should seek to repair any slow leaks that are pump supplied.

The study has highlighted that a rainwater supply system with a single speed pump may be more efficient than another of the same type, simply due to occupant usage habits and water end-use appliance flow rates being better aligned with the installed model of the pump. The subject of future research could be to determine the lowest pump power that could supply all required flow rates for a given home, as this would logically represent the most efficient pump for a system.

The overwhelming majority of systems were found to contain single speed pumps. The findings indicated that toilet half-flush events had the highest variability of energy intensity values between homes (1.05 to 3.32 Wh/l) and also the highest energy intensity at average 1.88 Wh/l. Full-flush toilet events had a tighter range (1.02 to 2.42 Wh/l) and slightly lower energy intensity than half-flush events, with an average 1.61 Wh/l. Toilet flushing had high energy intensities mainly due to the short duration of these events and the flow rate of cistern filling being considerably lower than the optimal pumping flow rate of the single speed pumps. Clothes washer energy intensity values were quite variable (0.90 to 2.73 Wh/l), but on average were lower than the toilet flushing with an average of 1.32 Wh/l. Finally, irrigation events had a wide range (0.65 to 2.5 Wh/l), but the lowest average energy intensity of 1.10 Wh/l. Irrigation event energy intensities were lower, since they typically operated within the optimal operating capacity of installed pumps.

The major underlying factor determining energy intensity was identified as being the rainwater flow rate. Put simply, high flow rate events are more efficient than low flow rate events, with single speed pumps incapable of adjusting their energy consumption in accordance with water demand. The findings of this study will help to refine design guidelines for future IPRWT systems to ensure they function efficiently across all micro-components or end-uses connected to them.

Taking a wider perspective, the energy intensities that were measured are between 2–6 times higher than the current energy intensity for centrally supplied water in SEQ. However, IPRWT energy intensity values compare better to those for RO desalination. This highlights that on an energy intensity basis, IPRWT pumps can supply water more efficiently than some alternative water supply systems in Australia, but not traditional surface water supply through normal distribution networks. Additionally, policymakers should be aware of the vast range of water end-use and mean system energy intensities identified in this 19 home sample. This study has shown that IPRWT systems can operate efficiently under certain ideal conditions, but those conditions are not present for the majority of end-use events. Better IPRWT design guidelines need to be established to ensure that pumps are better aligned to a household's water use characteristics. This will help to ensure that systems installed in the future are operating as efficiently as they can in order to reduce their energy footprint and associated greenhouse gas emissions.

REFERENCES

Aquacraft (1997). Trace Wizard. Boulder, Colorado.

Australian Bureau of Statistics (2007). Environmental Issues: People's Views and Practices. Volume 13. Canberra, Australia.

Beal C., Hood B., Gardner T., Christiansen C. and Lane J. (2008). Energy and water metabolism of a sustainable subdivision in South East Queensland: a reality check. *Proceedings of the Enviro'08 Conference*, Melbourne, Australia.

Beal C., Stewart R., Huang T. and Rey E. (2011). SEQ residential end use study. *Water: Journal of the Australian Water Association*, **38**, 80–84.

Campling P., Nocker L. D., Schiettecatte W., Iacovides A. L., Dworak T., Arenas M., Pozo C., Mat O., Mattheiß V. and Kervarec F. (2008) Assessment of the risks and impact of four alternative water supply options. European Commission – DG Environment, http://www.ecologic.eu/sites/files/project/2013/task_1_report_March_2009.pdf. (accessed 1 February 2014).

Chiu Y., Liaw C. and Chen L. (2009). Optimising rainwater harvesting systems as an innovative approach to saving energy in hilly communities. *Renewable Energy*, **34**, 492–498.

Coombes P., Kuczera G. and Kalma J. (2003). Economic, water quantity and quality impacts from the use of a rainwater tank in the inner city. *Australian Journal of Water Resources*, **7**, 111–120.

CSIRO (2011). Climate Change: Science and Solutions for Australia. CSIRO Publishing, Collingwood.

Cunio L. N. and Sproul A. B. (2009). Low Energy Pumping Systems for Rainwater Tanks. Proceedings of Solar'09, the 47th ANZES Annual Conference, Townsville, Australia.

de Haas D., Lane J. and Lant P. (2011). Life cycle assessment of the gold coast urban water system. Urban Water Security Research Alliance Technical Report No. 52.

Department of Infrastructure and Planning (2009). South East Queensland Regional Plan 2009–2031. Department of Infrastructure and Planning, Queensland Government.

Ferguson M. (2012). A 12-month rainwater tank water savings and energy use study for 52 real life installations. Proceedings of the Ozwater'12 Conference, Sydney, Australia.

Flower D. J. M., Mitchell V. G. and Codner G. (2007). Urban Water Systems: Drivers of Climate Change? Institute for Sustainable Water Resources and Department of Civil Engineering, Monash University, Melbourne, Australia.

Gardner A. (2009). Domestic rainwater tanks: usage and maintenance patterns in south east Queensland. *Water*, **36**, 73–77.

Ghisi E. and de Oliveira S. (2007). Potential for potable water savings by combining the use of rainwater and greywater in houses in southern Brazil. *Building and Environment*, **42**, 1731–1742.

Hall M. R., West J., Sherman B., Lane J. and de Haas D. (2011). Long-term trends and opportunities for managing regional water supply and wastewater greenhouse gas emissions. *Environmental Science and Technology*, **45**(12), 5434–5440.

Hallman M., Grant T. and Alsop N. (2003). Life Cycle Assessment and Life Cycle Costing of Water Tanks as a Supplement to Mains Water Supply. http://www.yvw.com.au/yvw/groups/public/documents/document/yvw1001682.pdf (accessed 1 February 2014).

Hauber-Davidson G. and Shortt J. (2011). Energy consumption of domestic rainwater tanks – why supplying rainwater uses more energy than it should. *Water: Journal of the Australian Water Association*, **38**, 72–76.

Hoang M., Bolto B., Haskard C., Barron O., Gray S. and Leslie G. (2009). Desalination in Australia. Water for a Healthy Country Flagship Report series ISSN: 1835-095X. CSIRO, Clayton, Victoria.

Hood B., Gardner E., Barton R., Gardiner R., Beal C., Hyde R. and Walton C. (2010). Decentralised development: the ecovillage at Currumbin. *Water: Journal of the Australian Water Association*, **37**, 37–43.

Kenway S. (2008). Energy use in the provision and consumption of urban water in Australia and New Zealand. *Water for a Healthy Country National Research Flagship*. CSIRO, Clayton, Victoria.

Marsden Jacob Associates (2007). The Economics of Rainwater Tanks and Alternate Water Supply Options. *Waterlines*. a Report Prepared for ACF, Nature Conservation Council and Environment Australia, Canberra, Australia.

MathWorks (2012). MATLAB. Natick, Massachusetts, U.S.A.

Microsoft Corporation (2010). Microsoft 2010 Professional Suite. Albuquerque, New Mexico.

Parkes C., Kershaw H., Hart J., Sibille R. and Grant Z. (2010). Energy and Carbon Implications of Rainwater Harvesting and Greywater Recycling. Environment Agency, United Kingdom.

Pink B. (2010). 2009–10 Year Book Australia. Australian Bureau of Statistics, Canberra.

Queensland Government (2008). Queensland Development Code Part MP 4.2 Water savings targets. Department of Infrastructure and Planning, Queensland, Australia.

Retamal M., Glassmire J., Abeysuriya K., Turner A. and White S. (2009). The Water-Energy Nexus: Investigation into the Energy Implications of Household Rainwater Systems. Institute for Sustainable Futures, University of Technology, Sydney.

SEWL (2008). Energy Consumption of Domestic Rainwater Harvesting. A Report Prepared by Water Conservation Group for South East Water Limited, Australia.

Stewart R. (2011). Verifying the End Use Potable Water Savings from Contemporary Residential Water Supply Schemes. National Water Commission *Waterlines* Report No. 61, Australian Government, Canberra.

Talebpour M. R., Stewart R. A., Beal C., Dowling B., Sharma A. and Fane S. (2011). Rainwater tank end usage and energy demand: a pilot study. *Water: Journal of the Australian Water Association*, **38**, 97–101.

Tam V., Tam T. and Zeng S. (2010). Cost effectiveness and tradeoff on the use of rainwater tank: an empirical study in Australian residential decision-making. *Resources, Conservation and Recycling*, **54**, 174–186.

Tjandraatmadja G., Pollard C., Sharma A. and Gardner T. (2011). Dissecting rainwater pump energy use in urban households. Science forum and stakeholder engagement: building linkages, collaboration and science quality. Urban Water Security Research Alliance. Brisbane, Queensland.

Umapathi S., Chongb M. N. and Sharma A. K. (2013). Evaluation of plumbed rainwater tanks in households for sustainable water resource management: a real-time monitoring study. *Journal of Cleaner Production*, **42**, 204–214.

Ward S., Butler D. and Memon F. A. (2012). Benchmarking energy consumption and CO_2 emissions from rainwater-harvesting systems: an improved method by proxy. *Water and Environment Journal*, **26**, 184–190.

White I. W. (2009). Decentralised Environmental Technology Adoption: The Household Experience with Rainwater Harvesting. Unpublished PhD, Griffith University, Australia.

Willis R., Stewart R., Panuwatwanich K., Capati B. and Guirco D. (2009). Gold coast domestic water end use study. *Water: Journal of the Australian Water Association*, **36**, 79–85.

World Water Assessment Programme (2012). The United Nations World Water Development Report 4: Managing Water under Uncertainty and Risk. UNESCO, Paris.

Chapter 3

The verification of a behavioural model for simulating the hydraulic performance of rainwater harvesting systems

Alan Fewkes

3.1 INTRODUCTION

In the UK, the majority of the population receives water via a mains network and disposes of wastewater via a piped sewerage system. A number of problems have been linked to centralised systems of water supply and disposal (Pratt, 1995), these include: increasing water demand, resources not located in areas of high demand, increased surface water runoff volumes and high discharge rates due to urban and highway development. The traditional solution to these problems has been the development of new water supplies, distribution networks and flood alleviation schemes.

An alternative and potentially more sustainable strategy is the use of decentralised technologies. For example, the use of planted or green roofs and landscaping results in partial water retention and reduced peak runoff flows into the stormwater sewer network. Stormwater sewer connections may be eliminated completely if techniques of onsite infiltration are used. Rain or stormwater runoff collected from roofs can be used for non-potable applications, potentially reducing the utilisation of potable water. The major application for rainwater utilisation is for WC flushing and garden watering. The benefits include: conservation of water resources, relief of demand on public water supplies and potential attenuation of peak runoff into the storm sewer network (Butler & Memon, 2010).

The concept of rainwater utilisation is not new, but previously social and economic factors have prevented its development and integration within the traditional water supply and disposal system. However, attitudes from industry, government and the public are changing towards demand management measures, which include rainwater utilisation. These changes in attitude are perhaps more evident in other European countries such as Germany, where government subsidies have been used to encourage rainwater utilisation (Konig, 1999).

The contribution rainwater utilisation can make to water conservation is related to non-potable water demand, household occupancy and housing density. The average domestic water consumption in the UK is approximately 150 litres/person/day. WC flushing accounts for about 33% and garden watering 3% of potable water supplied to domestic households (Griggs *et al.* 1996). The volume of runoff depends upon the roof area, which is related to architectural style and density of housing. Based upon an average rainfall of 580 mm/annum and housing densities ranging from 8 to 35 houses/hectare, Pratt (1995) predicts an average runoff of between 59 and 269 litres/house/day. Therefore rainwater utilisation could contribute significantly to the demand for WC flushing and garden watering. A storage tank is required to collect the rainwater runoff because rainfall events occur more erratically than WC flushing and garden watering demands. The capacity of the rainwater storage tank is important both economically and operationally.

This chapter describes the field testing of a rainwater harvesting (RWH) system in the UK and the verification of a model, which simulates its performance as a water conservation device. The model developed in this chapter can also be used to evaluate RWH systems as a method of stormwater control, but this application is not considered in detail in this chapter. The approaches to incorporate a stormwater attenuation allowance within RWH systems are discussed in Chapter 4. The objectives of the field tests were fourfold. Firstly, to monitor and record the rainfall, wind speed and wind direction at each Test Site. The wind conditions were monitored to investigate if there was any correlation between these and the rainwater runoff from the roof. Secondly, to monitor and record the inflows and outflows from the RWH system at each Test Site to determine the volume of mains water conserved per annum. Thirdly, to use the data collected to verify and refine a model that simulates the operation and hydraulic performance of a RWH system. Finally, the sensitivity of the model to the time interval of the input data time series and the method of modelling the rainfall losses is investigated.

3.2 THE RAINWATER HARVESTING SYSTEM AND INSTRUMENTATION

A RWH system with a storage tank of 2032 litres was installed successively into three domestic UK properties and its performance monitored at each location for periods of twelve, eight and six months. The collected rainwater was used only for WC flushing in each of the test properties. Each of the Test Sites was located within the East Midlands area of the UK. The properties at Test Sites 1 and 2 were two storey houses, while at Test Site 3 the property was a bungalow. All of the properties had pitched roofs covered with profiled, granular faced concrete tiles. Rainwater was collected from the whole roof area at each site. The projected roof plan areas were 85 m^2 (Site 1), 57 m^2 (Site 2) and 56 m^2 (Site 3). All of the properties were fitted with 9 litre dual flush WCs.

The system that was monitored is available commercially and uses a pump and accumulator (pump & acc.) to distribute water to the WC (Figure 3.1). Rainwater is collected from the house roof by gravity via a 100 mm diameter downpipe into a polythene tank. A coarse filter fitted into the downpipe ensures debris, such as leaves, does not collect in the tank. An overflow is fitted to the storage tank, which discharges into the household's surface water drain. Water is supplied under pressure using a pump in conjunction with an accumulator. When there is insufficient rainwater, a float switch fitted near to the bottom of the tank activates a magnetic valve, which allows approximately 250 litres of mains water to flow into the collector via a funnel.

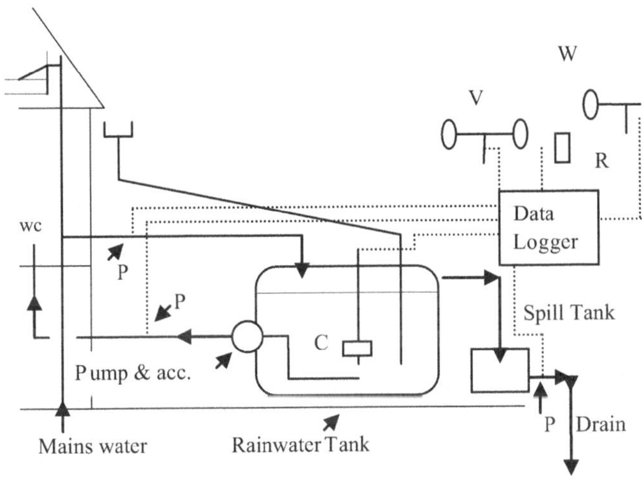

Key
C = Pressure transducer
P = Positive displacement flow meter
R = Tipping bucket rain gauge
V = Vane anemometer
W = Wind direction indicator

Figure 3.1 Rainwater harvesting system and instrumentation.

A schematic diagram of the instrumentation system is included in Figure 3.1. The water flow rate from the RWH system to the WC was measured using a positive displacement flow meter. A data logger recorded the total flow at time intervals of 1 minute at Site 1 and 1 hour at Sites 2 and 3. The inflow of mains make-up water was monitored using the same method. The volume of rainwater inflow from the roof was determined by measuring the level of water in the collection tank at time intervals of 1 minute at Site 1 and 1 hour at Sites 2 and 3 using a pressure transducer. Water overflowing from the system was collected in a 250 litre spill tank. Discharge into the drain was via a 25 mm pipe fitted with a positive displacement flow meter. The data collected was used to determine the percentage of WC flushing water conserved each month.

At each Test Site a weather station was installed to monitor rainfall, wind speed and direction. The weather station was used to quantify runoff losses due to wind effects and absorption by the roofs at each Test Site. A detailed description of the instrumentation and the justifications for the techniques adopted is reported elsewhere (Fewkes, 2004).

3.3 FIELD TESTING RESULTS AND DISCUSSION

The variables measured during the study are identified in Figure 3.2. The performance of a rainwater collector is described by its water saving efficiency *(E_T)*. Water saving efficiency is a measure of how much mains water has been conserved in comparison to the overall demand of the WC and is given by Equation 3.1.

$$E_T = \frac{\sum_{t=1}^{n} Y_t}{\sum_{t=1}^{n} D_t} \times 100 \qquad (3.1)$$

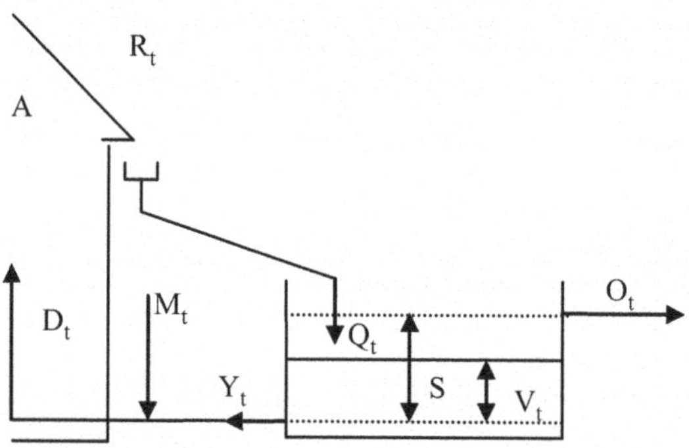

Key

R_t = *Rainfall (mm) during time interval, t*

Q_t = *Rainfall runoff (litres) during time interval, t*

M_t = *Mains make up (litres) during time interval, t*

O_t = *Overflow (litres) during time interval, t*

V_t = *Volume in store (litres) during time interval, t*

Y_t = *Yield from store (litres) during time interval, t*

D_t = *Demand (litres) during time interval, t*

S = *Rainwater tank capacity (litres)*

A = *Roof Area (m^2)*

Figure 3.2 System variables.

where:

t = duration of time interval, for example, minute, hour or a day.
$T = t_1 + t_2 + \text{-----------------} + t_n$ = time period under consideration, for example, a month or a year.

The results for Test Sites 1, 2 and 3 are given in Tables 3.1, 3.2 and 3.3, respectively. (Note: for Site 1, $t = 1$ minute and $T = 1$ month and for Sites 2 and 3, $t = 1$ hour and $T = 1$ month). The performance of the system in terms of its water saving efficiency is given for each of the months the system was monitored. The selection of a monthly monitoring period was arbitrary. The water saving efficiency at Site 1 ranged from 4% for June to 100% for September and February. At Site 2 the minimum water saving efficiency occurred in February with a value of 37% and maximums of 100% occurred during January, March and April. A maximum saving of 100% was achieved at Site 3 during October with the minimum efficiency of 59% being recorded during March. The WC demand was fairly constant at each site. The average WC usage at Site 1 was 6.5 flushes per day per person and the corresponding values for Sites 2 and 3 were 2.7 and 3.8 flushes per day per person.

Table 3.1 Monthly performance indicators including water saving efficiencies (Site 1).

	Monthly total							
Month	**Final vol. (V_T)* (litres)**	**Overflow vol. (O_T) (litres)**	**WC demand (D_T) (litres)**	**Make-up vol. (M_T) (litres)**	**Store yield (Y_T) (litres)**	**Rainwater runoff (Q_T) (litres)**	**Rainfall (R_T) (mm)**	**Water saving efficiency (E_T) (%)**
July	430	0	4951	4298	653	776	16	13.19
August	908	0	5650	4372	1278	1756	26.8	22.62
September	1255	3500	4949	0	4949	8796	110.2	100.00
October	942	673	5071	1537	3534	3894	49.8	69.69
November	441	2746	5134	975	4159	6404	75.4	81.01
December	1153	1903	5970	502	5468	8083	94.8	91.59
January	1830	2244	5417	787	4630	7551	87.4	85.47
February	1020	0	4856	0	4856	4046	50.6	100.00
March	504	0	5493	2070	3423	2907	39.8	62.32
April	323	0	6515	5637	878	697	11	13.48
May	392	0	4778	2299	2479	2548	33.2	51.88
June	485	0	4998	4810	188	281	6	3.76
Total		11,066	63,782	27,287	36,495	47,739	601	57.22

*Volume in storage at end of time period, for example, month.

Table 3.2 Monthly performance indicators including water saving efficiencies (Site 2).

Month	Final vol. (V_T)* (litres)	Overflow vol. (O_T) (litres)	WC demand (D_T) (litres)	Make-up vol. (M_T) (litres)	Store yield (Y_T) (litres)	Rainwater runoff (Q_T) (litres)	Rainfall (R_T) (mm)	Water saving efficiency (E_T) (%)
	Monthly total							
January	822	16.5	2140	0	2140	1308.5	27	100.00
February	387	0	2809	1777	1032	597	8.6	36.74
March	694	0	3228	0	3228	3535	61.6	100.00
April	1766	1942	1767	0	1767	4781	95.3	100.00
May	425	0	3731	505	3226	1885	30	86.46
June	1598	1440	3076	256	2820	5433	105.8	91.68
Total		3398.5	16,751	2538	14,213	17,539.5	328.3	84.85

*Volume in storage at end of time period, for example, month.

Table 3.3 Monthly performance indicators including water saving efficiencies (Site 3).

Month	Final vol. (V_T)* (litres)	Overflow vol. (O_T) (litres)	WC demand (D_T) (litres)	Make-up vol. (M_T) (litres)	Store yield (Y_T) (litres)	Rainwater runoff (Q_T) (litres)	Rainfall (R_T) (mm)	Water saving efficiency (E_T) (%)
	Monthly total							
January	482	223.5	2981.5	260	2721.5	1753	36.6	91.28
February	631	85.5	3196.5	249.5	2947	3181.5	59.4	92.19
March	390	0	3106	1265.5	1840.5	1599.5	30.8	59.26
April	498	0	3575.5	1002.5	2573	2681	50.6	71.96
May	722	432.1	2949	762.5	2186.5	2842.6	62.4	74.14
September	1386	0	3120.5	249	2871.5	2943.5	70	92.02
October	1562	1224.5	3357.5	0	3357.5	4758	97.6	100.00
November	382	0	2786	784.5	2001.5	821.5	23.4	71.84
Total		1965.6	25,072.5	4573.5	20,499	20,580.6	430.8	81.76

*Volume in storage at end of time period, for example, month.

Domestic water usage in the UK has been researched by various academics, for example, Thackray *et al.* (1978) and Butler (1993). Butler's survey estimated the average WC usage in a household was 3.7 flushes per day per person, which is in close agreement with Thackray's figure of 3.3 flushes per day per person. WC usage at Test Site 1 was higher than expected. The high usage rate may in part be attributable to the downstairs WC, which usually required at least two flushes to clear the WC pan.

In terms of losses, rainfall loss during collection occurs due to absorption by the roofing material and wind effects around the roof. The rainfall loss was modelled using an initial depression storage loss (E) with a runoff coefficient (C_f) (Pratt & Parkar, 1987). The model is of the general form:

$$Q_{Ti} = \sum_{t=1}^{n} Q_t = \left(\sum_{t=1}^{n} R_t \cdot A \cdot C_f \right) - E \tag{3.2}$$

where:

$T_i = t_1 + t_2 + \text{------------------} + t_n$ = time period for rainfall event, i.

Therefore

$$Q_{Ti} = (R_{Ti} \cdot A \cdot C_f) - E \tag{3.3}$$

where Q_{Ti} is rainwater runoff volume during rainfall event i (litres), T_i is duration of rainfall event i (minutes), E is depression storage loss (litres), C_f is runoff coefficient, R_{Ti} is rainfall during rainfall event i (mm) and A is the projected plan roof area (m^2). Other variables are as previously defined. It is worth noting that E can also be expressed in mm of rainfall by dividing the depression loss by the collection area.

Linear regression analysis was used to produce the rainfall loss parameters for each Test Site and the results are summarised in Table 3.4. The values of E were 0.21, 0.12 and 0.21 for Sites 1, 2 and 3 respectively.

Table 3.4 Rainfall loss parameters.

Test site	Number in data set	Coeff. of determination, r^2	Depression storage loss, E (mm)	Runoff coefficient, C_f
1	34	0.995	0.21	1.04
2	22	0.969	0.12	0.95
3	34	0.96	0.21	0.93

At Site 1 the value of C_f was 1.04, which was high as compared to values of 0.95 and 0.93 for Sites 2 and 3 respectively. The high value was probably attributable to an area of vertical walling adjacent to a single storey construction covered with a mono pitched roof that abutted the front elevation of the property. Pratt and Parkar (1987) obtained a runoff coefficient of 0.987 and a depression storage loss of 0.32 mm for a roof sub-catchment of five bungalows. The runoff coefficients and depression storage losses for the present study are comparable with Pratt and Parkar's values. An alternative approach is to use an overall runoff coefficient, which is estimated using the relationship:

$$C_{fo} = \frac{Q_T}{R_T A} \tag{3.4}$$

where Q_T, R_T and A are as previously defined.

The value of Q_T was equated to the total volume of roof rainwater runoff and R_T to the total rainfall during the trial period at each respective site. The overall runoff coefficients (C_{f0}) for Sites 1, 2 and 3 were 0.93, 0.93 and 0.86 respectively.

The correlation between rainwater runoff and both wind speed and direction was also investigated. Data collected from both the weather station and the collection system were analysed. The correlation between rainwater runoff, wind speed and direction was very weak with values of the coefficient of determination (r^2) ranging between 0.041 and 0.243. Consequently, it can be concluded that at each Test Site the wind speed and direction did not significantly influence the amount of rainwater collected.

3.4 MODELLING SYSTEM PERFORMANCE

McMahon and Mein (1978) identified three general types of reservoir sizing models, namely: critical period, Moran and behavioural models. Critical period methods identify and use sequences of flows where demand exceeds supply to determine the storage capacity. The sequences of flows or time series used in this method are usually derived from historical data. Moran-related methods are a development of Moran's theory of storage (Moran, 1959). A system of simultaneous equations is used with this method to relate reservoir capacity, demand and supply. The analysis is based upon queuing theory, which models or predicts queue lengths and waiting times for a particular service, for example the length of the queue and waiting time for a bus. Moran applied this theory to predict the likely volume of water in a store or reservoir during any time interval. Behavioural models simulate the operation of the reservoir with respect to time by routing simulated mass flows through an algorithm that describes the operation of the reservoir. The operation of the rainwater collector will usually be simulated over a period of years. The input data, which is in time series form, is used to simulate the mass flows through the model and will be based upon a time interval of either a minute, hour, day or month. A behavioural model was used to simulate the performance of the RWH system reported in this chapter because of its inherent adaptability.

The data collected (refer to the previous section) was used to assess the desirable characteristics of a RWH sizing model. The derived RWH sizing model consists of two parts:

- Provision of rainwater supply and WC demand patterns or time series;
- Simulation of system operation.

The rainfall and WC usage data collected during the monitoring periods were used as input into the system simulation model. The algorithm for the model used a yield after spillage (YAS) operating rule (Jenkins *et al.* 1978):

$$Y_t = \min(D_t, V_{t-1}) \qquad (3.5)$$

$$V_t = \min(V_{t-1} + Q_t, S) - Y_t \qquad (3.6)$$

The variables are as previously defined.

The YAS operating rule assigns the yield as either the volume of rainwater in storage from the preceding time interval or the demand in the current time interval, whichever is the smaller. The rainwater runoff in the current time interval is then added to the volume of rainwater in storage from the preceding time interval with any excess spilling via the overflow and then subtracts the yield.

The sensitivity of the rainwater collection sizing model to: i) the time interval of the rainfall and WC time series; ii) the magnitude of the runoff coefficient; and iii) the nature of the WC usage time series is investigated in subsequent sections of this chapter.

3.5 VERIFICATION OF THE RAINWATER HARVESTING SYSTEM MODEL

The correlations between the monthly modelled values of E_T, V_T, O_T and M_T and the corresponding measured values at each site were determined. For example, the predicted values of E_T at Site 1 were plotted against the respective measured values. A straight line was fitted to the data points using linear regression. The intercept of the straight line was arbitrarily set to zero before determining the gradient *(m)* of the line and the coefficient of determination *(r^2)*. The values of m and r^2 for E_T, V_T, O_T and M_T at Sites 1, 2 and 3 are given in Tables 3.5, 3.6, and 3.7, respectively.

Table 3.5 Correlations between monitored and modelled values of performance indicators at Site 1.

Model time interval	Model runoff coefficient	Percentage conserved		Final vols.		Overflow		Make-up	
		m	*r*²	*m*	*r*²	*m*	*r*²	*m*	*r*²
Hourly	$y = 1.04x - 0.21$	1	0.98	0.99	0.89	1.03	0.99	0.95	0.98
Hourly	$y = 0.93x$	1.01	0.97	1	0.87	0.88	0.99	0.95	0.98
Daily	$y = 1.04x - 0.21$	1.01	0.98	0.89	0.93	1.02	0.99	0.95	0.98
Daily	$y = 0.93x$	1.02	0.97	0.92	0.91	0.85	0.98	0.94	0.98
Daily	$y = 0.93x$ & Av Flush	0.88	0.7	1.03	0.95	0.83	0.95	0.83	0.92

Note: y = Rainfall runoff (mm)
x = Rainfall (mm)

The values of m and r^2 for E_T range between 0.98–1 and 0.83–0.98, respectively. The largest range of m and r^2 values are between 0.95–1.14 and 0.86–0.98, respectively and are associated with M_T. The lowest value of r^2 is linked to Site 2 and the modelled value of M_T (Tables 3.5–3.7). The results of this analysis indicate a YAS model based on an hourly time interval using an initial depression storage loss with a runoff coefficient accurately simulates the performance of the field tested 2032 litre rainwater collection system (e.g., Model Time Interval 'Hourly' and Model Runoff Coefficient '$y = 1.04x - 0.21$' in Table 3.5).

Table 3.6 Correlations between monitored and modelled values of performance indicators at Site 2.

Model time interval	Model runoff coefficient	Percentage conserved		Final vols.		Overflow		Make-up	
		m	r^2	*m*	r^2	*m*	r^2	*m*	r^2
Hourly	$y = 0.95x - 0.12$	0.98	0.91	0.98	0.96	1.02	1	1.04	0.86
Hourly	$y = 0.93x$	0.99	0.98	0.99	0.98	1.11	1	1.02	0.95
Daily	$y = 0.95x - 0.12$	0.97	0.95	0.94	0.95	1.05	0.99	0.98	0.99
Daily	$y = 0.93x$	0.97	0.98	0.94	0.96	1.06	1	1.02	0.95
Daily	$y = 0.93$ & Av Flush	1.05	0.93	0.99	0.98	0.82	0.91	0.77	0.89

Note: y = Rainfall runoff (mm)
x = Rainfall (mm)

Table 3.7 Correlations between monitored and modelled values of performance indicators at Site 3.

Model time interval	Model runoff coefficient	Percentage conserved		Final vols.		Overflow		Make-up	
		m	r^2	*m*	r^2	*m*	r^2	*m*	r^2
Hourly	$y = 0.93x - 0.21$	1	0.83	1.01	0.96	1.09	0.99	1.14	0.87
Hourly	$y = 0.87x$	1.02	0.83	1.07	0.96	1.23	0.98	1	0.84
Daily	$y = 0.93x - 0.21$	0.99	0.82	0.96	0.93	1.03	0.96	1.19	0.88
Daily	$y = 0.87x$	1.01	0.81	1.07	0.96	1.17	0.99	1	0.84
Daily	$y = 0.87x$ & AvFlush	1.02	0.83	0.96	0.9	1.05	0.99	1.18	0.81

Note: y = Rainfall runoff (mm)
x = Rainfall (mm)

3.5.1 Time interval sensitivity

The sensitivity of the RWH system to the time interval of the input time series has been investigated by other researchers. For example, Heggen (1993) demonstrated that daily time series' result in more accurate simulation of system performance than either weekly or monthly time series. More recently, Coombes and Barry (2007) used a sub-hourly time interval to simulate the performance of RWH systems located in various parts of Australia.

In the present study, the accuracy of models using daily time intervals compared to hourly time intervals was investigated. The sensitivity of the model to the time interval (t) of the input WC time series was investigated using a daily time interval YAS operating algorithm. The loss variables E and C_f were set to the same values as used in the hourly model (those being Model Time Interval 'Daily' and Model Runoff Coefficient '$y = 1.04x - 0.21$' in Table 3.5). The correlations between the monthly modelled values of the performance indicators (E_T, V_T, O_T, and M_T) and the corresponding measured values at Sites 1, 2 and 3 are given in Tables 3.5, 3.6 and 3.7, respectively.

The values of m and r^2 at Site 1 range between 0.89–1.01 and 0.98–0.93, respectively (Table 3.5). The lowest values of m and r^2 are associated with the final volumes. At Site 2, m varies between 0.94–1.05, whilst the limits of r^2 are 0.95–0.99. Again the lowest values are associated with V_T (Table 3.6). The ranges of m and r^2 at Site 3 are 0.96–1.19 and 0.82–0.96, respectively. The high value of m is related to M_T and the low value of r^2 to E_T (Table 3.7). These results indicate a YAS model with a time interval of a day produces results comparable to the hourly model and accurately simulates the performance of the field tested 2032 litre RWH system.

3.5.2 Rainfall loss sensitivity

The sensitivity of both the hourly and daily time interval models to rainfall losses was investigated by using an overall runoff coefficient (Equation 3.4) as opposed to an initial depression loss with runoff coefficient. The values of the overall runoff coefficients were 0.93 (Site 1), 0.93 (Site 2) and 0.87 (Site 3) (Section 3.3). The correlations between the monthly predicted values of the performance indicators and the corresponding measured values are given in Tables 3.5, 3.6 and 3.7. The values of m and r^2 for the daily model, range between 0.85–1.17 and 0.81–1.0, respectively. The ranges of m and r^2 for the hourly model are 0.88–1.23 and 0.83–1.0, respectively. Generally the correlation analysis indicated that the values of E_T and V_T are more accurately modelled than M_T and O_T. The use of an overall runoff coefficient appears justified in either the hourly or daily time interval models.

3.5.3 WC demand sensitivity

The daily WC demand time series used as input data in the respective models for each site were replaced with an appropriate average daily WC demand. The average demands used were 175.51 litres/household/day (Site 1), 97.2 litres/household/day (Site 2) and 102.64 litres/household/day (Site 3) in conjunction with overall rainfall coefficients of 0.93, 0.93 and 0.87 for Sites 1, 2 and 3, respectively.

The correlations between the modelled performance indicators and the measured values are given in Tables 3.5, 3.6 and 3.7 for Sites 1, 2 and 3, respectively. The values of m are between 0.77–1.18, whilst the range of r^2 is 0.7–0.99. Compared to the other models, the incorporation of average constant demand patterns and overall rainfall coefficients results in the least accurate modelling of the performance indicators. However, the correlation analysis does indicate that the overall integrity of this model type has been retained and could be used for the sizing of RWH systems.

3.6 DESIGN CURVES

The behavioural model was used to assess the performance of a RWH system in terms of its water saving efficiency. Average daily flushing demand data and fifteen years of historical daily rainfall data for eleven different UK locations were used as input time series to the system simulation model described in the previous section. A set of RWH system performance curves for each of the geographic locations was developed. From the location-specific curves, a set of average curves were determined, which have been shown to be sufficiently accurate for estimating RWH system performance in the UK. These curves are illustrated in Figure 3.3 (Fewkes & Warm, 2000).

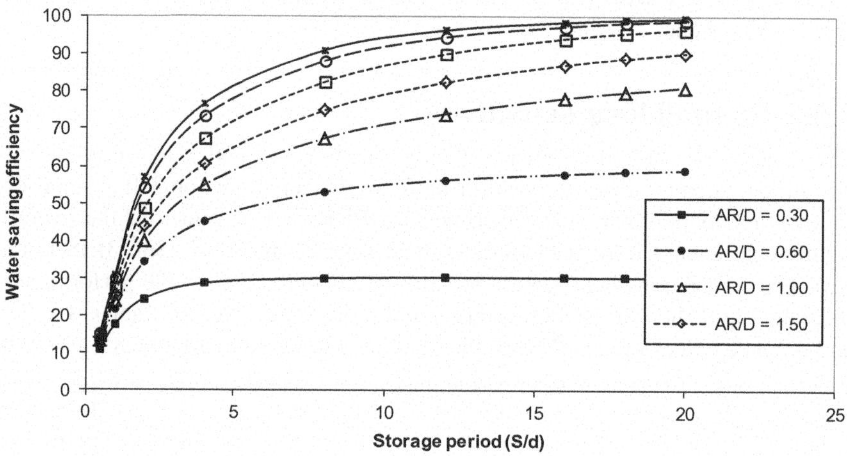

Figure 3.3 Average water saving efficiency versus storage period (AR/D = 0.3–100) for RWH systems in the UK.

The water saving efficiency for a particular combination of roof area, rainfall and demand can be determined from Figure 3.3 using the demand fraction (AR/D) and the storage period (S/d). The demand fraction is a dimensionless ratio and the storage period is expressed in days where *d* is the average daily demand in litres, all other variables areas previously defined (Section 3.3). The curves are a powerful design tool that are used to determine the storage capacity required to achieve a desired level of water conservation. From a practical viewpoint, when using the curves to determine the storage volume of a RWH system, if the system design point falls on the near horizontal portion of any of the curves the tank size can often be reduced, resulting in cost savings, but only a small decrease in system effectiveness.

Conversely, if the system design point lies on any of the steeper parts of the curves, a small increase in storage volume results in a large increase in system performance. The curves provide a valuable design aid for the accurate and therefore economic sizing of RWH systems. For example, a four-person household with a roof area of 100 m^2 situated in a relatively low rainfall zone (600 mm per annum) could conserve approximately 65% of their non-potable water demand using a 2000 litre storage tank. The same household in a relatively high rainfall zone (1200 mm per annum) could conserve approximately 80% of their non-potable water demand using a 2000 litre tank.

3.7 DISCUSSION

The collection of rainwater from roofs, its storage and subsequent use is a simple method of reducing the demand on both public water supplies and waste treatment facilities. The capacity of the rainwater storage tank is important because it affects both system and installation costs. This chapter has described the field testing of a commercially available RWH system and the verification of a model that simulates system performance.

The performance of a RWH system was successfully monitored in three UK properties for periods ranging from six to twelve months. The flows of both rainwater and mains make-up water into and out of the system were measured and logged. A weather station adjacent to each Test Site was used to monitor rainfall, wind speed and wind direction. The average water saving efficiency at Sites 1, 2 and 3 was 57%, 85% and 82%, respectively. The lower system efficiency at Site 1 was attributable mainly to the high WC flushing demand at this test property. The average WC flushing demands at Sites 1, 2 and 3 were 6.5, 2.7 and 3.8 flushes/person/day, respectively. Previous research (Thackray *et al.* 1978 and Butler, 1993) estimated average WC usage between 3.3–3.7 flushes/person/day. The WC demand at Site 3 was in agreement with previous studies, the demand at Site 2 was low and at Site 1 the demand was high. The average monthly rainfall levels were 50.1 mm/month (Site 1), 54.7 mm/month (Site 2) and 53.9 mm/month (Site 3). The fifty year average monthly rainfall for the East Midlands area within which the test properties were situated is 52.8 mm/month.

The rainfall loss parameters for each roof at the Test Sites were investigated using two modelling approaches. Firstly using an initial depression storage loss with a runoff coefficient and secondly, using an overall runoff coefficient. The values of the initial depression storage loss were 0.21 mm (Site 1), 0.12 mm (Site 2) and 0.21 mm (Site 3) and the corresponding runoff coefficients were 1.04 (Site 1), 0.95 (Site 2) and 0.93 (Site 3). These values are in general agreement with the rainfall loss parameters determined by Pratt and Parkar (1987), who determined a runoff coefficient of 0.987 and a depression storage loss of 0.32 mm for a roof sub-catchment of five bungalows. The values of the overall runoff coefficient determined in this study were 0.93 (Site 1), 0.93 (Site 2) and 0.86 (Site 3).

The model verified in this chapter is a behavioural model, which simulates the operation of the RWH system's storage tank with respect to time by routing simulated mass flows through an algorithm that describes the operation of the store. The input data in time series form is used to simulate the mass flow through the model based upon a time interval of either an hour or day. The model was used to predict system performance for different combinations of roof area, demand, storage volume and rainfall level. A sensitivity analysis was used to identify the essential characteristics of a RWH sizing model. Similarly, rainfall losses during collection were quantified and incorporated into the model. Finally, a series of curves were presented based upon the verified model, which enables the performance of RWH systems to be predicted in the UK.

3.8 CONCLUSIONS

The results from the field tests described in this chapter indicate a model using a daily time interval time series can be used to accurately predict rainwater harvesting system performance. The use of hourly time series is not necessary to determine the percentage of WC flushing water conserved. The daily RWH sizing model with a YAS operating rule can be used as a basis against which other models can be evaluated. The form of the WC demand time series does not have to be defined for accurate modelling; average usage data is satisfactory. However, this observation may not be universally applicable to all RWH systems. Demand patterns that exhibit significant daily variance will potentially require more precise modelling.

The incorporation of rainfall losses into a RWH sizing model is also necessary if the systems' performance is to be accurately assessed. The rainfall loss parameters for the collection areas in this study were modelled using an initial depression storage with constant proportional loss model. A simplified model using only a constant proportional loss or runoff coefficient was demonstrated to produce acceptable results. Finally, the amount of rainwater collected was not found to be significantly affected by wind speed and direction, for this particular study.

REFERENCES

Butler D. (1993). The influence of dwelling occupancy and day of the week on domestic appliance wastewater discharges. *Building and Environment*, **28**(1), 73–79.

Butler D. and Memon F. A. (2006). Water Demand Management. IWA Publishing, London, ISBN 1843390787.

Coombes P. J. and Barry M. E. (2007). The effect of selection of time steps and average assumptions on the continuous simulation of rainwater harvesting strategies. *Water Science and Technology*, **55**(4), 125–133.

Fewkes A. (2004). The modelling and Testing of a Rainwater Catchment System in the UK. Unpublished PhD Thesis, Nottingham Trent University, UK.

Fewkes A. and Warm P. (2000). Method of modelling the performance of rainwater collection systems in the United Kingdom. *Building Services Engineering Research and Technology*, **21**(4) 257–265.

Griggs J. C., Shouler M. C. and Hall J. (1996). Water conservation and the built Environment. In: 21AD: Water Architectural Digest for the 21st Century. Oxford Brooks University, UK.

Heggen R. J. (1993). Value of daily data for rainwater catchment. *Proceedings of the 6th IRCSA Conference*, 1–6 August, Nairobi, Kenya.

Jenkins D., Pearson F., Moore E., Sun J. K. and Valentine R. (1978). Feasibility of rainwater collection systems in California. Contribution No. 173 (Californian Water Resources Centre), University of California.

Konig K. W. (1999). Rainwater in cities: a note on ecology and practice. In: T. Inoguchi, E. Newman and G. Paoletto (eds), Cities and the environment: New approaches for Eco-Societies. United Nations University Press, Tokyo, pp. 203–215.

McMahon T. A. and Mein R. G. (1978). Reservoir Capacity and Yield. Developments in Water Science 9, Elsevier, Amsterdam.

Moran P. A. P. (1959). The Theory of Storage. Methuen, London.

Pratt C. J. (1995). Use of porous pavements for water storage. *Proceedings of the 7th IRCSA Conference,* June 21 – 25, Beijing, China.

Pratt C. J. and Parkar M. A. (1987). Rainfall loss estimation on experimental surfaces. *Proceedings of the 4th International Conference on Urban Storm Drainage.* Lausanne, Switzerland.

Thackray J. E., Cocker V. and Archibald G. (1978). The Malvern and Mansfield studies of water usage. *Proceedings of the ICE*, **64**(1), 37–61.

Chapter 4

Rainwater harvesting for domestic water demand and stormwater management

Richard Kellagher and Juan Gutierrez Andres

4.1 INTRODUCTION

It is now widely recognised that water is a scarce and precious resource in most places in the world. The impact of climate change, along with population growth, will continue to make this an increasing problem. Rainwater harvesting (RWH) for domestic use has been in practice for centuries. Nowadays, it is still promoted and widely used in situations where the infrastructure for water supply is poor or does not exist. However, where potable water networks exist and provide a reliable supply, RWH systems are generally seen as offering marginal benefits and therefore there is limited emphasis in promoting their use in the developed world.

The general position taken by drainage design professionals is that RWH tanks cannot be assumed to have sufficient storage available during an extreme rainfall event to contribute meaningfully to the management of stormwater runoff. However, with the growing awareness that stormwater is a potentially valuable resource (if it can be shown that it can also contribute significantly to flood protection and pollution reduction of streams and rivers), RWH will be seen in a completely new light. This chapter aims to demonstrate just this; that RWH cannot only save water, but that it also has the potential for controlling stormwater runoff. This chapter provides an overview of:

- The importance of understanding the relationship between the runoff yield and the demand for non-potable water in using RWH for stormwater management;
- The uncertainties associated with estimating yield, demand and the performance of the storage tanks in meeting the stormwater control requirements;

- A design methodology for providing stormwater control through the use of RWH tanks;
- Application of the methodology to a pilot study to demonstrate the effectiveness of the stormwater control design method; and
- The benefits of using an actively managed RWH system.

4.1.1 Types of RWH

Figure 4.1 summarises the different types of RWH systems that can be designed. The main distinction between the different systems (defined in Figure 4.1 as 'RWH objective') is whether they are intended for water saving only, or for water saving and stormwater control. Water saving systems are effectively standard RWH systems that are usually based on simple design rules. The most common of these is that the tank is sized based on the smaller of two values: 5% of the annual demand or 5% of the annual yield. There is little difference in the design approach whether it is sized for one or more than one property. These systems are not discussed further in this chapter.

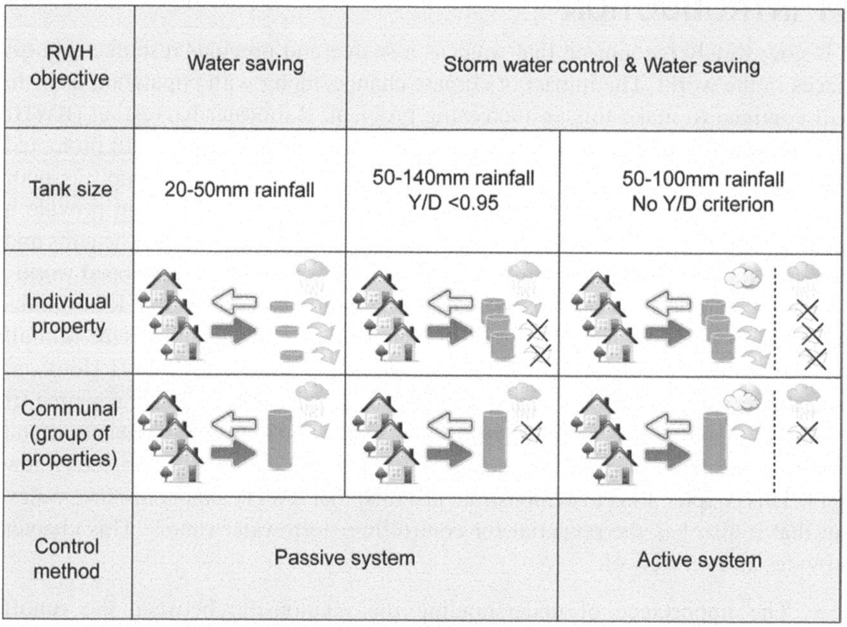

Figure 4.1 Types of RWH system.

Stormwater control RWH systems are designed specifically for control of stormwater runoff, but have the same water saving capability. They are sized to specifically address a certain storm rainfall depth. These stormwater control systems (defined in Figure 4.1 as 'Control method') can be 'passive' or 'active'. The second row in Figure 4.1 provides an indication of the basis for sizing tanks depending on the rainwater objective and control method. The third row of Figure 4.1 shows the performance of RWH systems that are provided for individual properties. Water saving systems are not designed to control stormwater runoff, so a large rainfall event will probably result in the overflow coming into operation. The passive RWH stormwater control system is designed to prevent overflow occurring for an extreme event of a certain size. However, a proportion of houses cannot be assumed to control stormwater runoff, which has implications for the design of the drainage system serving the development (i.e., explained further, later in the chapter). In contrast, active control systems can prevent runoff from all properties. The fourth row of Figure 4.1 shows schematically the same aspects on overflow performance for a single RWH system serving a group of houses. These RWH systems operate in the same way except the passive control RWH systems can be designed with the assumption that the overflow will not come into operation when storing runoff from the design storm (this is explained later in the chapter).

Nearly all RWH systems are built to work on a 'passive' basis; where the water level in the storage tank is purely a function of the demand and the runoff yield. However, passive stormwater control systems need to meet certain design criteria in order to control the runoff. There are two key criteria:

(i) The demand must be regular and fairly well quantified;
(ii) The demand (D) must be greater than the average yield (Y) from the collection surface, or the tank will often be full. In practical terms this is expressed as:

$$\frac{Y}{D} < 0.95 \tag{4.1}$$

An active system, as opposed to a passive system, actively manages the volume of water within the storage tank so that the design storm event can be stored when it takes place. In this case, the two criteria that need to be complied with for passive systems do not need to apply. This type of system and its advantages are discussed at the end of the chapter (Section 4.6).

The final distinction between systems is whether the RWH system is designed to serve an individual property or a group of properties. A RWH system for a group of properties is designed using the same criteria as for one property. However, it has significant advantages in terms of its stormwater control performance, which is explained later in Section 4.4.5.

4.1.2 The background research

This chapter is based on research into RWH systems conducted by HR Wallingford (2012). Further information on the research can be found in the report.

4.2 UNCERTAINTIES ASSOCIATED WITH DESIGNING RWH TANKS FOR STORMWATER CONTROL

The water retained in the tank of a RWH system is a function of its recharge rate and the water demand. The recharge rate is a function of the rainfall events through the year and the contributing surface area. The demand for the water is a function of the frequency of use of non-potable water-based appliances, which is strongly linked to the occupancy in a dwelling. The effect of these two aspects and their relationship (the Yield/Demand ratio) is illustrated in Figure 4.2. *Y/D* for a range of storage tanks is illustrated in the figure, from a low *Y/D* ratio that shows significant spare storage is available for most of the time, to a high *Y/D* value (greater than 1.0), where there is virtually no spare storage at any time. The Yield and Demand values can be assessed on an annual average basis, as long as the average yield and demand through the year does not vary significantly. This means that where the *Y/D* ratio is lower than 1.0, there is normally spare storage available in the tank.

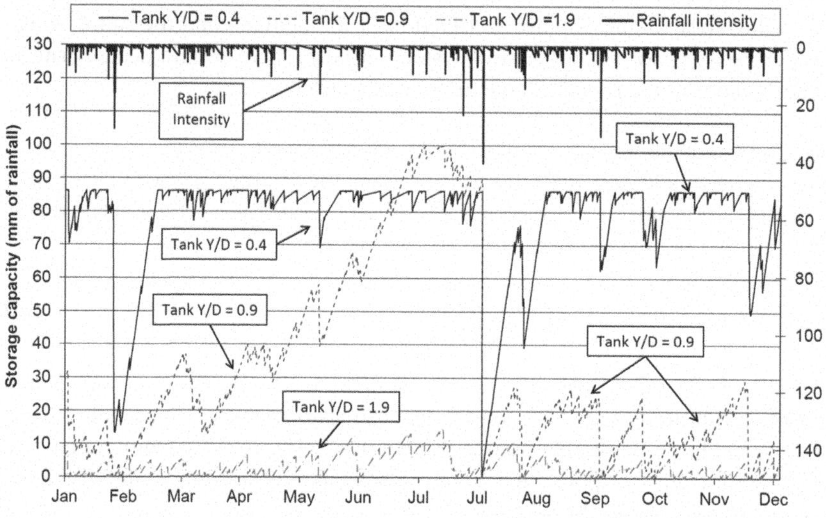

Figure 4.2 Typical storage availability through the year (shown as depth of rainfall) for 3 RWH systems with *Y/D* ratio ranging from 0.4 to 1.9 (HR Wallingford, 2012).

Yield can be calculated relatively accurately, as the roof area being drained is known, therefore uncertainty is a function of the random nature of rainfall events in size and frequency (unless the collection surface is unusual, such as a green roof). However, there is generally much greater uncertainty associated with Demand. The use of rainwater for internal domestic application is normally limited to toilet flushing and washing machines to avoid the potential health risks associated with RWH systems. The water use is generally assumed to be closely related to the number of people in a property and their habits, along with the hydraulic characteristics of the appliances used. Consequently water use can vary significantly. Even more important is the uncertainty associated with the number of people in a house, which will vary due to both ownership changes and the working and leisure activities of the occupants. The basis by which the occupancy can be estimated best is by using the number of bedrooms together with supporting statistics on the average occupancy for each category of house type. These two uncertainties (of Yield and Demand) are addressed in two different ways:

(i) The Yield over a period of time is a function of the variability of event size and inter-event dry period. This uncertainty is catered for by increasing the available storage in the tank and this amount has to increase as the *Y/D* ratio rises towards 1.0;
(ii) Demand varies the *Y/D* ratio based on the occupancy of the property. This means that the statistics of the mean and standard deviation of occupancy have to be used in sizing the systems and also calculating the effectiveness of the stormwater control achieved.

The probability distribution of the occupancy is assumed to be a binomial distribution (Royle, 2004), assuming a minimum occupancy of one person in a property and an upper bound constraint of 2 people per bedroom. This assumption has not been tested and requires confirmation. Whether this distribution is correct or not, it is important to utilise the occupancy distribution statistics to establish the likelihood of the number of people in a property of each category. The variability of water consumption due to variation in use of water by individuals has not been taken into account, although this could be added, based on the same principle.

4.3 THE STORMWATER SIZING METHODOLOGY

The formula for sizing a RWH tank is shown in Equation 4.2. A detailed explanation can be found in the British Standard for RWH (BS 8515: 2009) and the HR Wallingford (2012) report.

$$R_d = sP_{50} - A_d + \frac{(V_{YR} - 1)}{A} \cdot CP_{50} \cdot 1000 \qquad (4.2)$$

where R_d is the net rainfall (measured in millimetres of rainfall depth) passing into the tank from the design storm event. The calculation of R_d should take into consideration any loss elements that occur in generating runoff as well as other processes such as filtration; sP_{50} is the average amount of storage available in a 1 m^3 storage tank for 50% of the time (measured in millimetres of rainfall depth).

The value for sP_{50} provides an estimate of the rainfall depth that could be catered for by providing a tank of 1 m^3; it is a function of *Y/D*. CP_{50} is a coefficient that accounts for the effective proportion of the storage provided for a tank larger than 1 m^3; it is a function of *Y/D*. The sP_{50} and CP_{50} values are based on storage available for 50% of the time. Therefore there needs to be an adjustment to the volume of storage to provide a higher level of certainty for storing all the runoff of an event equal to the design depth. A_d is an additional rainfall depth allowance to cater for the uncertainty of storage availability for the design storm event (a function of *Y/D*; measured in millimetres of rainfall depth). V_{YR} is the stormwater control tank size (measured in m^3), and A is the collection area (m^2) – normally the roof plan area.

The storage volume is sized based on the design depth of rainfall to be controlled. However, the research showed that the effectiveness of the storage volume provided is a function of the *Y/D* ratio. As *Y/D* increases towards 1.0, greater provision for storage is needed. In addition, more storage is needed to take account of the stochastic randomness of rainfall events, and again this increases as *Y/D* increases (HR Wallingford, 2012).

4.4 THE PILOT STUDY – HANWELL FIELDS (BANBURY, UK)

This section describes a pilot study carried out by HR Wallingford (2012), based on survey field data collected by Inch (2010), to demonstrate and test the methodology described in Section 4.3. The objective was to size tanks for stormwater control for a design rainfall event of 60 mm. The pilot study site comprised a mix of 66 properties ranging in roof area size and numbers of bedrooms. The property breakdown by number of bedrooms and mean occupancy is shown in Table 4.1.

Table 4.1 Number of properties in the Hanwell Fields development by number of bedrooms (development data).

Number of bedrooms	Number of properties
1	3
2	13
2.5	3
3	40
4	7

In addition, a survey was carried out to establish the actual occupancy in each property. Questionnaire returns were incomplete, but 34 of the 66 households did provide this information (Inch, 2010). This information was used to carry out an analysis based on a detailed simulation model built in InfoWorks CS (Innovyze, 2014). The model represented all the dwellings individually and was run with a 100 years of continuous stochastic rainfall data (HR Wallingford, 2012). Three different models were constructed for the pilot study:

(i) Model 1 – Tanks sized for *individual properties*, with the occupancy of each property assumed to be the mean for each property category using regional occupancy statistics (Section 4.4.3);
(ii) Model 2 – Tanks sized as for Model 1 for individual properties, but demand based on *actual occupancy* (from information based on field survey data (Inch, 2010) (Section 4.4.4);
(iii) Model 3 – *Communal tanks* sized with demand based on total population using mean occupancy from regional occupancy statistics (Section 4.4.5).

The main reasons for building and running these three different models were:

- To assess both the theoretical performance of the tank system (Model 1) and its actual performance based on the known population (Model 2) and to check the validity of using regional occupancy statistics when designing RWH tanks. As the actual occupancy rate of properties will not be known for a new development when the houses are built, sizing the RWH tanks has to rely on the use of average occupancy statistics. Theoretically, this applies in all situations (even if occupancy is known at the time of putting in a RWH system), as the occupancy of all dwellings will change in time;
- To identify the advantages (in terms of stormwater control performance) of a RWH system for a group of properties (Model 3) provides a comparison to the use of individual RWH tank systems (Model 1).

4.4.1 Design of individual tanks (models 1 & 2)

The statistical mean occupancy from regional statistics for each category of property by the number of bedrooms used for the designing of the RWH tanks for both models 1 and 2 is shown in Table 4.2. Based on the information on property bedrooms and roof areas, along with an assumed demand of 40 litres per person per day (lpd) and the annual rainfall for Banbury, the following *Y/D* ratios were identified (Table 4.3) to size the storage tanks. The use of 40 lpd was based on detailed investigation into the frequency of use and water consumption of modern toilets and washing machines (DCLG, 2007; Inch, 2010; Waterwise, 2010). Traditionally, 50 lpd is assumed, but as the *Y/D* ratio is so important, an over-estimation of Demand would result in under-estimation of tank sizes and possibly include some properties that had ratios greater than 0.95.

Table 4.2 Mean occupancies for each type of property based on regional statistics.

Number of bedrooms	House occupancy
1	1.40
2	1.74
2.5	2.08*
3	2.41
4	3.02

*This number was obtained by linear interpolation.
Source: ONS (2004); OCC (2009); Inch (2010).

Table 4.3 Number of properties by *Y/D* ratios (values used for sizing the RWH tanks).

Y/D	Number of properties
$Y/D \geq 0.95$	11
$0.95 > Y/D \geq 0.90$	10
$0.90 > Y/D \geq 0.85$	4
$0.85 > Y/D \geq 0.80$	3
$0.80 > Y/D \geq 0.75$	19
$Y/D < 0.75$	19

A key outcome of this element of the investigation is that toilet flushing on its own (due to a very significant increase in water efficiency measures) provides insufficient Demand to ensure stormwater management capability for systems serving individual properties, as *Y/D* will normally be greater than 1.0. It was therefore assumed that both toilet and washing machine use would constitute the Demand for non-potable water use. Other supply aspects (car washing, gardening) are not included as these cannot be regarded as being a regular daily demand throughout the year.

The figures in Table 4.3 show that 11 of the 66 properties had ratios higher than 0.95 and, in accordance with the criterion expressed in Equation 4.1, these properties were excluded from being provided with RWH systems (Section 4.1.1), resulting in a final sample size of 55 houses. The properties that had difficulty in complying with the *Y/D* ratio criterion were the smaller 1 and 2 bedroom properties. Three bedroom properties were found to be the most efficient in terms of minimising the *Y/D* ratio.

4.4.2 The importance of actual vs. assumed occupancy for the performance of RWH stormwater control systems

In the field survey (Inch, 2010), actual occupancies were obtained for 34 of the 66 properties in the pilot study area. Of these 34 properties, 31 had been assumed to warrant the use of stormwater control RWH systems based on property type average occupancy statistics. Table 4.4 summarises the actual *Y/D* ratios for the 34 properties where the occupancy was known, as well as the assumed mean occupancy based on statistics.

Table 4.4 *Y/D* ratios for 34 properties based on actual occupancy (using survey data) and mean occupancy (using statistical data).

	Number of properties in *Y/D* band based on actual occupancy using survey data	Number of properties in *Y/D* band based on mean occupancy using regional statistics
Total number of properties	34	31
Properties with $Y/D < 0.75$	13	12
Properties with $Y/D < 0.80$	15	21
Properties with $Y/D < 0.85$	21	23
Properties with $Y/D < 0.90$	22	25
Properties with $Y/D < 0.95$	24	31

Table 4.4 shows that of the 34 properties, only 24 actually complied with the *Y/D* criterion (Equation 4.1) and 31 properties were assumed to comply with the *Y/D* criterion based on mean occupancy statistics. This means that 7 of the 34 properties actually had *Y/D* ratios greater than 0.95. These 7 properties would have tanks that would often be full and would be very unlikely to have the storage available to store a large rainfall event when it occurred. As the results demonstrate (described in subsequent sections), these dwellings effectively failed to provide any useful storage for controlling large stormwater events.

Table 4.4 also shows that certain properties would have been provided with slightly less storage than that which would have been provided if the occupancy was known. It can be seen that only 15 properties had actual *Y/D* ratios of less than 0.80, while it had been assumed that 21 properties had ratios less than 0.80. However, the opposite is also true, in that some properties were provided with more storage due to a higher calculated *Y/D* ratio than existed in practice. The effect of these differences was found to be much less important in the performance of the tanks, but it should be noted that too much storage serving one property does not compensate for reduced storage in other properties, as those with under-sized storage would be slightly less effective at retaining all the runoff from the design event storm.

4.4.3 Model 1 – performance of the design scenario: Tanks for individual properties with occupancy levels based on mean occupancy statistics

Model 1 shows the performance of the methodology outlined in Section 4.3 in dealing with the stochastic variability of rainfall, as the occupancy is assumed to be known from statistical data. The design depth of rainfall to be captured was set at 60 mm. Results were obtained for individual properties and the whole site for:

- The proportion of events in each rainfall depth band that had a spill[1] equivalent of more than 1 mm rainfall for each property;
- The proportion of events in each rainfall depth band that had a spill equivalent of more than 1 mm rainfall on average for all the properties;
- An assessment of the spill depth and depth retained for each extreme event (24 events larger than 50 mm and 54 events over 40 mm).

Examination of the stochastically generated rainfall characteristics (based on a 6-hour inter-event dry period) showed that the vast majority of events were of 5 mm or less. Figure 4.3 shows that there is roughly only one event per year that is greater than 30 mm and only one event every two years larger than 40 mm.

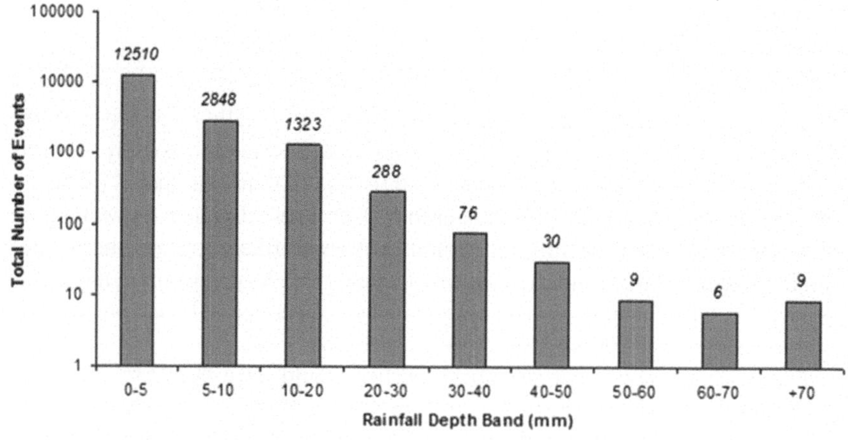

Figure 4.3 Rainfall events in the 100 year rainfall time series by depth band (HR Wallingford, 2012).

An analysis of all events was made by grouping events into 10 mm rainfall depth bands and recording a 'failure' as taking place for any property that had a spill equivalent of more than 1 mm of rainfall. Figure 4.4 shows the proportion of events for which there was a spill from each of the tanks (ranked in terms of *Y/D*). In Figure 4.4, for each rainfall depth band, all 55 properties are plotted in rank order based on *Y/D*. As all 55 properties are plotted, it is difficult to see individual results, but the trend of low to high values of *Y/D* shows that properties with a low *Y/D* perform better (even though the tanks are smaller in size) than those with high *Y/D* ratios. The analysis showed that the higher the value of *Y/D*, the less successful each property is

[1]Overflow from the rainwater harvesting tank to the drainage/sewerage system.

in retaining all the rainfall events, whatever the size of the event. Below a value of *Y/D* of 0.8 the performance is good, while higher values do less well. However, it is also worth noting that even for events greater than 70 mm, around 50% of events still do not spill from the majority of the tanks. It is also evident for these large events that there is less distinction between *Y/D* ratios and this is because there is a relatively limited allowance for extra storage for tanks with low *Y/D* ratios. Consequently, for a property with a low *Y/D* ratio a tank is sized such that it cannot retain an event that is much greater than 60 mm, even if it is empty at the start of the event.

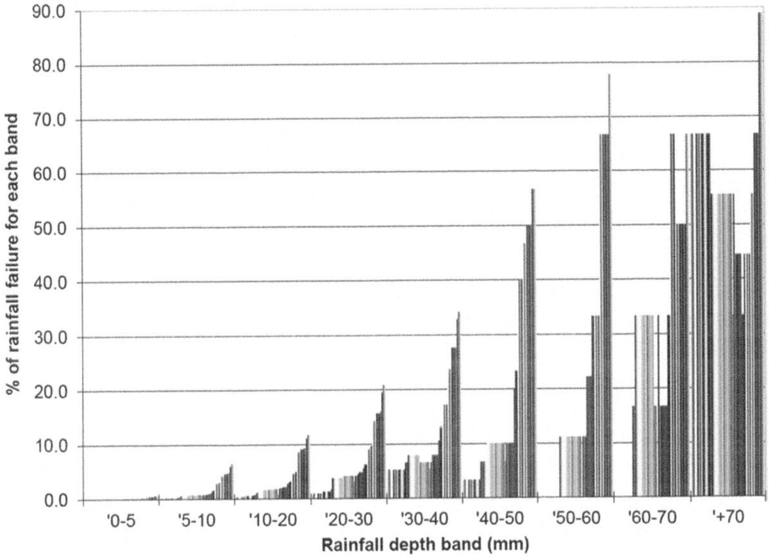

Figure 4.4 Proportion of events with 1 mm or more of rainfall spilling from each tank in 10 mm rainfall depth bands (the 55 properties are plotted in rank order based on *Y/D* ratio in each rainfall depth band) (HR Wallingford, 2012).

Figure 4.5 shows the proportion of events for which there was a spill of more than 1 mm (of rainfall) from the overflow pipe serving all 55 properties with RWH tanks. Around 67% of events in the range of 50 mm to 70 mm generate a spill. This graph also shows the number of events in each rainfall depth band. Additionally, Figure 4.6 illustrates an analysis of the amount of water spilled during the large rainfall events. It shows that, on average, the spilled depth for the 50–60 mm group of events, although quite variable, is only 4 mm. It also shows that for events that are larger than the design depth of 60 mm, the tanks retain most of the runoff. Only 3.3 mm spills for the 60–70 mm group and on average the storage system retains 64 mm of effective rainfall. The results demonstrate that although many small spills do take place for all events, the storage tanks are generally quite effective in retaining the design storm rainfall depth for extreme events.

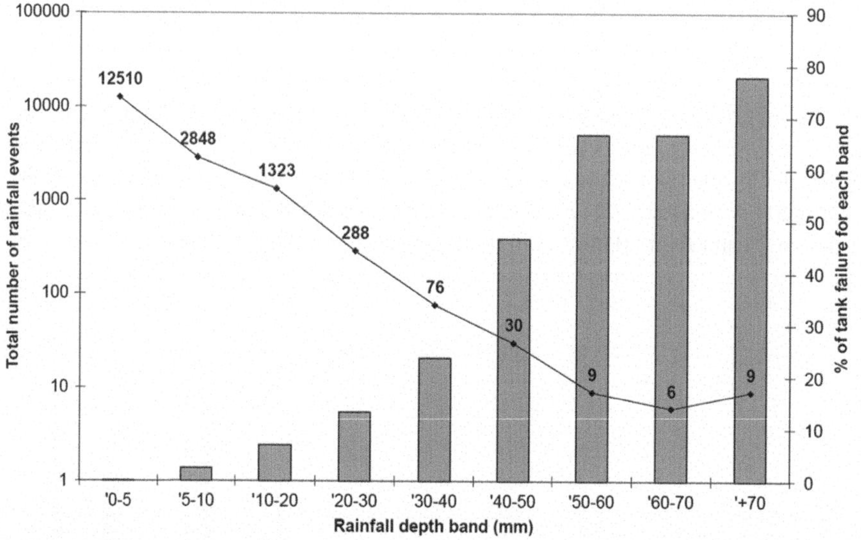

Figure 4.5 Proportion of events with an average of more than 1 mm of rainfall spilling from all tanks, also showing number of events by rainfall depth ranges (HR Wallingford, 2012).

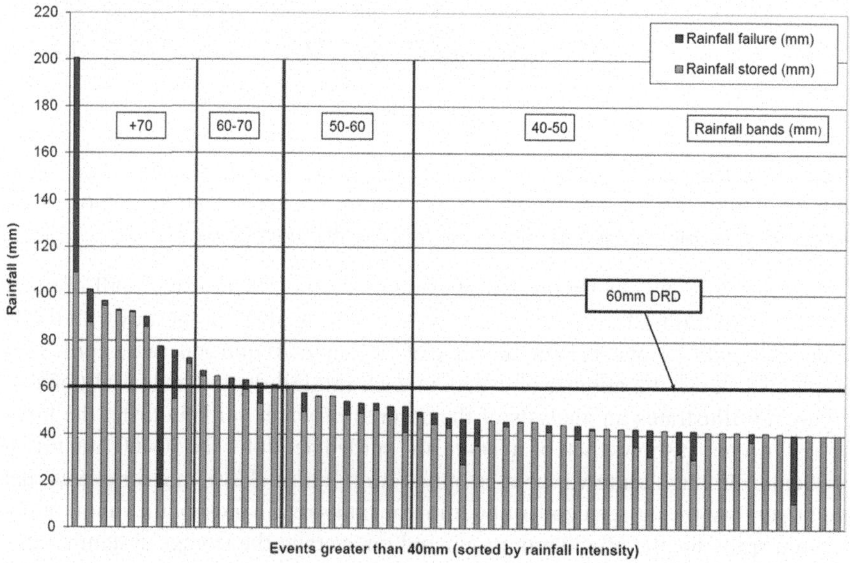

Figure 4.6 Volumes stored and spilled for each of the 54 events larger than 40 mm, based on the assumed occupancy of dwellings (HR Wallingford, 2012).

A detailed examination of event spill performance was also conducted and the results are illustrated in Figure 4.7. Figure 4.7 is a cloud plot of points representing all events of the 100 year series for a property with a *Y/D* of 0.65. The horizontal axis of the figure shows the rainfall depth for each event. On the vertical, axis the following variables are shown:

- Spare capacity in the tank or the spilled volume (both measured in millimetres of rainfall);
- Spare capacity of the tank at the end of the event (if the tank did not spill). This is represented in the negative part of the axis (i.e., how much more rainfall could have been stored);
- Total spill volume for the event if the tank did not store the total rainfall event depth.

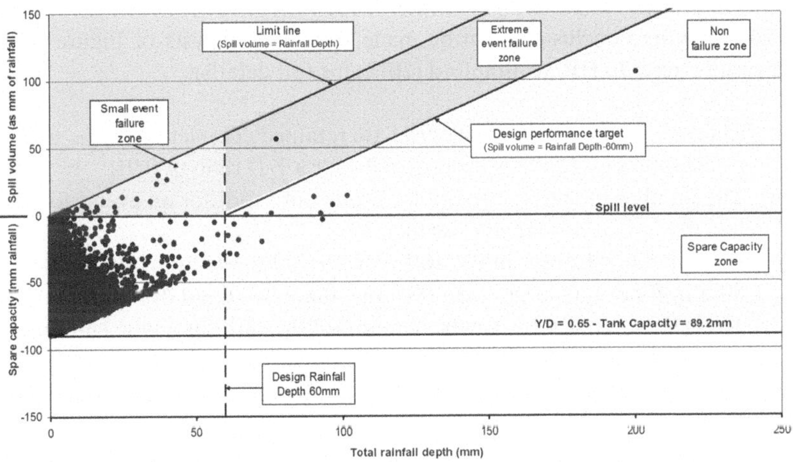

Figure 4.7 RWH tank spill performance for all rainfall events for the 100 year rainfall series for a *Y/D* ratio of 0.65 (HR Wallingford, 2012).

Figure 4.7 is divided into five zones by four different lines:

- The spill level line: a horizontal line that divides the events that spilled and the events that were completely stored by the tank;
- The limit line: a 45 degree line where the total rainfall depth is equal to spill volume (that being when the tank would be completely full);
- The design rainfall depth line: a vertical line defined by the design rainfall depth to be stored in the tank (60 mm in this case);
- The design performance target line: a 45 degree line where the spill volume is equal to the total rainfall minus the design event rainfall depth. (that being a spill from the tank for an event that is larger than design event, but that has stored the design depth before the spill commences).

Therefore the five areas represent:

- The area above the limit line, where no results are possible;
- The spare capacity zone: the area below the spill level line, which includes all the events that are completely stored by the tank;
- The non-failure zone: the area that includes all the events in which the spill occurs after the tank has stored a rainfall depth in excess of the design rainfall depth;
- The small event failure zone: those events less than the design rainfall that result in a spill;
- The extreme event failure zone: those events greater than the design rainfall that result in a spill and where the tank did not manage to store the design rainfall depth from an event greater than the design event.

The following conclusions can be made from the analysis of Figure 4.7 and other results (refer to HR Wallingford (2012) for full details):

(i) The vast majority of events (99%) are retained completely by the tanks for a *Y/D* ratio of 0.65. This drops to 97% for a *Y/D* ratio of 0.91;

(ii) The number of spills in the failure zone (any spill for an event) for events less than 60 mm is small (less than 1%);

(iii) There is only 1 event in the 100 year series for which the design depth of 60 mm for events larger than 60 mm cannot be stored (for a property with *Y/D* = 0.91, there are 5 events, but most of the spills are quite small). These are the events above the 'design performance target' line;

(iv) There are a number of events that are greater than the design rainfall depth, but for which at least 60 mm is retained before spilling takes place;

(v) Although the tank for the *Y/D* ratio of 0.91 performs slightly worse than that for the *Y/D* of 0.65, for the number of events that can be constituted as failures, it stores more water for extreme events larger than the design event and has fewer spills for these events.

These conclusions indicate that the design methodology devised is generally successful in achieving control of stormwater runoff for situations where the actual population is known. However, as *Y/D* gets close to 0.95 the reliability reduces a little, even though more storage is provided.

An analysis of the seasonal performance was also carried out (HR Wallingford, 2012). This is not discussed in this chapter, but it is important to consider this issue in the context of the required performance of the drainage system in relation to potential impacts on a receiving water body. This is because the requirements for protecting receiving water bodies from non-point pollution (as may be found in surface water runoff) may vary due to seasonal variations related to the potential level of dilution, temperature and other factors.

4.4.4 Model 2 – performance of the actual scenario for individual tanks

As discussed in Section 4.4.2, there were 7 properties with tanks with a *Y/D* ratio greater than 0.95 that were provided with RWH systems. These properties have a significantly worse performance than other properties with a ratio less than 0.95 and this influences the performance of the storage systems when examined as a whole. Figure 4.8 shows the retained and spilled performance for the tanks during all the large rainfall events. It shows that the tanks fail to store the rainfall volume effectively for all storm events less than or equal to the design depth, with a fairly uniform proportion of failure increasing from 15% for the smaller events through to 25% for the biggest events. This proportion can be compared to the percentage of properties that had a *Y/D* ratio >0.95, which was 7 in 31; approximately 23%. This is very useful in that it shows that if the proportion of properties can be determined where actual occupancy rates do not comply with $Y/D < 0.95$, then the 'failure' proportion of any rainfall can be determined and incorporated when designing the drainage system serving the development.

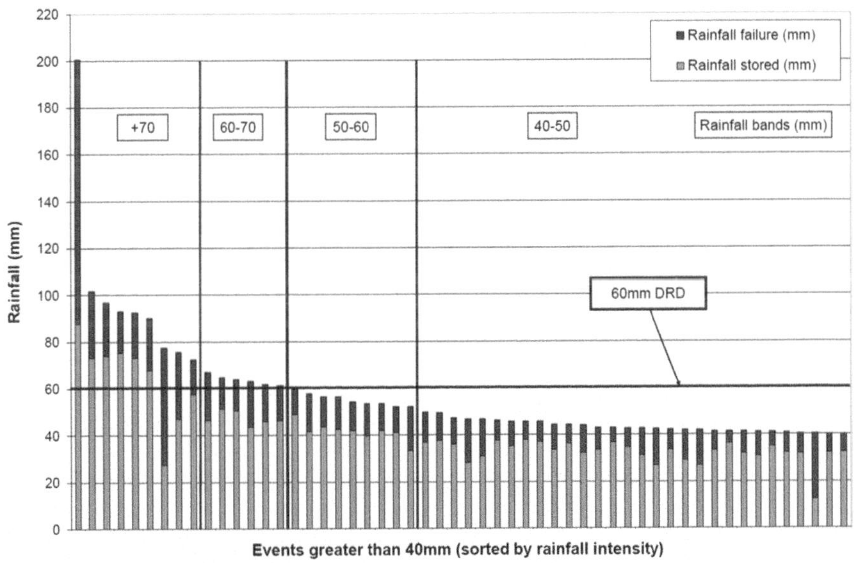

Figure 4.8 Volumes stored and spilled for each of the 54 events larger than 40 mm, based on the actual occupancy of dwellings for 31 properties (HR Wallingford, 2012).

To demonstrate the effect of the *Y/D* ratio and the threshold set at 0.95, Figures 4.9 and 4.10 show the overflow pipe performance for extreme events for the 7 properties with $Y/D > 0.95$ and the 24 properties with $Y/D < 0.95$. It can be seen that the storage provided by the properties failing to comply with *Y/D* of 0.95 retain, on average, around 8 mm of rainfall for these events. However, for those properties with a *Y/D* ratio <0.95, the results are slightly better than that obtained from the assumed statistical population (Section 4.4.2). These results clearly show that the properties with *Y/D* > 0.95 are effectively useless in providing storage for stormwater control, but those that do comply do provide very effective runoff control. It should also be noted that the tank sizing mechanism, due to the need to cater for the stochastic uncertainty of rainfall events by providing some additional storage, does provide even greater benefits for those storms that are significantly greater than the design storm depth.

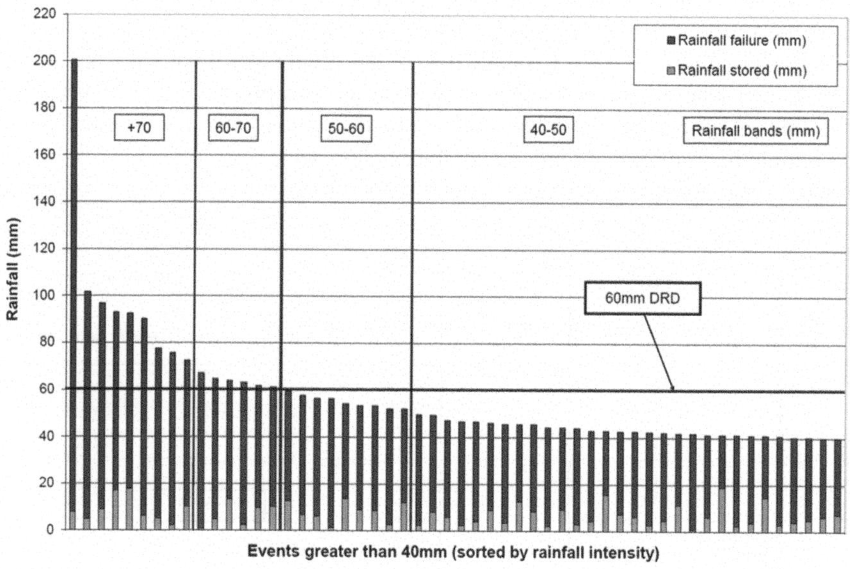

Figure 4.9 Volumes stored and spilled for each of the 54 events larger than 40 mm, based on the actual occupancy of dwellings for 7 properties with *Y/D* >0.95 (HR Wallingford, 2012).

4.4.5 Model 3 – performance of the design scenario for a communal tank

Providing individual properties with RWH tanks and then making provision for a proportion of them failing to retain the design depth of rainfall is clearly an expensive option if a more efficient solution can be devised. An alternative solution is the use of communal RWH using a single, large tank. This is because the total population in a group of houses will converge on the statistical mean. The mean *Y/D* ratio for the 31 properties of actual occupancy served with tanks is 0.76, which is virtually the same as that obtained for the 55 properties provided with tanks based on the assumed mean population, having *Y/D* ratio (0.77). This means that as more houses are served by a single tank, there is greater assurance that the *Y/D* ratio calculated will be approximately correct.

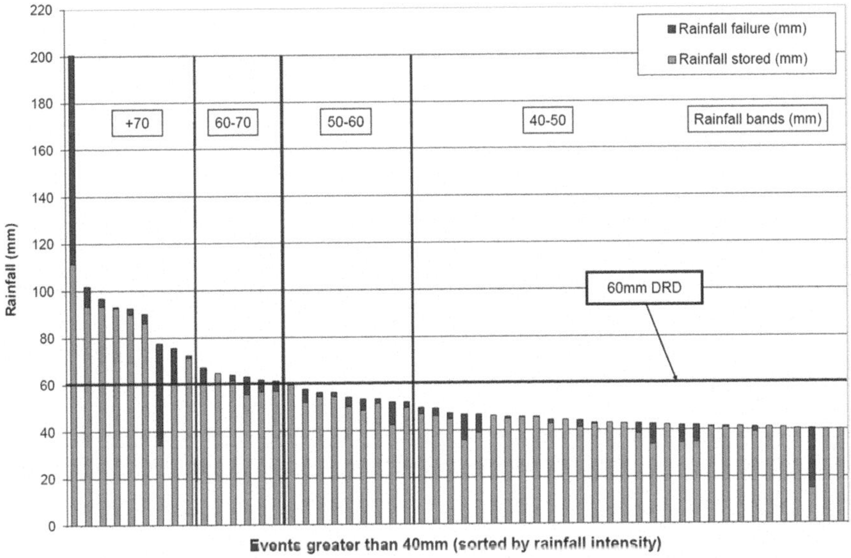

Figure 4.10 Volumes stored and spilled for each of the 54 events larger than 40 mm, based on the actual occupancy of dwellings for 24 properties, with *Y/D* <0.95 (HR Wallingford, 2012).

Based on a *Y/D* ratio of 0.76, the storage tank volume for using a communal tank to serve all 55 properties is 235 m^3. This compares to the 256 m^3 of storage provided for the sum of all property systems supplied with individual tanks. The performance of the communal tank (Figure 4.11), even though the storage is less, results in a better performance than the tanks serving each of the 55 properties. For the 53 events (excluding the 211 mm event), only 11 events have any spill and the mean spill volume for events between 60 and 70 mm is 2 mm. The mean volume retained for rainfall events greater than 60 mm is 71 mm; significantly greater than the design event. These results clearly show that there is a distinct advantage in terms of hydraulic performance, in providing communal RWH storage for stormwater management. The question of what constitutes a communal system can be examined statistically in the same way as can the assessment of the proportion of single property failures (when the actual mean population is likely to be a close approximation to the statistical mean, such that it can be reasonably confidently assumed that *Y/D* is less than 0.95) (Section 4.5).

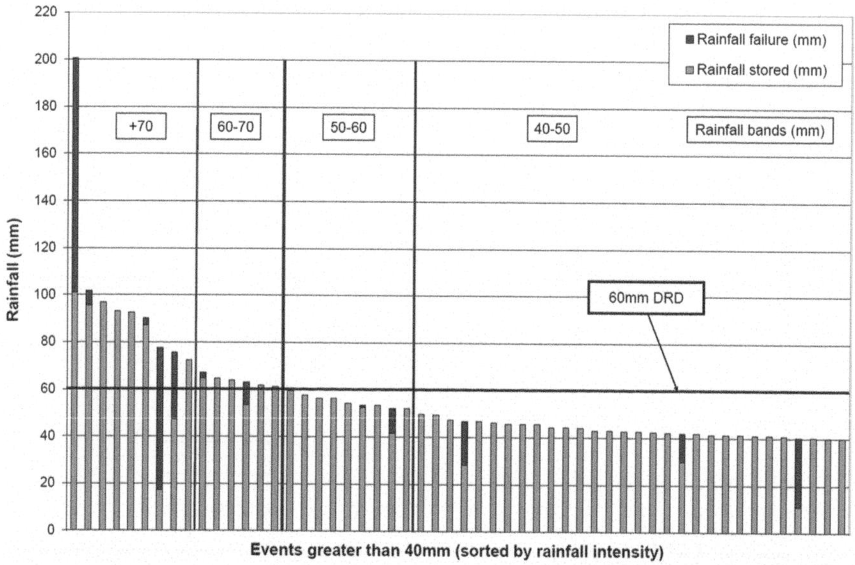

Figure 4.11 Volumes retained and stored for each event for a communal tank designed to retain 60 mm rainfall: 54 events larger than 40 mm (HR Wallingford, 2012).

4.5 A METHODOLOGY FOR ASSESSING UNCERTAINTY OF PROPERTY OCCUPANCY

It is clear that compliance with the *Y*/*D* ratio is essential. Therefore a methodology for estimating the proportion of properties that will be non-compliant based on actual occupancy is important, if the design of stormwater systems should take into account the runoff from properties that do not retain stormwater runoff. This can be done subject to some assumptions. A probabilistic approach can be taken if the statistical distribution (mean and standard deviation) of property occupancy is known. Where the roof area is not known for each property type, then this can also be included in the analysis in the same way if the mean area and standard deviation are known, although in most cases the actual roof area is likely to be known.

The assumption made is that occupancy is based on a binomial distribution with the minimum occupancy of a property being one person, with an upper-bound occupancy of twice the number of bedrooms. If roof areas are to be included in the analysis then a normal distribution is assumed. An example of applying the method to a 2-bedroom house with a specific known roof area is provided here. The average occupancy is assumed to be 1.72 people with a standard deviation of 0.73. The roof area is 42 m^2. This gives a *Y*/*D* ratio of 0.86 and therefore complies with the *Y/D* rule. The probability is, however, that there is a 44% chance of failure (*Y/D* > 0.9 in this case). This is represented in Figure 4.12 as the dark shaded area on the probability mass function of occupancy population. It should be noted that if the roof area had been 45 m^2, although this would have a higher *Y/D* that is compliant with the criterion, the probability of failure would still be the same. This is because the failures are associated with the chance of having only 1 person occupying the property.

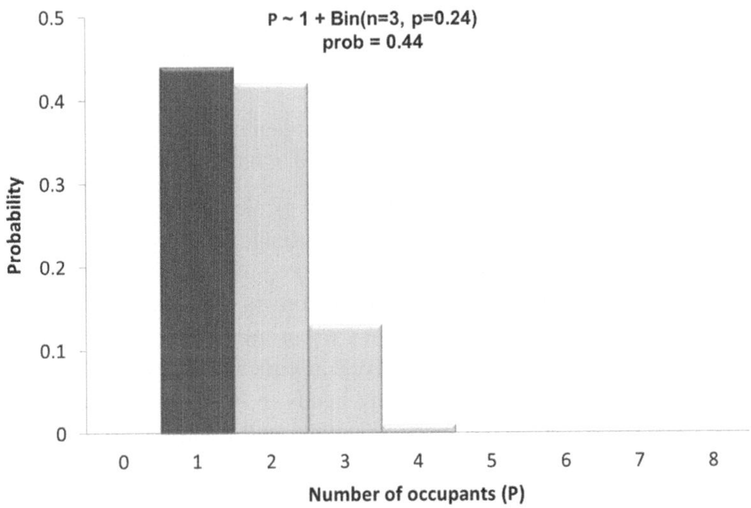

Figure 4.12 Binomial distribution for a 2-bedroom property occupancy showing the probability of compliance with a *Y/D* ratio of > 0.90 for a roof of 42 m^2 (HR Wallingford, 2012).

4.6 ACTIVE MANAGEMENT OF RWH SYSTEMS

The analysis presented in the previous sections has shown that the stochastic nature of rainfall has to be catered for specifically in the sizing of tanks. Additionally, only properties that can comply with a *Y/D* ratio less than 0.95 will provide stormwater control. This may exclude quite a large number of properties, especially in the wetter areas of a country. Furthermore, estimated tank sizes become relatively large when *Y/D* ratios increase above 0.75 to address this uncertainty. The only occasion where active management would not provide any benefit over a passive system, is where *Y/D* ratios were known to be less than around 0.7. In this situation, the tank sizes would be fairly similar and they would normally be close to empty for most of the time. Active management of the water level in the tank is therefore an alternative option so that stormwater storage is always available. This not only provides confidence in achieving stormwater control, it also means that all properties will comply irrespective of the *Y/D* ratio. It should also be noted that no less water is likely to be used by the household when using an active system. The drawdown pump operation will only be frequent when the *Y/D* ratio is above 1.0 and in this situation the householder, by definition, is not able to consume all the water being supplied.

4.6.1 Active control decision rules

Active control of the storage in a tank means that storage has to be maintained at a level where there is capacity for runoff from a large rainfall event. However, if a large storm event is in progress, it is critical that a pump does not start emptying the tank. Rules therefore have to be in place to try and avoid this happening. This means that the tank must not be emptied when:

1. A significant rainfall event is likely to happen in the near future; or
2. A significant rainfall event is taking place or has very recently taken place.

In the first case, unless systems are linked to meteorological forecasting, a decision to empty cannot be linked to future conditions. Although this is technically possible, it is an unlikely solution to be applied and there are minimal additional benefits compared to applying the second option. In fact, there is a significant disadvantage in that many rainwater tanks in an area would probably all start emptying at the same time, which may have consequences downstream.

In the case of the second option, there needs to be a time delay introduced between a water level threshold being exceeded and pumping commencing, as the threshold trip may occur due to a large event taking place. It is suggested that a 2 day delay would be sufficient to allow for any downstream drainage system to have coped with the runoff from a major event. This means that the discharge can then take place safely and storage maintained in the tank for the next major event.

4.7 CONCLUSIONS

This chapter has presented a methodology for providing stormwater control using RWH storage tanks. The methodology presented was applied to a pilot study of 66 houses, 55 of which were suitable for implementing RWH. The methodology for sizing a tank does not depend on knowing the household occupancy, but on the number of bedrooms, the statistics on average occupancy for that type of property in that region and the size of the roof.

- RWH can be used to control the runoff from large stormwater events;
- The ratio of *Y/D* is very important and it is an essential design parameter unless active control is provided to maintain a stormwater storage volume;
- The use of annual rainfall is a very simple way of estimating average yield, but if seasonal issues are important (either in terms of demand or yield) then the analysis should be based on seasonal information;

- National and local statistics on property occupancy are available to enable calculations to be made for the *Y/D* ratio and to assess the probability of non-compliance for individual properties. Mean and standard deviation information on occupancy is available for all types of properties based on numbers of bedrooms (DCLG, 2007).
- The use of RWH for a group of houses (a communal system) is significantly more cost effective in the control of the stormwater runoff in terms of effective storage volume;
- Where RWH is provided to individual houses, a statistical analysis needs to be conducted to estimate the proportion of properties that will not provide stormwater control of a large stormwater event. This is due to the variability associated with the property occupancy, which has a direct effect on the consumption of water from the tank;
- A similar statistical analysis can be carried out for communal systems to determine the uncertainty range of *Y/D* for the group of properties being served, particularly where the number of properties is small. For large numbers of properties, reasonable confidence can be placed in the *Y/D* value using the statistical mean population.

REFERENCES

BSI (2009). BS8515:2009. Rainwater Harvesting Systems: Code of Practice. British Standards Institute, London, UK.

DCLG (2007). Live tables on household characteristics – Table 810: Household characteristics: mean household size by type of accommodation and by number of bedrooms. http://www.communities.gov.uk/housing/housingresearch/housingstatistics/housingstatisticsby/householdcharacteristics/livetables/ (accessed 1 June 2010).

HR Wallingford (2012). SR736_Developing Stormwater Management using Rainwater Harvesting. Testing the Kellagher/ Gerolin methodology on a pilot study. Ver. 2.

Inch D. (2010). Rainwater Harvesting as a Potential Control for Urban Storm Flooding. Unpublished MSc Thesis, Coventry University.

Innovyze (2014). InfoWorks CS. (http://www.innovyze.com/products/infoworks_cs/ (accessed 25 August 2014).

OCC (2009). Survey of New Housing 2008: Summary of Results. Oxfordshire County Council Policy Unit, Oxford, UK.

ONS (2004). Number of People Living in Households (UV51). Office for National Statistics http://www.neighbourhood.statistics.gov.uk/dissemination/LeadTableView.do?a=3&b=277085&c=Cherwell&d=13&e=7&g=479953&i= 1001x1003x1004&m=0&=1&s=1281100205272&enc=1&dsFamilyId=143 (accessed 1 June 2010).

Royle J. A. (2004). N-mixture models for estimating population size from spatially replicated counts. *Biometrics*, **60**, 108–115.

Waterwise (2010). Saving water at home. http://www.waterwise.org.uk/reducing_water_wastage_in_the_uk/house_and_garden/save_water_at_home.html (accessed 1 June 2010).

Chapter 5

Rainwater harvesting for toilet flushing in UK schools: Opportunities for combining with water efficiency education

Cath Hassell and Judith Thornton

5.1 INTRODUCTION

As discussed by Russell and Fielding (2010), water conservation can be subdivided into *efficiency behaviours* (one-off technological changes, such as controls for urinal flushing, reducing flush volumes from WCs, replacing taps, rainwater harvesting (RWH)) and *curtailment behaviours* (individual actions such as turning the tap off when brushing teeth, only using the washing machine for a full load). The underlying drivers for these two distinct behaviour types have been argued to be quite different (Gardner & Stern, 1996). In a more general sense, whilst it is regularly postulated that take-up of one sustainable behaviour can lead to the take up of another via a catalysing or spillover effect (Austin *et al.* 2011), there is relatively little evidence supporting this in practice. Neither is there evidence supporting the idea that an *efficiency behaviour* can influence a *curtailment behaviour.* Despite this, the two behaviours are regularly linked by policy makers; '*Implementing water efficiency measures may also provide an excellent opportunity for schools to educate students in the need to conserve water – a key component of sustainability*' (Duggin & Read, 2006).

Installing RWH systems into non-residential buildings is one such efficiency behaviour. In buildings, where the predominant uses are WC and urinal flushing (i.e., daytime use in buildings such as schools and offices), there is the potential to save considerable amounts of mains water, particularly where roof areas are large. In the UK, policy measures such as the *Schools for a Future* programme, and the requirement to meet higher levels of BREEAM[1] (Barlow, 2011) have stimulated the

[1]BREEAM (Building Research Establishment Environmental Assessment Methodology) is a sustainability certification scheme for buildings in the UK. It includes aspects such as thermal performance, material choice, generation of energy, and use of water.

uptake of RWH in new school buildings and it is also a recommendation in standard guidance for water management in schools (Duggin & Read, 2006). There are also examples where RWH has been retrofitted into existing schools, and these have tended to be one-off pilot projects funded by local authorities or the local water companies to 'demonstrate best practice' with regards to reducing the demand for mains water (e.g., Retrofitting RWH in London Schools 2012 (Hammersmith & Fulham, 2014), as described in this chapter).

The desire to make the link between *efficiency behaviours* and *curtailment behaviours* stems from the magnitude of the scope for water savings that could result if the link exists, as illustrated by the following theoretical example. A typical large primary school (i.e., catering for Year 1 to Year 6 pupils) in the UK with two form entry (i.e., two classes per year group) and the standard 30 pupils per class will have 360 pupils and nearly 50/50 split between girls and boys. For this notional school, the total water demand for WC and urinal flushing (two applications where harvested rainwater can be used) is estimated as 788 m^3 (Table 5.1). This is based on the assumption that the school was fitted with 9 litre WCs and has 8 urinals (flushed every hour with a maximum allowable flushing rate of 7.5 litres/hour during the school hours). The financial cost to the school for this volume of water is US$2575–7144 (£1568–£4350)[2] depending on where in the UK the school is situated.[3]

Table 5.1 Estimation of water demand for flushing WCs and urinals in the notional school.

Parameter	Value
Number of girls	180
Number of boys	180
Number of school days	200
Water use per WC flush (litres)	9
WC uses per day by each girl pupil (ech$_2$o, 2012)	2
Number of urinals	8
Water required to flush each urinal (litres/hour)[a]	7.5 litres
School term days	200
Number of hours in a school day	10
Annual water demand for flushing WCs – girls (m^3)	*(180 × 9 × 2 × 200)/1000 = 648*
Annual water demand for flushing urinals – boys (m^3)	*(8 × 7.5 × 10 × 200)/1000 = 120*
Annual water demand for flushing WCs – boys[b] (m^3)	*(5.5 × 9 × 2 × 200)/1000 = 20*
Total water demand to be met by RWH system	*788 m^3*

[a]Water Regulations (1990).
[b]WC flushing for boys based on an assumed 3% of boys using the toilet instead of the urinal.

[2]Rate of exchange 1 UK pound = 1.64 US dollar. Rate correct on 15th January 2014.

[3]Water costs vary widely across the UK. Of the main water supply and sewage companies, Thames Water is cheapest at US$3.27/m3 (£1.99/m^3) (2013–21014 prices) (Thames Water, 2013) and South West Water is the most expensive at US$9.06/m^3 (£5.52/m^3) (2013–21014 prices) (South West Water, 2013).

Using the Intermediate Method, Equation 5.1, from the British Standard for RWH for calculating storage requirements (BS 8515, British Standards Institute, 2009), it can be estimated that a roof area of 1650 m^2 and rainfall of 620 mm a year could provide all of the required rainwater for WC and urinal flushing (this level of rainfall is typical for the East of England), as could a school roof of 1030 m^2 coupled with rainfall of 1000 mm a year (the UK average rainfall).[4] This figure of 788 m^3 is equivalent to each pupil saving 6 litres of water a day at home throughout the year.[5]

This is a very modest water saving, and the question therefore, is what types of behaviour change might be required domestically to save an equivalent volume of water, or indeed, what additional water savings could be realised by coupling a behavioural change project with a technological change in the school environment? In contrast to the school environment, where water uses are relatively inelastic in relation to function (i.e., they are urinal and WC related) behavioural change interventions at home could comprise a variety of changes which are inherently more personal and adaptable; shallower baths, shorter showers or washing up using a bowl are all potential solutions. In addition, since these behavioural changes may also involve hot water savings, water, energy and CO_2 are saved, leading to an overall greater environmental benefit.

This chapter first introduces water use and RWH in UK schools, and then investigates how a potential target of 6 litres/pupil/day is achievable, using data collected from two projects (Section 5.6 and Section 5.7) delivered in schools in London, West Sussex and Hampshire.

5.2 WATER USE IN SCHOOLS

Section 83 of the Water Act 2003 requires public authorities (which is defined as any public body and includes schools) to take into account, where relevant, the desirability of conserving water supplied or to be supplied to premises. Local water companies have a legal duty to promote water efficiency to their customers (Defra, 2013). Whilst neither of these pieces of legislation specifically refer to RWH, they are both drivers to reduce water consumption in schools, and school-specific guidance in BREEAM indicates how savings might be achieved. Data from DfES (2003) indicates that the average UK primary school (without a pool) uses 3.8 m^3 of water/pupil/year, with a best practice school using 2.7 m^3 of water/pupil/year. The corresponding figures for secondary schools are 3.9 m^3 and 2.7 m^3 of water/pupil/year. Disaggregation of this data into micro-components is not available, but the typical water uses will include WC flushing, urinal flushing, hand washing, caretaker cleaning, catering requirements (dishwashing and food preparation) and grounds watering (both primary and secondary schools). In primary schools, there

[4]Assuming a roof drainage factor of 0.85, a filter efficiency of 0.9 and adequate storage. Required roof area to meet whole demand is unlikely to be available in an existing school.

[5]360 pupils × 6 litres × 365 days = 788,400 litres = 788.4 m^3.

is often high usage at classroom sinks (e.g., drinking water, hand washing, paint-brush cleaning). In secondary schools, usage includes science labs and cooking classrooms and shower use. The contribution of leakage to these figures is unknown. Non-potable water requirements (WC flushing and urinal flushing) are thought to be a relatively high proportion of the total, and the use of rainwater is therefore often considered to be appropriate for these purposes, from both environmental and economic perspectives. Table 5.2 gives an indication of current and suggested best practice consumption values in schools.

Table 5.2 Benchmarks for water use in schools.

Type of school	Typical practice (m^3/pupil/year)	Best practice (m^3/pupil/year)
Primary School (no pool)	3.8	2.7
Primary School (with pool)	4.3	3.1
Secondary School (no pool)	3.9	2.7
Secondary School (with pool)	5.1	3.6

Note: From a sample of 14,330 schools in the UK.
Source: (DfES, 2003)

5.3 CONFIGURATION OF RWH SYSTEMS IN UK SCHOOL BUILDINGS

There are three basic configurations for RWH systems in the UK, illustrated in Figure 5.1 (a, b and c). They are classified in BS 8515 as:

(a) water collected in storage tank(s) and pumped directly to the points of use;
(b) water collected in storage tank(s) and fed by gravity to the points of use;
(c) water collected in storage tank(s), pumped to an elevated cistern and fed by gravity to the points of use.

RWH systems installed in schools are usually type C, meaning that in the event of a pump breakdown the WCs can still be flushed and the school can remain open. Storage volumes are usually sized so that rainwater collected at weekends and during school holidays is available for use during normal building occupation.

In the UK, RWH systems generally incorporate filtration prior to storage, to minimise the presence of leaves and other detritus in the storage tank and first flush mechanisms are rarely used. Pumps are generally multi stage pressure pumps (housed in the rainwater storage tank itself), although suction pumps can be found in some systems. Pumps are automatically protected from dry running by an integrated control panel. The control panel will also operate a mains back up to automatically deliver water to ensure that WCs and urinals can still be flushed (detailed in BS 8515, British Standards Institute, 2009). Rainwater is classified as

Fluid Category 5 (Water Supply (and Fittings) Regulations, 1999). To conform to the Water Regulations, the rainwater must be isolated from the mains supply via a Type AA or type AB air gap to protect the mains from back siphonage (Water Supply (and Fittings) Regulations, 1999). Depending on system configuration, the mains back up is supplied to the rainwater storage tank, to a break tank in a the service area for the building or into an elevated cistern. When the mains back up is to the rainwater storage tank or into a break tank, then any mains back up requires pumping. A UV disinfection system may be incorporated.

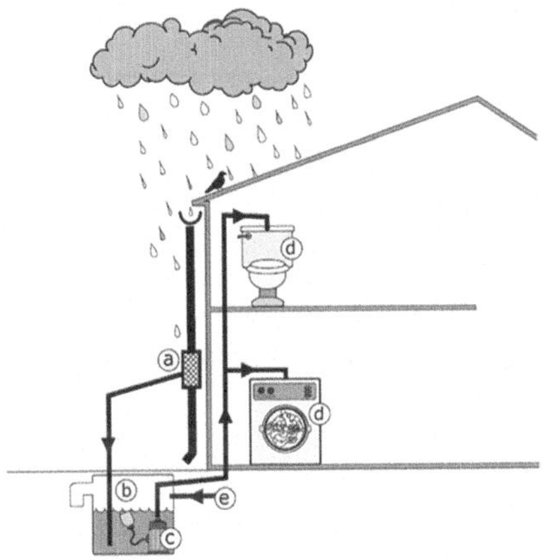

Figure 5.1a Directly pumped rainwater system. Water passes through the filter (a) and into the storage tank (b). It is then pumped (c) directly to the appliances (d). Mains water backup (e) is to the storage tank via a suitable air-gap. (Thornton, 2013).

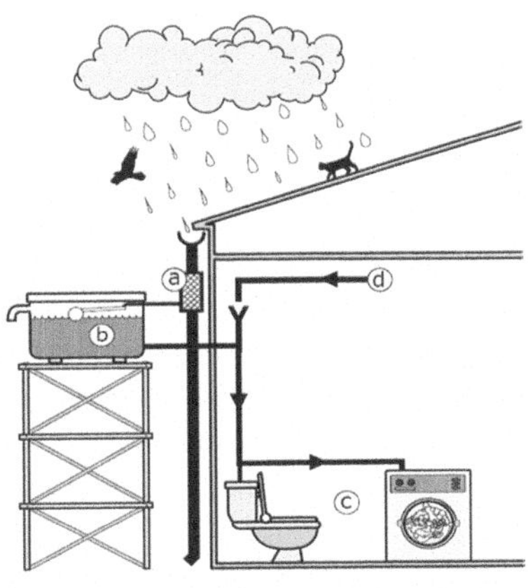

Figure 5.1b Gravity fed rainwater system. Water passes through the filter (a) and directly into a roof-level storage tank (b), from where it flows into the appliances (c). Mains water backup (d) is via a suitable air gap. (Thornton, 2013).

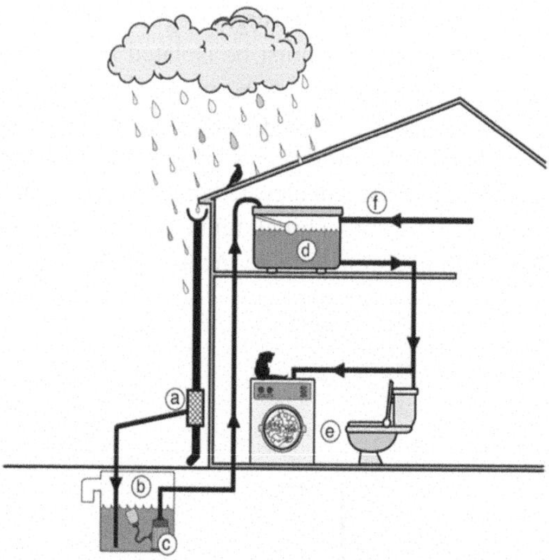

Figure 5.1c Indirectly pumped RWH system. Rainwater passes through a filter (a) and into the storage tank (b). Water is pumped (c) into a header tank (d), from where it flows to the appliances (e). Mains water backup (f) is provided to the header tank via a suitable air-gap. (Thornton, 2013).

The annual yield (Y) in litres from rainwater is generally calculated using the formula in BS 8515:2009 (British Standards Institute, 2009):

$$Y = A \cdot i \cdot D_f \cdot F_e \tag{5.1}$$

where,

A = available roof area in m^2
i = average yearly rainfall in mm
D_f = drainage factor of the roof
F_e = filter efficiency

The answer is divided by 1000 to get a yearly yield in m^3. Whilst this is accepted to be a simplification (given the variability of rainfall from the average), this remains probably the best available method for determining yield from a RWH system. The yield is dependent on storage being sized at 5% of the annual yield (BS 8515:2009). In order to account for uncertainties associated with rainfall and demand patterns and stormwater attenuation benefits, complex methods have been researched including approaches discussed in Chapter 4.

RWH systems can be difficult to retrofit into existing buildings, as new supply pipework will need to be installed, rainwater down pipes may need to be rerouted and it can be difficult to bury tanks due to existing infrastructure. RWH systems also require ongoing maintenance by external professionals.

5.4 BENEFITS OF RWH IN THE UK CONTEXT

Whilst RWH is a standard technology in many countries with relatively undeveloped mains water infrastructure, it remains unusual in the UK, with an estimated 80,000 (UKRHA, 2014) systems installed across all sectors, with 21% of installed systems classified as commercial units.[6] Although no separate figures are kept for schools, industry sources estimate that between 600–750 new schools were built with RWH systems fitted. Most of these schools were built under the *Building Schools for a Future* programme which ran from 2005 to 2010. The harvested rainwater is typically used for non-potable uses such as WC and/or urinal flushing (Thornton, 2013), and in addition to offsetting the use of mains water, the amount of stormwater reaching the sewers is reduced. In many UK cities, rainwater and foul water drainage systems are not separate, and are conveyed jointly to the sewage treatment plant (Butler & Davies, 2010). During intense rainfall events, these combined sewers become overloaded and raw sewage is discharged directly into rivers and seas via a Combined Sewer Overflow (CSO). In addition to separating rain and foul water drainage systems, approaches that minimise the amount of rainfall leaving site are often considered; these systems are generally referred to as Sustainable Drainage System (SuDS) solutions (Woods-Ballard *et al.* 2007). Where possible, these systems infiltrate a proportion of rainwater to ground, and where soil porosity or water table does not permit this, the focus is on attenuating peak flows, generally via specific attenuation tanks. Whilst the volumes of water requiring attenuation are generally far in excess of that normally incorporated into RWH systems, nevertheless, RWH is considered as an important tool in reducing volumes of rainwater runoff from a site (Department of Communities and Local Government, 2010). Stormwater attenuation aspects related to RWH are discussed in Chapter 4.

5.5 ENGAGING WITH PUPILS TO ENCOURAGE WATER EFFICIENT BEHAVIOUR

Average domestic water use in the UK is generally stated to be around 150 litres/person/day (e.g., Market Transformation Programme, 2008; Defra, 2008), and as reviewed by Memon and Butler (2006), most studies have found that the underlying frequency distribution curve has a positive skew (i.e., that median water use is lower than mean, and that a relatively small number of people are very high water users). However, as discussed by Parker and Wilby (2013), very few rigorous studies on domestic water use are conducted, and *'household water use is notoriously difficult to infer because it is shaped by local political, social, economic and meteorological factors; by changes in population, uptake of demand reduction measures, and technology, by price elasticity of consumption linked to household size; and by interplay between these drivers'*.

[6]UKRHA figures from 31st May 2006 to 31st August 2013.

Despite the lack of understanding of domestic water use, water efficiency campaigns are regularly carried out by water companies, with varying degrees of effectiveness, as discussed by Omambala *et al.* (2011), in the Waterwise Evidence Base project (a compilation of UK based water efficiency studies). It remains unusual for such campaigns to be based on any underlying theory of behaviour, or to draw on academic literature on environmental psychology, pro-environmental behaviour, or practice based approaches. 'Practice-based' approaches to water efficiency start from the perspective that water consumption is a consequence of the service provided, such as cleanliness, leisure and comfort (e.g., Browne *et al.* 2013). Understanding the service and the role that water plays in this service is therefore important, and Browne *et al.* (2013) argue that *'there is too much water in water demand research'*. The 'Patterns of Water' project (Pullinger *et al.* 2013), demonstrated an enormous range of practices in relation to a service (such as laundry, personal hygiene), and describe the difficulty of clustering people into behavioural groups. As discussed by Pearce *et al.* (2012), grounded theory could well be used to generate theories of water using behaviours, and undertaking sufficiently rigorous studies of existing behaviours may be a prerequisite for eliciting changes in those behaviours. It is certainly the case that water efficiency interventions stand in contrast to wider society, where messages and products are marketed at very specific groups of people, and the product itself is part of an aspirational vision based on generating an emotion within the potential consumer of the product, as opposed to any rational reaction.

Environmental psychologists subdivide conservation behaviours in relation to a number of underlying causes (Stern, 2000). In the context of household water use, family dynamics will have a strong influence via several underlying causes:

- habits and routines that are normal within the family unit – these are *'executed without deliberate consideration, and result from automatic processes, as opposed to controlled processes like consciously made decisions'* (Verplanken & Holland, 2002).
- via a more obvious and stated belief such as concern for the environment, or a more specific belief in relation to water.
- attitudinal factors in relation to a particular behaviour and whether or not it is perceived as beneficial.

Clearly, any daily household consumption may also mask variation of water uses within households; diary based studies, focus groups and interviews are all possible approaches, but even then, collecting data on water use by children compared to adults is problematic. For the purposes of the current study, the idea of encouraging children to consider a target saving of 6 litres/pupil/day as a curtailment behaviour was therefore based entirely on an equivalence with the predicted savings from a RWH system, as opposed to any data from actual household water uses.

More sophisticated approaches to understanding water using practices are clearly needed, given the lack of effectiveness of water efficiency campaigns as discussed above. However, these approaches stand in direct contrast to the very rational approach of standard educational practice. The approach to pupil engagement in the current study therefore followed a relatively conventional approach, as discussed in Bunn (2006). The engagement needs to be relevant, relate to actual situation, age appropriate, culturally sensitive with achievable targets that are clearly explained. Educational input should start with how the school is currently performing with regard to their water consumption, highlighting where they are doing well. It should cover the environmental reasons to reduce water use, and the advantages and workings of the technological upgrades. The engagement is related to wider environmental aspects of the National Curriculum where possible. School pupils are often set challenges during the school day. Effective behaviour change can make use of this fact by setting other challenges, for example, to save a certain amount of water at home. Behaviour change is more likely if people understand their own behaviour first and how that actually impacts on their own water use. By personalising water consumption and patterns of use, the most effective savings can be highlighted for each pupil. Learning materials should be designed to be taken home so that behaviour change within the family is also influenced. It is recognised that, just as a technological solution requires maintenance to keep it performing at its optimum, that the message around behaviour change should be repeated to maintain the changes achieved.

Given the importance of combatting climate change, and the fact that children study it as part of the National Curriculum, the relationship between water use and CO_2 emissions formed part of the educational projects described below, and is therefore reviewed here. In the UK it takes 1.2 kWh of mostly electrical energy to supply 1 m^3 of cold water to a building and to clean the resultant 1 m^3 of wastewater, and given the carbon intensity of the UK grid (0.57 $kgCO_2$/kWh), this means that 1 kg of CO_2e is produced for every cubic metre of water supplied and treated. This comprises 0.6% of total UK emissions (Environment Agency, 2008).

However, when water is heated for showers, baths etc, its carbon load increases greatly[7]; CO_2 emissions from domestic hot water are 6% of total UK emissions (Environment Agency, 2008). As shown in Figure 5.2 for a typical new home with gas central heating and hot water, whilst 46% of the CO_2 emissions are from water heated by gas (showers, baths, kitchen and basin taps), the volumes of water that are heated electrically (and therefore with a higher carbon intensity) for the washing machine and dishwasher have a disproportionately high impact.

[7]Unless the property heats its hot water by solar thermal or a biomass boiler or a wood stove with a back burner.

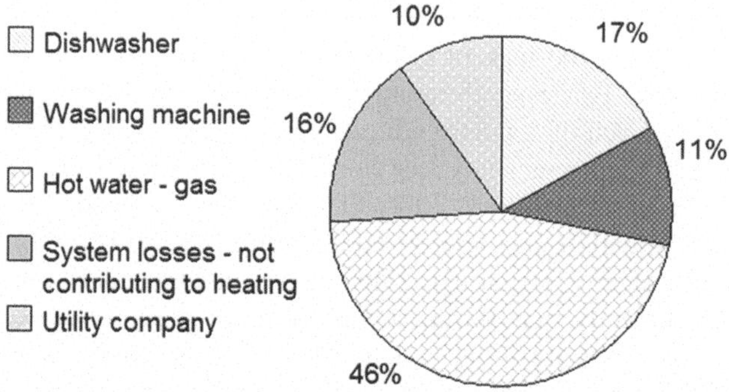

Figure 5.2 CO_2 emissions from domestic water use, showing both household emissions and those relating to the supply and treatment of water and wastewater (utility company). Example assumes a new build property with gas condensing boiler, occupancy of 2.4 and standard appliances. Data from Clarke *et al.* (2009).

5.6 RETROFITTING RWH SYSTEMS INTO LONDON SCHOOLS

As an illustration of the scope for mains water savings from RWH, the projected savings from four installations (two primary schools and two secondary schools) are given below.

Project details: RWH systems were retrofitted into four London schools in 2012, in a project funded by the Greater London Authority (GLA), the Environment Agency (EA) and Thames Water (TW). TW identified a series of schools in London that had higher than usual water usage, of which four were chosen to take part in the pilot project. Water consumption per pupil in the chosen schools ranged from 6.6 to 33.5 m^3 per pupil per year (Table 5.3), far higher than in a typical school which is 3.9 m^3/pupil/year for secondary schools and 3.8 m^3/pupil/year for primary schools (as shown earlier in Table 5.2).[8] Therefore there was considerable scope for water saving measures within these schools and, as part of the project, the reasons for the high consumption were identified and later rectified.

As all the RWH systems were retrofitted, the roof area from which rainwater could be collected was lower than the total roof area of the buildings (Table 5.4). Projected yearly yield was calculated using Equation 5.1 and is based on the roof area as stated, an average rainfall of 600 mm a year, a drainage factor of 0.85 (pitched roof) and a filter efficiency of 0.9. Tanks were sized by the company providing the RWH system to match demand with yield as far as possible, with the exception of PS1, where attenuating stormwater runoff into a local stream

[8]The high water consumption in the two secondary schools was due to a combination of factors, including leakage, uncontrolled urinals and low pupil numbers. Wastage of water in schools with smaller pupil numbers always impacts greatly on benchmark figures.

was a priority and therefore a larger tank was installed. As all storage tanks were above ground, tank volumes were lower than if underground and in three of the four schools did not meet the 5% of yield calculated according to the intermediate sizing formula recommended in BS8515:2009.

Table 5.3 Benchmarking School's Water Consumption – from Retrofitting RWH Systems into London Schools.

School	Pupils in school[a]	Actual water consumption in m^3 (2011)	Cost to school in 2011 at \$3.08/m^3	Cost to school in 2011 at £1.88/m^3	Water consumption (m^3/pupil/year)
SS1	80	814	2507	1530	10.2
SS2	153	5121	15,773	9627	33.5
PS1	400	2646	8150	4974	6.6
PS2	238	1559	4802	2931	6.6

[a]The secondary schools in this project were for students with special educational needs and pupil numbers in those schools are far smaller than in mainstream secondary schools.

Table 5.4 Calculated potential savings from RWH in the schools.

School	Total roof area (m^2)	Roof area collected from (m^2)	Tank volume (m^3)	Projected annual yield (m^3)	Projected annual savings from water bill (at \$3.08/m^3)	Annual CO_2 savings in ($kgCO_2e$[a])
SS1	1019	350	5	161	496	161
SS2	1250	475	5	218	671	218
PS1	1100	350	10	161	496	161
PS2	1634	280	3	129	397	129
Total				669	2061	669
Average per school				167	514	

[a]The CO_2 savings do not take into account any CO_2 produced by pumping the collected rainwater

Table 5.5 takes data from both Tables 5.3 and 5.4 to show that the RWH systems retrofitted in these schools had the potential to save between 0.4 and 2.0 m^3 of water per pupil per year. As the table shows, water usage at each school after this intervention was still far higher than in a typical UK school, and that retrofitting RWH is unlikely to be the priority solution in a school with high water consumption. The requirement for leaks rectification, appliance upgrades and behaviour change to reduce usage to the typical 3.9 m^3/pupil/year (secondary schools) or 3.8 m^3/pupil/year (primary schools) was high in all of these schools.

Table 5.5 Comparing water savings from retrofitting RWH systems against total water consumption.

School	Pupils in school	Water consumption (m^3/pupil/ year)	Projected annual yield from rainwater (m^3)	Projected mains water saving per pupil/ year (m^3)	Water consumption after RWH retrofitted (m^3/pupil/ year)
SS1	80	10.2	161	2.0	8.2
SS2	153	33.5	218	1.4	32.1
PS1	400	6.6	161	0.4	6.2
PS2	238	6.6	129	0.5	6.1

5.7 BE A WATER DETECTIVE

5.7.1 Project background and context

A UK Housing Association (with over 57,000 homes in England) wanted to support their tenants, many of whom have difficulty paying bills and are in fuel poverty, by demonstrating the savings that could be made on water and energy bills through simple behaviour change. The Housing Association was also concerned that tenants would struggle to cope with metered water bills following the universal metering programmes that were being implemented in many of the areas where they held property. Many families who change to paying for water through a meter (as opposed to the rateable value) find that their bills increase. At the same time, they wanted to raise awareness of the environmental cost of water. The Housing Association envisaged that engaging with their tenants might be easier to achieve via school age children in a household. Furthermore, by working in local schools the wider community could benefit; many of the pupils may not live in homes belonging to the Housing Association, but they and their families could still benefit from the advice given.

The Housing Association contracted ech$_2$o consultants to carry out the pupils engagement work. ech$_2$o have designed a pupil and family behaviour change programme called *Be Water Aware – Be a Water Detective*. The programme has been delivered in a series of schools across the UK. One of the core elements of this programme is that pupils go homes and ask family members how they use water. The workshops have been designed following many years work with both adults and children around sustainable water use. ech$_2$o have had many comments of how once a person started to think about water, they realised they were using more than they needed for a particular task and so changed their behaviour and cut down on excess water use. These anecdotal comments were a particularly strong driver in the design of the programme.

ech$_2$o built on the *Be Water Aware – Be a Water Detective* programme to deliver the pupil engagement programme for the Housing Association. For pupils, the message was primarily focussed on environmental reasons to save water, such as reducing the pressure on local rivers, and the reduction in CO_2 emissions from using less hot water. A series of assemblies and workshops were delivered to pupils in all the schools to emphasis this message, followed by a challenge to save 6 litres of water per day, with advice on how that figure could be achieved. For households, the emphasis was on the monetary savings from using less water. Hot water adds \$374 (£228), approximately 16%, to the average annual combined energy bill (Energy Saving Trust, 2013). In new, small, well insulated flats, the £228 required for hot water use can be as much as is required to heat the dwelling.

Schools were identified by the Housing Association in areas where Southern Water and South East Water were implementing universal metering, or in areas where the percentage of the Housing Association's homes among the school catchment area was particularly high. Five schools were chosen to take part in the project. The schools are identified by their initials and whether primary or junior schools. In each of the five schools, the following approaches and activities were used for behavioural change interventions:

- Water use benchmarking and discussion with teachers/facilities staff
- School assembly
- Leaflet
- Water audit

1284 pupils attended a *'Be Water Aware'* assembly and took home a leaflet, and 213 pupils attended *a 'Be a Water Detective'* workshop.

5.7.2 Water use benchmarking and discussion with teachers/facilities staff

Where possible, water consumption in each school was benchmarked against other UK schools.[9] Table 5.6 shows benchmarked results for four of the five schools involved in the *'Be a Water Detective'* project. Water consumption ranged from 3.4–10.5 m^3 of water/pupil/year against 3.8 m^3/pupil/year for the typical UK primary or junior school. Information was provided to the head and bursar in each school to show how much the school spent a year on water and how efficient the school was compared to other UK schools. Where water consumption was high, the schools were guided in identifying where excessive use could be occurring and advice was given on the most cost effective ways to rectify the excessive use. For example, ech$_2$o noted that PS2 had three uncontrolled urinals that were wasting 152 m^2/water/year. They also have an outdoor swimming pool, which requires periodic draining down and re-chlorination, adding greatly to the water load. ech$_2$o provided a written report for the school which recommended and priced new urinal controls and an insulated pool cover.

[9]Data could not be obtained for one of the schools.

Table 5.6 Benchmarking schools' water consumption – from *Be a Water Detective* project.

School	Pupils in school	Water consumption in m³ (2011)	Cost to school of 1 m³ of water (2012)[a] ($)	Cost to school (2012 prices) ($)	Water consumption m³/pupil/year
JS1	308	no data	2.71	n/a	n/a
JS2	313	1064	5.94	6320	3.4
PS1	273	1514	5.94	8993	5.5
PS2	180	1896	4.46	8456	10.5
PS3	210	832	5.94	4942	4.0

[a]Water for JS2, PS1 and PS3 schools is supplied by South East Water. Water for PS2 School is supplied by Portsmouth Water.

5.7.3 *Be Water Aware* school assembly

The assembly consisted of three main sections. After an initial introduction by the teacher, the ech$_2$o facilitator showed the school's water use in a simple graphic form to emphasise how much water the school used over the previous years. Any notable changes were highlighted and discussed as shown in Figure 5.3. Pupils were also shown how their school performs compared to a typical school and a water efficient school (Figure 5.4). The pupils were asked to consider how the school could become more water efficient. The link between how much water their school used with the environmental pressures on local rivers due to over abstraction, and CO_2 emissions from heating hot water with its effect on climate change was highlighted.

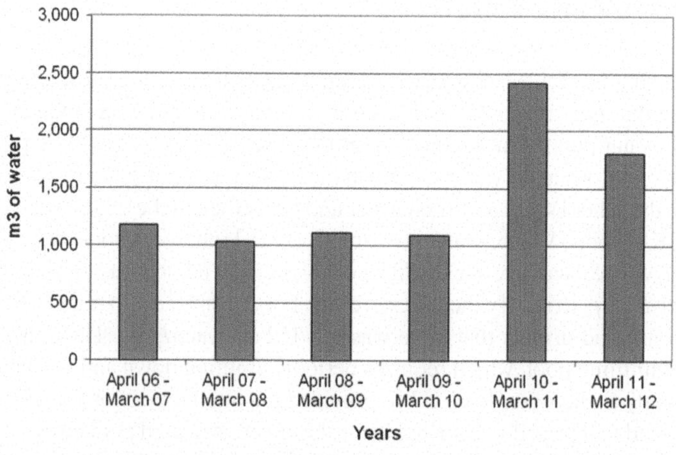

Figure 5.3 Annual water consumption of PS2.

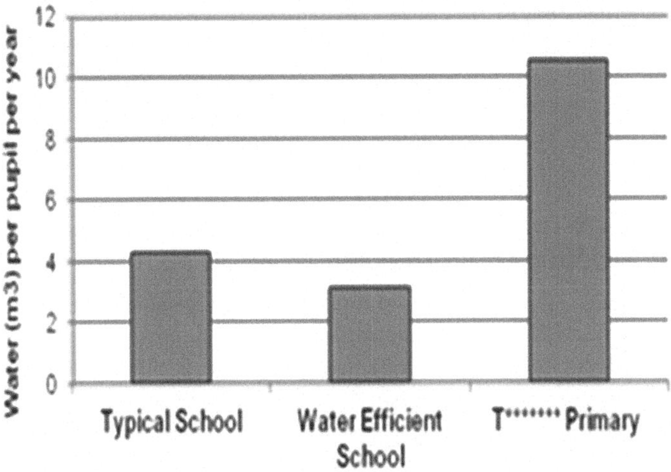

Figure 5.4 Benchmarking water consumption of PS2.

The second section of the assembly asked pupils where they thought people usually used most hot water and to identify simple ways to save water (both hot and cold) at home and at school. The assembly concluded with a challenge for all pupils to save 6 litres of water a day coupled with advice about which measures would achieve such a saving, and the cumulative savings (Table 5.7) if all the pupils met this simple target, showing the power of collective action.

Table 5.7 Yearly savings by meeting the 6 litres/day challenge.

School	Number of pupils in school	Yearly savings if all pupils meet the 6 litres saving a day challenge (m^3)
M. Junior School	308	675
M. Primary School	273	598
S. Junior School	313	685
T. Primary School	180	394
W. P. Primary School	210	460
Total	1284	2822

5.7.4 Leaflet

Information was delivered to over 1000 households about the cost of a unit (m^3) of water and how much money they could save on their water bills by taking shorter showers or shallower baths. The information was in the form of a leaflet designed by ech$_2$o that the pupils took home in their book bags. The leaflet also contained information about support for households who were having difficulty paying bills for example, Watersure (a UK scheme that caps the bills of vulnerable customers regardless of water use) (OFWAT, 2014) and any local initiatives. All monetary savings were based on the cost of water in the area where the school was (as compared to 'the average UK water price' which is usually used and is unhelpful as actual water costs vary significantly across the UK).

5.7.5 *Be a Water Detective* Water audit

ech$_2$o has devised a water audit programme that encourages school pupils to be a water detective with their family members.

Pupils involved in the interactive workshop filled out a water audit form about how they used water and also identified the type of appliances they had at home. The pupils took the forms home to collect the same information from other household members. Everyone who answered the audit had to answer the following questions:

- How old are you? Over 18, 12–18, or up to 12?
- Are you a male or female?
- How many times do you bath or shower a week?
- Do you have a bucket bath, shallow, medium or deep bath?
- How long is your average shower time in minutes?
- Do you wash up in a bowl, with a soapy sponge, in a dishwasher or under running water?
- Do you turn off the tap when brushing your teeth?
- How often do you use the toilet or urinal per day when you are at school? (For ages 5–18 only)

The pupils were also asked to identify the following information about appliances in their homes:

- Do you have a water meter?
- Do you have a dual flush toilet?
- What type of shower do you have? A shower connected to bath taps, an electric shower, a mixer shower or a power shower?

The pupils filled out the form about their own behaviour and information about the appliances in their homes with help from the ech$_2$o workshop facilitators. The workshop facilitators were careful to remain neutral on how many times a person showers or baths, whether that is twice a week or 14 times a week. The focus is on saving water *when* showering or bathing. This also helps to ensure that pupils enter their actual behaviour not what they think is the right or wrong answer. It also means that when asking family members the message of how do you *actually* use water not how *should* you use water is carried across when the pupils are being water detectives at home. By filling out information about the appliances in their home in the classroom with the workshop facilitators also means that the appliances can be correctly identified. ech$_2$o have found that it is best to deliver the *Be a Water Detective* workshop with primary school pupils from Years 4, 5 and 6 (i.e., pupils aged 8 to 11). The value of getting the pupils to go home and ask all family members is that people have to actually think about how they use water.

In the final part of the workshop, pupils are asked to pledge one thing they will do to cut their water use. Two core messages are delivered for bath usage. When you are in the bath, unless you already have a shallow bath, fill it 2.5 cm less full.[10] A discussion is also held about whether those who have a deep bath could cut down to a medium bath. The core message for shower use to the pupils is to see how many pupils already meet the four minute shower challenge and set the challenge for those who do not. All pupils get a four-minute shower timer. If they do not have a shower at home, the workshop facilitators deliver a four minute bath challenge - to run the bath taps for no more than four minutes. The workshop facilitators talk about how they used to use water and the changes they have made and whether that was easy or hard. The role of the teacher is also important as they represent another adult who can identify their own potential for water saving and state their intention to change.

The results of the water audit carried out under '*Be a Water Detective*' programme are discussed in the section below.

5.8 THE WATER AUDIT

Section 5.7.5 provides details of the '*Be a Water Detective*' programme. In this section the results obtained from the water audit are discussed.

5.8.1 Behaviour

213 audit forms were taken home and 114 were returned. Overall this was a 55% return. The rate of return of the audit sheets between schools differed greatly; one school returned just 28% of audit sheets that were taken home, whereas in another school 76% of audit sheets taken home were returned.[11] Data was collected and analysed from 471 people across 114 households. People answering the audit are identified by gender and divided into three age groups (as shown in Figure 5.5). Average size of household was 4.1 people with a range of 2–9 household members.

[10]2.5 cm less from a full standard size bath is a saving of 19 litres (1.5 m × 0.025 m × 0.5 m × 1000 = 19 litres).

[11]Teachers stated that the rate of return of the water audit forms was similar to, or slightly higher than, the usual return rate for normal homework.

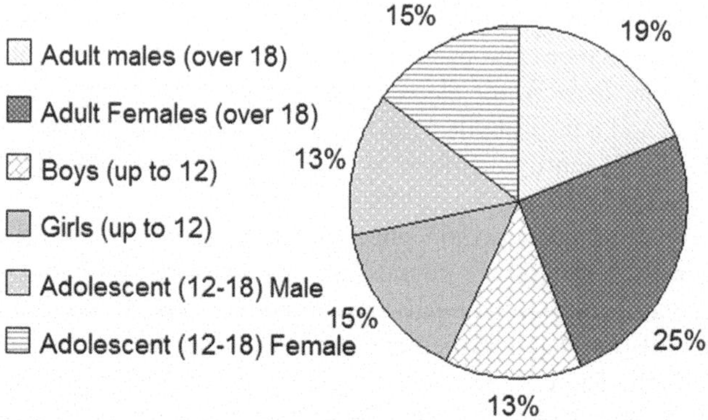

Figure 5.5 People who took part in the survey.

45% of respondents bath or shower once a day, 47% do so less than once a day, and 8% bath or shower more than once a day, as shown in Figure 5.6. 303 people who answered the survey, regularly have a bath. Of these, 105 only have a bath. The average number of baths is three times a week, with a range of 1–10. Of the 103 who regularly have a bath, 74 have a bath every day, (41 adults, 4 adolescents and 27 children). Of the 217 who bath less than once a day, 63 are adults, 19 are adolescents and 135 are children. 4 people have more than 1 bath a day. They are all adults. As Figure 5.7 shows, most people have a medium bath (57%). More people have a deep bath (29%) than have a shallow bath (11%). 3% of people have a bucket bath.

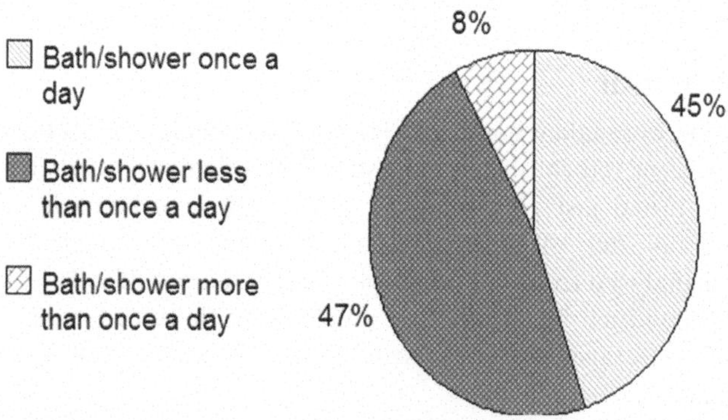

Figure 5.6 Bathing habits.

366 people regularly have a shower. 168 only have showers. The average shower time is 9 minutes, with a range from 2 to 48 minutes. The most popular

length of time in the shower is 5 minutes; the second most popular is 10 minutes (Figure 5.8). Average shower frequency amongst the respondents is 5 times a week, with a range of 2–14 times a week. Of those people who regularly have a shower, 122 have a shower every day. This breaks down as 80 adults, 14 adolescents and 28 children. 225 have a shower less than once a day, (78 adults, 24 adolescents and 123 children). 19 have eight or more showers a week. This is 13 adults, 4 adolescents and 2 children.

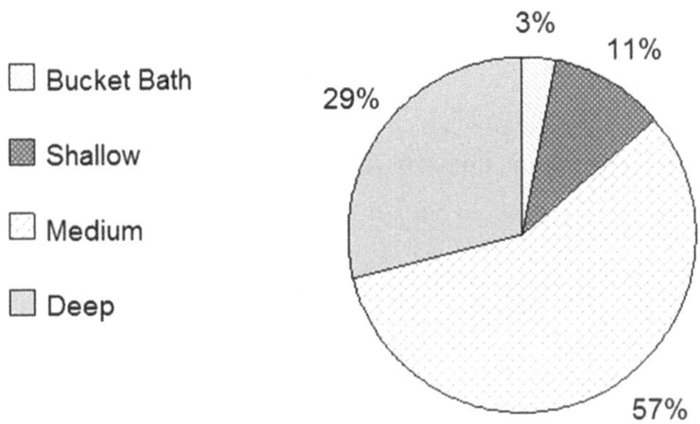

Figure 5.7 Responses to question; 'How deep is your normal bath?'.

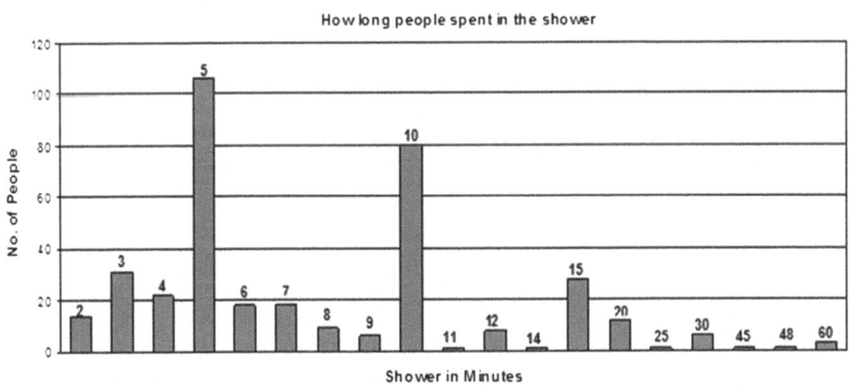

Figure 5.8 How long people spend in the shower.

Most people brush their teeth twice a day, every day, (97% in this survey). 15% of respondents leave the tap running when they are brushing their teeth as shown in Figure 5.9. Figure 5.10 shows most people in this survey have water efficient washing up habits with just 5% washing up under running water.

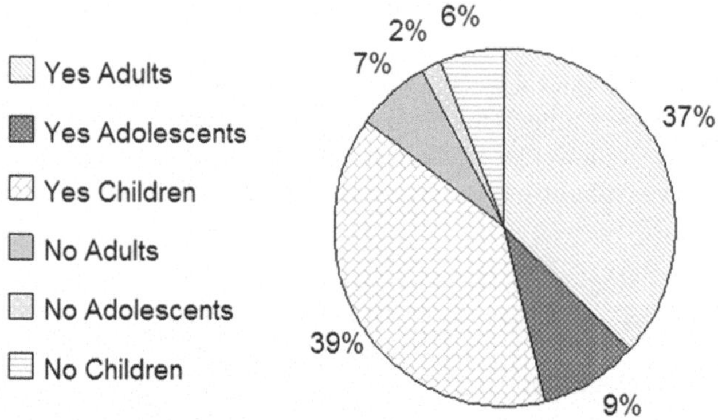

Figure 5.9 Responses to the question: 'Do you turn off the tap when brushing your teeth?'.

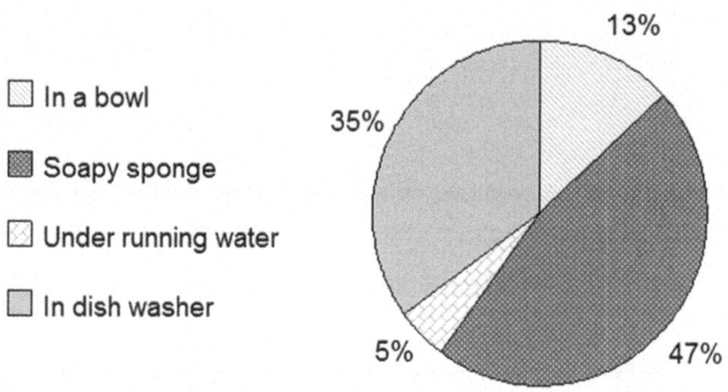

Figure 5.10 How people wash up.

5.8.2 Appliances

Information was also collected about some of the household appliances. Respondents were asked whether they had a dual flush WC, whether they were on a meter, and what type of shower they had. 46% of homes had at least one dual flush WC. 73% of households were on a meter, almost twice the UK average. However, this figure is not as surprising as it seems, given the fact that schools were targeted in areas where mass retrofitting of meters was occurring. As can be seen in Figure 5.11, 7% of households do not have a shower at all, and almost half of all households only have a shower connected to the bath taps. Just 17% have a thermostatic mixing shower and 23% have an electric shower. The schools in this study were in the catchment areas with a large amount of social housing which is likely to be the reason that shower ownership is less than the UK average.

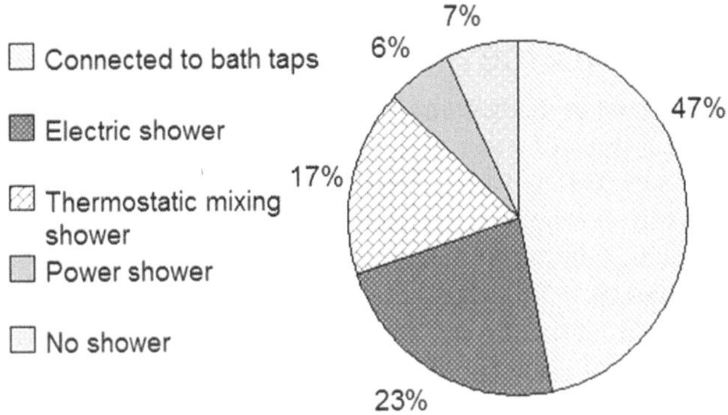

Figure 5.11 Types of shower in households.

5.8.3 Engagement and responsiveness

Whilst formal follow up surveys were not undertaken, at the end of every workshop pupils were asked via a show of hands whether they had a) learnt anything new about water, b) enjoyed the lesson, and c) would start saving water. Over 95% of all pupils replied yes to the questions. Comments from children were occasionally received on the returned water audit forms (although the forms did not ask for any feedback) with several children stating that they had met the four minute shower challenge. For example, '*Me and my dad done the 4 min challenge!!*' Boy pupil, Year 5 – PS1.

A lot of children commented during the lesson about other family members and how they used water. '*My sister needs the four minute challenge. She's in the shower for ages*', and '*My mum always has a deep bath*'. They also reflected on differing practices of visitors; '*My gran washes up under running water when she comes to stay*'.

Teachers reported via a feedback form that the workshops were useful and fun and they felt that the pupils engaged very well, for example '*The children really enjoyed the lesson and I am impressed by the number of returned audit forms*'.

Specific comments suggested that the teachers also expected that water savings would result:

> *"The children's response to your workshop has been very positive and there has been lots of talk about meeting the 4 minute shower challenge (including from the staff!)".*

> *"Thank you, the children really enjoyed the session. They will be pestering their families to save water".*

> *"The most useful part was for the children to start thinking about how much water they use and how easily they can save water".*

5.9 SAVING SIX LITRES OF WATER A DAY – WHAT DOES IT MEAN IN PRACTICE?

In the notional school at the beginning of this chapter (Section 5.1), pupils were required to save 6 litres of water a day to match the savings made from RWH. In the pilot schools in Retrofitting RWH Systems into London Schools project (Section 5.6), it was shown that the potential water savings from behaviour change by 670 pupils saving 6 litres of water a day is 1468 m^3 a year compared to 669 m^3 from the RWH systems that were installed, a factor of 2.2 times more water saved. If the 1284 pupils from the '*Be a Water Detective*' project also saved 6 litres of water a day, 2822 m^3 of water could be saved a year. Additional amounts saved at home by transference of such behaviours may increase this figure. So the question becomes, how likely that is to happen? How easy/difficult is to achieve a 6 litre/person/day reduction via various methods? Analysing the behavioural data in Section 5.8 enables a better understanding as to whether the six litres a day saving is likely. It is shown below that it is possible to make 6 litres a day saving from even a single water usage unless one is currently using water very efficiently. For example, if a person has a bucket bath or showers for two or three minutes under an electric shower, turns the tap off when they are brushing their teeth and washes up using a bowl, the potential for saving 6 litres of water a day is not high. However, as has been shown in Section 5.8 there are a great many people where it would be possible to save water and so this section considers what they would need to do to save the required 6 litres of water a day. As the workshop focussed on saving water from shorter showers and shallower baths these two uses are further discussed as below.

5.9.1 Saving 6 litres of water from a shallower bath

It is difficult for a householder to calculate savings from a bath as there is no measuring mechanism at the side of a bath and the amount of time spent in a bath is not the major factor as to how much water is used. '*Be a Water Detective*' data shows that average bath frequency among those surveyed is less than UK Government assumptions at 3 times a week compared to 4.76 (Market Transformation Programme, 2011a). Most people in the survey had a medium or deep bath, so saving 6 litres of water per bath is easy without a noticeable change in bath comfort. Saving 6 litres per bath from a shallow bath is far harder and from a bucket bath is virtually impossible.[12] To save an average of 6 litres per day (and assuming the person is not showering in the days when they are not having a bath), each person needs to save 14 litres each time they bathe. *A reduction in depth of 2.5 cm could provide this amount of saving in most baths even when the starting depth of the water is at the half way mark (medium bath).*

[12]Most people who take a bucket bath use one 10 litre bucket of water.

5.9.2 Saving six litres of water from a shorter shower

Shower flow rates vary, from 3 litres/minute (small electric showers) to over 20 litres/minute (power showers) and there are a range of flow rates within the separate shower types (Table 5.8, compiled from Clarke *et al.* 2009) and data measured by one of the authors).

Table 5.8 Shower flow rates.

Type of shower	Average flow rate litres/minute
Shower connected to bath taps	5–8
Electric shower	3–6
Thermostatic mixing shower	6–12
Power shower	15 and above

Note: Showers that are connected to bath taps are usually fed from a storage cistern in the loft, though can be fed from a combination boiler.

Therefore, saving 6 litres of water a day requires between 20 seconds to 2 minutes less in the shower, *if* showering once a day, which the UK Government considers is the norm (UK Government assumed frequency of 1.04 showers per person per day in 2010, rising to 1.12 in 2015, Market Transformation Programme, 2011c). Assuming that the average flow rate from the shower is 6 litres/minute, spending one minute less per shower will save the required 6 litres of water. The UK Government bases most of its calculations about shower use on the premise that the average length of a shower is five minutes. As Figure 5.8 showed, in this survey (and backed up by many other surveys, as summarised by Clarke *et al.* 2009) the average shower time is far greater (Figure 5.8). As stated earlier, for the respondents in this study, the average shower time is 9 minutes and the most popular length of time in the shower is 5 minutes with the second most popular being 10 minutes. Average shower frequency amongst the respondents is 5 times a week. Therefore, to save an average of 6 litres per day, each person needs to save 8.4 litres each time they shower. *This requires spending 1.4 minutes less in the shower, a 15% reduction from the average shower time of 9 minutes.* As part of the '*Be a Water Detective*' project, the four minute shower challenge produced an estimated average potential savings per pupil higher than 6 litres/day. As discussed by Shove and Walker (2010), shower using behaviour is socially constructed and people shower for many reasons. Consequently, there is a need to understand the underlying purposes and practices behind showering behaviour before considering what reductions might be realistic and how best to tailor water using behaviour discussions with individuals. Six basic clusters of washing/bathing behaviours were identified in Pullinger *et al.* (2013), and the scope for reducing shower duration will obviously differ between groups.

5.9.3 Saving six litres from brushing teeth

Turning off the tap when brushing your teeth saves 10 litres of water each time (assuming two minutes teeth brushing, tap flow rate of 6 litres/minute and allowing a generous 2 litres of water for wetting/rinsing etc.). However, as most people turn off the tap when brushing their teeth the overall potential of saving water is less than the water savings from baths and showers. In this survey, 15% of respondents leave the tap running when they are brushing their teeth. *If that relatively small number of respondents started to turn the tap off, they would save 20 litres each a day.*

5.9.4 Saving 6 litres of water from efficient washing up habits

Most of the respondents in this survey washed up very efficiently and so scope for saving 6 litres of water a day is relevant to a mere 5% of respondents, those who wash up under running water. For that small percentage, even one less minute with the tap running whilst washing up would save their required 6 litres.

5.9.5 Savings from WC flush

Working with pupils in in-depth workshops allows the demonstration of simple technological solutions for saving water in homes that will add to the savings from behaviour change.

Save-a-flush bags are silicon filled bags that when placed in a WC cistern swell up and displace one litre of water. They are designed for single flush WCs with a 7.5 or 9 litre flush volume, but can also work effectively in some 6 litre single flush WCs and are simple to fit. The Market Transformation Programme (2011b) states that average WC flush is 4.71 flushes per day at home and a save-a-flush bag can save up to 1 litre of water per flush. Assuming an average saving of 0.5 litres per flush, if a household does not have a save a flush bag and subsequently fits one, over 2 litres of water can be saved per person a day without any behaviour change required. In '*Be a Water Detective*', average household size is 4.1. *Therefore if a pupil takes a save-a-flush bag home and fits it, the savings attributed to that pupil can be calculated at 10 litres per day.* However, as 46% of households in this survey already had a dual flush WC, no savings can be made from reducing the WC flush in those properties.

Based on the discussion above, Table 5.9 presents a series of measures (changes in consumption behaviour) which can help to reduce per capita water consumption.

5.9.6 CO_2 savings

Assuming that the water saved from behaviour change is hot water for showers and baths, CO_2 savings will be considerable. It is difficult to calculate exactly how much, as showers may be heated by gas or electricity and hotter water is required for baths than for showers. As stated earlier, using data collected from '*Be a Water*

Detective' 105 people only had baths, 168 people only had showers and 198 people had a mixture of both. Assuming that people who have a mixture of both baths and showers make their 6 litre savings equally between the two, 43% of the savings come from saving bath water and 57% from saving shower water. Adding the information that 25% of those who had a shower had an electric shower, it can be seen from Table 5.10 that total CO_2 savings from the 360 pupils in the notional school (Section 5.1) would be 7 tonnes of CO_2 a year.

Table 5.9 Summary of how to achieve 6 litres of water saving per day from behaviour change.

Behaviour	Method	Litres saving achieved per use (to ensure a daily saving of 6 litres)
Taking a bath	Reduce bath depth by 2.5 cm	14
Taking a shower	Spend 1.4 minutes less in the shower	8
Brushing teeth	Turn tap off when brushing teeth	20
Washing up	Do not wash up under running water	6 litres for every minute tap is no longer running
Flushing the toilet	Fit a *Save-a-flush* bag in the WC cistern	10 for whole household

Table 5.10 CO_2 savings from 788 m^3 of hot water[13].

Savings from	Percentage of savings	Water saved (m^3)	Gas use related energy saved (kWh)	Electricity saved (kWh)	CO_2 saved (kg)
Electric shower	14	110	n/a	3420	1860
Bath	43	339	14,909	n/a	2758
Thermostatic mixing shower or shower connected to bath taps	43	339	13,215	n/a	2445
Total			28,124	3420	7063

[13] Assumptions: to heat 1 m^3 shower water requires 31 kWh electricity, or 39 kWh gas (Omambala *et al.* 2011). To heat 1 m^3 water for a bath requires 44 kWh gas (Omambala *et al.* 2011). The carbon contents of gas and electricity are taken as 0.185 $kgCO_2/kWh$ and 0.544 $kgCO_2/kWh$ respectively (both from Carbon Trust, 2011).

5.10 DISCUSSION

As discussed by Ward *et al.* (2012), very little published data exists on post-occupancy monitoring of the mains water savings from installing RWH systems, particularly in the UK context. Despite this, they are widely recommended in policy documents. In a study by Ward *et al.* (2012), a system in an office with a 1500 m^2 roof area and an 807 mm annual rainfall resulted in an 87% saving of mains water for WC flushing. However, the building was operating at approximately 1/3 of its design occupancy, so the storage tank was significantly oversized. Extrapolation to the predicted occupancy level resulted in the mains water saving decreasing to approximately 35%. This type of calculation is complicated by the sizing method used for the storage tank (three potential methods are indicated in BS 8515), and the way in which data is analysed (e.g., to account for periods when the system is not working, or to adjust for occupancy patterns). The financial value of the mains water saved was under £500/year, despite the building being located in the South West (which has the highest water and sewerage charges in England and Wales). Furthermore, as discussed by Roebuck *et al.* (2011), since factors such as maintenance and replacement costs, and the discount rate and discount period are often not considered when considering whole-life costing of RWH systems, it is far from clear that RWH systems represent a cost-effective approach to water demand management.

Post-occupancy monitoring data was not available for the 4 RWH systems in the schools described in Section 5.6, and in the absence of a dataset comparing projected savings with actual savings, it is inadvisable to extrapolate. Consequently, the notional mains water saving of 6 litres/person/day (Table 5.7) used as the basis for comparison with water efficiency measures in the current study is simply an estimate, but it is worth considering this in the context of how else a similar saving could be achieved. As shown in Table 5.9, similar savings could very easily be made via curtailment behaviours. Clearly, future studies should be designed to monitor these behaviours, although as already discussed; collecting domestic micro-component data is complicated. Furthermore, the effectiveness of water saving interventions is known to fade over time; Fielding *et al.* (2013) demonstrated that water use had returned to pre-intervention levels a year post-intervention in an Australian study (where householders were metered), although Omambala *et al.* (2011) report water savings are maintained 2–3 years post-intervention in four UK based studies.

Appliance ownership varies considerably with demographic group, and in general fewer thermostatic mixing showers are fitted in social housing. The Market Transformation Programme (2011c) assumes 42% ownership of mixer showers (standard or power) rising to 45% in 2015 and 39% ownership of electric showers in 2010, rising to 44% in 2015[14]. A survey carried out by a UK water

[14] Shower ownership is expected to increase even further, to 46% in 2020, 47% in 2025 and 48% in 2030. (Market Transformation Programme, 2011c)

company (Sutton & East Surrey Water, 2013) showed that 49% of their customers have a thermostatic mixing shower, 24% have a power shower, and 26% had an electric shower. As seen earlier, electric shower ownership in the current study is less than assumed by Defra at 23% and that the number of households that have a thermostatic mixer shower is far less at just 17%. 7% of households do not have a shower at all, and almost half of all households have a shower connected to the bath taps[15]. These differences serve to illustrate the importance of collecting data on appliance ownership and type in the population of interest, rather than using nationally averaged figures.

Whilst self-reported data on water using behaviours is notoriously inaccurate (Beal *et al.* 2011), some interesting points emerge. Firstly, in this population, the self-reported frequency of showering and bathing (Figure 5.6) is considerably lower than that reported in other studies; the Market Transformation Programme (2011a) assumes a bathing frequency (among households owning a bath) of 0.68 per person per day in 2010, falling to 0.66 in 2015[16]. Average showering frequencies are considered to be around twice those of bathing, (Herrington, 2006) and the Market Transformation Programme (2011c) assumes a shower frequency per person per day of 1.04 in 2010, rising to 1.12 in 2015[17]. Results from this survey show an average showering frequency of 5 times a week, or 0.7 times a day. These differences from national average figures may reflect the age or demographic of the population, and clearly there will be a link with appliance ownership (the population had a low penetration of power, thermostatic mixer, and electric showers compared to the wider UK population, Figure 5.11). There do not appear to be any other studies reporting showering/bathing frequency according to age to compare to our results, and simply note from the self-reported data presented here, that the frequency increases with age.

The accuracy of shower duration estimates is likely to be poor; the peaks of commonly understood numbers such as 5, 10, and 15 minute durations stated (as opposed to a minute longer or shorter) are very clearly demonstrated in Figure 5.8. Nevertheless, the positive skew on the distribution is consistent with that found in studies conducted with micro-component monitoring (e.g., Waylen *et al.* 2007), which demonstrated that median shower durations are a more appropriate summary statistic than mean.

It was not regarded as realistic to ask those surveyed to estimate an actual bath volume, but note that the number of deep baths seems relatively high, in contrast to the commonly used assumptions of 40% volume to overflow, based on Chambers

[15] The schools in this study were in catchment areas with a large amount of social housing which is likely to be the reason that shower ownership is less than the UK average.

[16] Bath frequency is expected to decrease even further to 0.63 in 2020, 0.61 in 2025 and 0.58 in 2030 (Market Transformation Programme, 2011a).

[17] Shower frequency is expected to increase even further 1.21 in 2020, 1.27 in 2025 and 1.33 in 2030 (Market Transformation Programme, 2011c).

et al. (2005). The study also confirms that the 'bucket bath' is a practice that exists in the UK today, usually by people from countries such as India, Pakistan and Bangladesh in circumstances where there is no access to a shower.

5.11 FINAL REMARKS

Whilst RWH could potentially meet WC and urinal flushing demand in many schools in the UK, it is problematic to retrofit and therefore a costly solution to reduce the demand for mains water and remove stormwater from the drainage system. In situations where there is a need to reduce stormwater runoff (such as in school PS1 in the current study) and at schools that drain surface water into combined sewers, lower technology approaches than RWH for WC flushing are preferable. Rain gardens, or collecting water for a school allotment are possible, and could easily be linked to educational activities. With regard to water savings, the wide range of mains water uses per pupil in the schools reported here demonstrates that in most instances it is likely that there are more cost effective measures to save water on the school estate than the installation of RWH. Beyond the school environment, a 6 litre/person/day reduction in water use would be strikingly easy to achieve in many ways, and as demonstrated, could result in significant energy and CO_2 savings if it was via curtailing a hot water using behaviour. Regardless of the efficiency behaviour chosen by the school, the potential for combining it with measures to increase curtailment behaviours, including beyond the school environment should not be neglected. Whilst there is as yet little evidence for the effectiveness of these approaches, it is a low cost intervention that shows considerable scope for community engagement and outreach, particularly for Housing Association tenants, who represent a group at high risk of water and fuel poverty.

REFERENCES

Austin A., Cox J., Barnett J. and Thomas C. (2011). Exploring Catalyst Behaviours. Full Report, Department for Environment, Food and Rural Affairs (Defra), UK. http://randd.defra.gov.uk/Default.aspx?Menu=Menu&Module=More&Location=None&Completed=0&ProjectID=16324 (accessed 23 May 2014).

Barlow S. (2011). Guide to BREEAM. RIBA, London, UK. ISBN: 9781859464250.

Beal C., Stewart R. A., Huang T. T. and Rey E. (2011). SEQ residential end use study. *Water: Journal of the Australian Water Association*, **38**, 92–96.

Browne A. L., Medd W. and Anderson B. (2013). Developing novel approaches to tracking domestic water demand under uncertainty. A reflection on the 'up scaling' of social science approaches in the United Kingdom. *Water Resource Management*, **27**, 1013–1035.

British Standards Institute (2009). BS 8515:2009 + A1:2013. RWH systems – Code of practice.

Bunn (2006). Sustainable Schools – Getting It Right. British Council for School Environments (BCSE), London, UK.

Butler D. and Davies J. W. (2010). Urban Drainage. CRC Press, Florida, USA, ISBN: 978-0415455268

Carbon Trust (2011). Conversion Factors, CTL153. www.carbontrust.com/media/18223/ctl153_conversion_factors.pdf (accessed 23 May 2014).

Chambers C., Glennie K. and Marshallsay D. (2005). Increasing the Value of Domestic Water Use data for Demand Management. Final Report, WRc P6832, Swindon, UK.

Clarke A., Grant N. and Thornton J. (2009). Quantifying the Energy and Carbon Effects of Water Saving. Final report for Energy Savings Trust and Environment Agency, London, UK. http://www.energysavingtrust.org.uk/Publications2/Housing-professionals/Heating-systems/Quantifying-the-energy-and-carbon-effects-of-water-saving-summary-report (accessed 23 May 2014).

Department of Communities and Local Government (2010). Code for Sustainable Homes – Technical Guide. November 2010, London, UK, ISBN 9781859463314.

Defra (2008). Future Water: The Government's Water Strategy for England. Defra, London, UK. Ref: PB13562. https://www.gov.uk/government/publications/future-water-the-government-s-water-strategy-for-england (accessed 23 May 2014).

Defra (2013). The Water Bill. Briefing notes overview. Defra, London, UK. https://www.gov.uk/government/collections/water-bill-briefing-notes (accessed 23 May 2014).

DfES (2003). Energy and Water Benchmarks for Maintained Schools in England: 2002–03. Statistical First Release, London, UK. http://webarchive.nationalarchives.gov.uk/20130401151655/http://www.education.gov.uk/researchandstatistics/statistics/allstatistics/a00194239/energy-and-water-benchmarks-for-maintained-schools (accessed on 23 May 2014).

Duggin J. and Reed J. (2006). Sustainable Water Management in Schools. W12. Based on Research Project RP716. Construction Industry Research and Information Association (CIRIA). London, UK. ISBN 0860176304.

ech$_2$o (2012). Reducing WC Flush Volumes in Schools – Are we Overestimating the Water Savings? ech$_2$o, London, UK, http://www.ech2o.co.uk/downloads/WC%20use%20in%20schools%20-%20an%20ech2o%20report%202012.pdf (accessed 23 May 2014).

Energy Saving Trust (2013). At Home with Water. Energy Saving Trust Report, http://www.energysavingtrust.org.uk/About-us/The-Foundation/At-Home-with-Water (accessed 23 May 2014).

Environment Agency (2008). Greenhouse gas emissions of water supply and demand management options. Science Report – SC070010. Environment Agency, Bristol, UK.

Fielding K. S., Spinks A., Russell S., McCrea R., Stewart R. and Gardner J. (2013). An experimental test of voluntary strategies to promote urban water demand management. *Journal of Environmental Management*, **114**, 343–351.

Gardner G. T. and Stern P. C. (1996). Environmental Problems and Human Behavior. Allyn and Bacon, Boston, USA, ISBN 9780205156054.

Hammersmith and Fulham (2014). School Case Studies 2011 and 2012. http://www.lbhf.gov.uk/Directory/Environment_and_Planning/Carbon_reduction/Green_schools/182389_School_case_studies_2011_and_2012.asp (accessed 02 September 2014).

Herrington P. R. (2006). The economics of water demand management. Chapter 10, In: Water Demand Management, D. Butler and F. A. Memon (eds), International Water Association, London, UK, ISBN 1843390787.

Market Transformation Programme (2008). BNWAT28: Water Consumption in New and Existing Homes. Defra, London, UK. http://efficient-products.ghkint.eu/cms/product-strategies/subsector/domestic-water-using-products.html#viewlist (accessed 23 May 2014).

Market Transformation Programme (2011a). BNWAT03: Baths: Market Projections and Product Details. Defra, London, UK. http://efficient-products.ghkint.eu/cms/product-strategies/subsector/domestic-water-using-products.html#viewlist (accessed 23 May 2014).

Market Transformation Programme (2011b). BNWAT01 WCs: Market Projections and Product Details. Defra, London, UK. http://efficient-products.ghkint.eu/cms/product-strategies/subsector/domestic-water-using-products.html#viewlist (accessed 23 May 2014).

Market Transformation Programme (2011c). BNWAT02: Showers: Market Projections and Product Details. Defra, London, UK. http://efficient-products.ghkint.eu/cms/product-strategies/subsector/domestic-water-using-products.html#viewlist (accessed 23 May 2014).

Memon F. A. and Butler D. (2006). Water consumption trends and demand forecasting techniques. Chapter 1, In: Water Demand Management, D. Butler and F. A. Memon (eds), International Water Association, London, UK, ISBN 1843390787

OFWAT (2014). WaterSure (vulnerable groups scheme). http://www.ofwat.gov.uk/consumerissues/assistance/watersure/ (accessed 02 September 2014).

Omambala I., Russell N., Tompkins J., Bremner S., Millar R., Rathouse K. and Cooper M. (2011). Evidence Base for Large-Scale Water Efficiency. Phase II Final Report. Waterwise, London, UK. http://www.waterwise.org.uk/resources.php/12/evidence-base-for-large-scale-water-efficiency-phase-ii-final-report (accessed 23 May 2014).

Parker J. M. and Wilby R. L. (2013). Quantifying household water demand management. *Water Resource Management*, **27**, 981–1011.

Pearce R., Dessai S. and Barr S. (2012). Re-framing environmental social science research for sustainable water management in a changing climate. *Water Resource Management*, **27**, 959–979.

Pullinger M., Browne A., Anderson B. and Medd W. (2013). The Water Related Practices of Households in Southern England, and their Influence on Water Consumption and Demand Management. Final report of the ARCC-Water/SPRG Patterns of Water projects. Available from http://www.sprg.ac.uk/projects-fellowships/patterns-of-water (accessed 6 January 2014).

Russell S. and Fielding K. (2010). Water demand management research: A psychological perspective. *Water Resources Research*, **46**(5), W05302.

Shove E. and Walker G. (2010). Governing transitions in the sustainability of everyday life. *Research Policy*, **39**, 471–476.

South West Water (2013). Charges Tariff Summary. https://www.southwestwater.co.uk/media/pdf/8/n/charges_tariff_summary.pdf (accessed 15 January 2014).

Stern P. C. (2000). Towards a coherent theory of environmentally significant behaviour. *Journal of Societal Issues*, **56**(3), 407–424.

Sutton and East Surrey Water (2013). Draft Water Resources Management Plan. http://www.waterplc.com/pages/about/publications/water-resources-and-drought-reports/draft-water-resources-management-plan-2013/ (accessed 15 January 2014).

Thames Water (2013). Metered Charges. Water and wastewater charges if you have a water meter. 2013–2014. http://www.thameswater.co.uk/tw/common/downloads/your-home/Metered_Charges_2013-2014.pdf (accessed January 15 2014).

Thornton J. (2013). Choosing Ecological Water Supply and Treatment. CAT Publications, Machynlleth, UK, ISBN: 978-1902175683.

UKRHA (2014). Figures supplied in an email from Terry Nash, Secretary of UKRHA. Personal communication to C Hassell on 3 January 2014.

Verplanken B. and Holland R. W. (2002). Motivated decision making: Effects of activation and self-centrality of values on choices and behavior. *Journal of Personal and Social Psychology*, **82**(3), 434–447.

Ward S., Memon F. A. and Butler D. (2012). Performance of a large building RWH system. *Water Research*, **46**, 5127–5134.

Water Supply (and Fittings) Regulations (1999). Statutory instruments. 1999 No. 1148. Water Industry, England and Wales. UK government, London, UK. http://www.legislation.gov.uk/uksi/1999/1148/contents/made (accessed 23 May 2014).

Waylen C., Bujnowicz A., Saddique S. and Kowalski D. (2007). Provision of Identiflow® Analysis for the Highland Park Water Efficiency and Seasonal Tariff Trial. UC7541, WRc plc, Swindon, UK.

Woods-Ballard B., Kellagher R., Martin P., Jefferies C., Bray R. and Shaffer P. (2007). The SuDS Manual. CIRIA, London, UK. ISBN 9780860176978.

Chapter 6

Community participation in decentralised rainwater systems: A Mexican case study

Ilan Adler, Luiza C. Campos and Sarah Bell

6.1 INTRODUCTION

Rainwater harvesting (RWH) is commonly discussed as a source of non-potable water to reduce demand for potable water from the mains network in cities. However, in remote and rural areas, which are not served by a mains drinking water network, rainwater is an important option for potable water supply. In such cases, the treatment and storage of the rainwater prior to use is critical to ensure that the water is safe for potable use. The public health risks of failure of these systems are much higher than for non-potable use.

The failure of water supply systems, both centralised and decentralised, can be attributed to many different causes, including social, technical and natural risks. In centralised urban water systems, the responsibility for the safe operation of the system and management of risk is delegated to the local water utility, which is usually able to employ highly skilled and specialised engineers and managers. In remote and rural locations, the responsibility for operating the water system often falls to local residents, who may not have specialist knowledge of water technology and management and who often have competing demands and responsibilities within their community. Community participation and technical capacity building are therefore vital in ensuring the success of remote water supplies, including rainwater harvesting systems.

This chapter presents a case study of the implementation of rainwater harvesting for potable water supply in rural communities in Mexico. It analyses the reasons for success and failure of systems implemented in schools, health centres and community halls in the San Miguel de Allende district, in Mexico, since 2007. A comparison of successful and failed systems shows that the key factors for success include the level of involvement of the end users in maintenance and operation, and the availability of technical support, training and replacement parts. In the case of RWH systems for communal buildings, rather than individual residences, knowledge transfer and the succession of responsibilities are also important factors determining success. Natural changes in the community, such as the election of new local government representatives, a change in parents at the school as children graduate, or staff turnover in community health centres, can undermine technical knowledge and responsibility for maintaining and operating the water system, contributing to system failure.

The chapter begins by describing the case study site and the technical design of the RWH systems that were implemented in 13 rural communities in San Miguel de Allende. It describes in detail cases of complete failure and abandonment of two of these systems and identifies factors contributing to these failures, as well as factors contributing to the short-term failure or poor performance of other systems in the programme. Community participation and leadership are analysed as the key success factors for community RWH systems, including the capacity for communities to deal with technical complexity. The chapter concludes with recommendations for community engagement and design of RWH systems for potable supply in rural areas.

6.2 BACKGROUND

6.2.1 Site description

The municipality of San Miguel de Allende is located 274 kilometres north of Mexico City, in the largely semi-arid State of Guanajuato. Its population is close to 160,000 inhabitants according to the 2010 census (INEGI, 2011), with 46% living in the main town of San Miguel and 54% living in smaller rural communities. Each rural community is represented in the local government by an elected 'delegate' (*Delegado*). Local government offices are based in the town of San Miguel, which is the main commercial and cultural hub of the area. The town of San Miguel de Allende was named as a UNESCO's World Heritage site in 2008. San Miguel's rich cultural and historical legacy attracts large numbers of tourists and foreign retirees all year round, driving up the prices of real estate around the city centre (Dixon *et al.* 2006). The main economic activities are tourism and agriculture (Garcia y Garcia, 2006), with minimal industrial activity.

In contrast to the prosperous main town, the surrounding rural communities live in conditions of abject poverty and receive far less public services. Some communities can only be accessed by dirt roads, which can become blocked during the rainy season, compounding the sense of isolation. Sanitation, garbage collection and water supply are dismal in many cases and tend to get increasingly worse the farther away the community is from the main town. San Miguel has one of the highest 'inequality rates' in Mexico (Székely *et al.* 2007). The link between water scarcity and poverty is inextricable and may be compounded by external factors such as climate change (Hemson *et al.* 2008; Stoddart, 2009).

Average precipitation in the region is 400–600 mm/year, with internal variations from one part of the municipality to another (SMN, 2010). The rainy season normally starts towards the end of May and finishes in early October, followed by a long dry period. These patterns, however, have been changing in recent years, with sudden heavy storms occurring in the middle of the dry season, or rains starting later than expected. In February 2010, at the start of fieldwork for the research presented in this chapter, 151 mm of rainfall were recorded in the State of Guanajuato, compared to a 6.5 mm average over the past 50 years (CAN, 2010b).

In 2007, the Ecology and Environment Department of the San Miguel de Allende municipality initiated a series of pilot RWH projects throughout the region, mostly in rural primary schools and a small number of health clinics. RWH was implemented solely as a means to provide drinking water, leaving well water strictly for non-potable uses (i.e., sanitation, washing, etc.), as much of the groundwater in these areas is contaminated with fluoride. Projects were funded mostly from municipal and state funding, with some support from local non-governmental organisations (NGOs) (ESF, 2006).

6.2.2 System design

Storage was constructed using easy to install geomembrane cisterns, with a geotextile underneath for greater protection. The initial systems installed in the project had the cistern buried underground, using beam and concrete covers, enlisting the help of local masons and builders (ESF, 2008). Subsequent systems were constructed using elevated tanks, with a lower installation cost. Larger cisterns were made of geomembrane (commercially known as 'quick tanks') while the smaller ones were purchased as pre-fabricated rigid plastic tanks (Table 6.1). The systems were designed to collect rainwater from rooftops using PVC guttering, with a 'first-flush' device or settling tank to remove larger debris and pollutants (Figures 6.1 and 6.2). For successful and safe operation, gutters and the settling tank need regular cleaning to avoid recontamination.

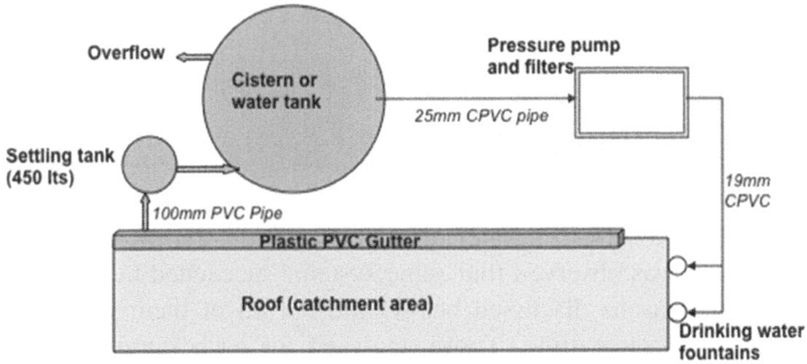

Figure 6.1 Typical rainwater harvesting (RWH) system (plan view).

Figure 6.2 Typical above-ground cistern/storage and settling/sedimentation tank.

Initially, conventional filtration and chlorination was used for disinfection. From 2009 onwards, silver ion devices were added to the systems in order to enhance the water quality and increase safety. The motivation for this was to find an alternative to chlorination as a disinfectant, the side effects of which are amply known (Xie, 2004). After an evaluation of the existing options, silver ionisation was chosen due to its durability, ease of maintenance and safety. Contact ionisation devices were purchased from a local Mexican supplier.

The responsibility for maintenance of the systems fell entirely on the community. A small financial incentive was initially offered by the municipality but did not receive the adequate follow-up and failed to materialise. Intermittent maintenance and supervision was performed by the contractors along with a local NGO (IRRI-Mexico), mostly on a voluntary basis. At the moment of writing, Engineers Without Borders (EWB-UCL) has taken the systems on board and is successfully working on their continued upkeep and improvement in collaboration with the local communities. By the end of summer 2014, most of the systems were fully operational.

6.3 SYSTEM EVALUATION

6.3.1 Water quality

In 2010 and 2012, a series of water quality evaluation studies were performed on the systems. It was observed that some systems succeeded better than others for a number of reasons discussed below, and not all of them were functional or available at the same time. Those analysed for each sampling period are listed in Table 6.1. The main results for these studies were published in Adler *et al.* (2011, 2013).

Table 6.1 RWH systems used for study.

ID	Community/sampling site	No. of Users	Cistern sampling			
			Size (m^3)	Type	2010	2012
1	Rancho Nuevo Guadalupe	39	5	TK	x	x
2	San Antonio de La Joya	51	17	OG	x	
3	Don Juan	12	7.5	OG	x	x
4	La Aurora	8	10	TK	x	x
5	San Miguel Viejo – Classroom	75	45	OG	x	x
6	San Miguel Viejo – Kitchen	75	17	UG	x	x
7	Augustin Gonzalez – Clinic	NA	45	UG	x	x
8	Augustin Gonzalez – School	104	80	UG	x	x
9	El Salitre	54	17	OG	x	x
10	Montecillo de Nieto	70	30	OG		x
11	Boca de la Cañada	54	17	OG		x

Note: Systems are all installed in schools in the respective communities, except for No. 7 (rural clinic). OG, Overground geomembrane; UG, buried/underground cistern with geomembrane liner; TK, pre-fabricated plastic tanks (5 m^3) with lids; NA, not available.

Although most systems complied with water quality standards, some notable deviations were observed. For instance, in 2010 the system at El Salitre (Site 9) completely failed to perform. Despite all the filtering and disinfection mechanisms, there were inordinately high coliform counts in the effluent, making the water unfit for consumption. Upon closer inspection, it was noted that the lid of the cistern (made of geomembrane) had collapsed and was partially torn. The water, exposed to the elements, was gradually contaminated with falling leaves and organic debris, becoming turbid and brownish. The settling tank was also heavily contaminated, having not been cleaned in several months. The role of the settling tank, which is normally meant to protect the cistern from the largest concentrations of pollutants, was reversed, becoming a focal point for recontamination.

After discussions with the community and several school parents, it emerged that the head teacher had been replaced recently, and the new one was unaware of the operation of the system. The training provided to the community in the previous year was largely lost, even though some parents (mainly mothers) knew about it and had attended the training. Without clear guidance, the system was simply left to perform by itself, with little or no maintenance and had fallen into disrepair.

The situation was resolved after much intervention and encouraging active participation from school parents and teachers. A new, albeit more informal, training was delivered, reviewing the main maintenance principles, as well as establishing an agenda for specific tasks. Within a few weeks the lid was repaired, a 'cleaning day' was scheduled, where several parents and staff attended to clear the roof of debris, prune the trees and so on and eventually the system was back on line. In the 2012 water quality evaluation, the system performed impeccably, with zero coliform counts at the drinking taps and overall good turbidity and water quality in the cistern.

6.3.2 Abandoned systems

Two of the systems, installed in Santuario de Atotonilco and Cruz del Palmar, deserve special mention as they were altogether abandoned, and therefore not included in the water quality studies. The reasons for this are complex and are discussed below, with valuable lessons to be learned from each case.

Santuario de Atotonilco: Installed at the community's main primary school, this system was controversial from the start. The municipality wanted one of the RWH systems installed here, as Atotonilco is an emblematic site. The Spanish colonial monastery attracts visitors from all over the globe and was named as a World Heritage site in 2008 by UNESCO, along with the main town of San Miguel de Allende. Furthermore, the community's water wells have a particularly high fluoride concentration (ESF, 2006), which was one of the main motivators for installing the systems in the first place.

However, the school staff and parents seemed to be less motivated here than those in other communities. One of the reasons cited by the head teacher was the sheer number of students (close to 100) and the lack of sufficient staff to keep an eye out for vandalism, which in her opinion could occur with an above-ground cistern. Due to insufficient funding, it was not possible to offer the option of a buried cistern. Local politics being a strong factor in this particular community, the fact that Augustin Gonzalez (Sites 7 and 8, Table 6.1) had large underground cisterns, created a potential source of friction and envy, even though these systems had been installed two years earlier with funding from different sources. Other communities which received cheaper above-ground instead of underground cisterns had similar concerns, perceiving it as 'unfair' that one location should get more attention or better technology than others.

The roof in the school chosen for rainwater harvesting belonged to a small classroom in very poor condition, badly in need of water-proofing and repair. Some community members related the RWH system to the actual state of the roof and thought that adding gutters or downpipes would somehow damage or put the structural capacity of the classroom at risk, which created further resistance. Other rooftops were not feasible as there was no space nearby to put the elevated cistern, so after much negotiation, the proposed site was agreed upon and construction initiated.

From the very beginning, community involvement was minimal and the project as a whole was received with much scepticism. This compares sharply with other sites, such as Rancho Nuevo, San Antonio de la Joya or San Miguel Viejo (Table 6.2), where parents, teachers and students helped out with the RWH system installation from day one, including donating food to workers and project managers. The system at Santuario de Atotonilco was not perceived as meeting any specific need or request of the community. It is not uncommon for communities in rural Mexico to be averse to ideas coming from the Government (or perceived as such). This is due mainly to a history of neglect and complex relationships between the Federal or Municipal authorities and local community leaders. Had the installation offered to refurbish the actual classroom building (an issue which seemed to concern the parents and teachers more), along with the RWH system, it is possible that it would have had a greater receptivity. The municipal government, however, was unable to provide this at the time from the allocated funding.

Once the system was installed and tested, it worked well for a short while before it was neglected. On a subsequent visit, the head teacher reported that '*children did not like the taste of the water*'. The lid of the cistern also collapsed due to vandalism (according to the same source), by children climbing onto its side. One of the drinking fountains was broken from a football and never replaced. After intense lobbying and fund-raising on behalf of EWB-UCL a buried plastic tank was procured, maintenance and training sessions were scheduled with the community, and the system is now back in operation.

Cruz del Palmar: Cruz del Palmar is the largest community in the Municipality of San Miguel, with over 1000 inhabitants (INEGI, 2011). It was established in 1516 and is also one of the farthest, geographically, from the main town. Until recently, like many of the other sites, it was accessible only via dirt roads, which became difficult to pass during the rainy season. However, a new highway has greatly facilitated connections and transport. The Municipality was keen to install a RWH system due to poor groundwater quality, occasional water scarcity and the higher social impact expected from a larger population.

The choice of the site was agreed with the *Delegado* (locally appointed leader or delegate), who was the main link between the community and the municipal authorities. Dealing with an elected representative, in contrast with a long-term and well established head teacher or community leader, presents important challenges, which were experienced at this site. The *Delegado* is usually elected every 3 years. The short period in office means that any commitments and follow-up to projects might not necessarily be honoured by a future *Delegado*. His/her effectiveness and long-term influence can also depend on kinship and political affiliations. The level of interest a project may receive can depend on how much time remains in the *Delegado's* term, how seriously committed is the representative to the community, or even the political advantage, if any, that can be gained from the project. The same challenges apply to State and Federal programmes (Adler, 2011), although in the latter case governing periods are usually longer (up to 6 years).

The site selected in consultation with the *Delegado* was a large communal space, known as a 'multiple-use hall', where weddings, meetings and local celebrations take place. The key to this space is usually in the hands of the *Delegado* and it is locked while not in use, since no staff work there on a regular basis. The plan was to install the cistern *inside* the hall with rainwater collected from the large metallic roof. The initial plan was for the purified water to be pumped across a small street to the local clinic, where two drinking taps were installed for public use. This technically challenging and expensive project was completed and delivered, along with a training session to the clinic staff and local authority, as scheduled. The training session, however, in contrast to other communities, had very low attendance. It was also difficult to know who was going to be in charge of the system, as the clinic staff had a high rotation and the *Delegado* was about to complete his 3-year term. After the installation the community protested that it was taking up too much space in the Hall and that it should be removed. The new *Delegado* seemed receptive to the RWH project and keen to restart the system and arranged a meeting with the town residents with the project at the top of the agenda. After several visits, it was agreed to move the system to the local secondary school. The head teacher there was highly interested, along with several of the parents and the entire system (including the 30 m^3 geomembrane tank, all the gutters, pipes and filters) was installed in the new location.

6.4 REASONS FOR FAILURE

The main reasons for a system failing to operate correctly can be divided between those related to maintenance and those linked to the actual system design. Problems relate to social and technical issues and often it is difficult to distinguish between the two. A vigilant community, actively engaged in the project, will be more likely to prevent or report potential problems before they get worse. On the other hand, even the most proactive stakeholders can do little about a pump or electrical failure, particularly if they do not have the spare parts or the know-how to repair it.

Technical failures that cannot be readily repaired can create frustration and eventual apathy towards the system, as was observed in some of the sites. If people do not feel that they can be a part of the solution to a problem, they start losing interest and simply leave it in the hands of others, creating in turn a larger probability of future failures, resulting in a vicious circle. The example of El Salitre (Site 9), clearly demonstrates this; without external intervention of some sort (from the authorities or contractors, for instance), it is likely that the system would have been abandoned.

The delicate balance between an intelligent, fail-proof design and active social engagement is not always easy to achieve, particularly in remote communities where regular inspection visits are not feasible and much is left to the community itself. Despite all the beneficiaries being left with telephone numbers and contact information should anything go wrong, very few calls were ever received, even when repairs were badly needed. A general tendency was observed, throughout all the sites, to not report problems but rather wait until someone came round for a visit or inspection. Complaints and concerns were therefore issued in person, informally, very rarely by phone and never in writing or by email.

The list below details some of the major causes for systems failing, grouped according to the most likely causal factors: design, maintenance or external factors. Some instances of failure fall into more than one category. For instance, a broken lid that happened for external reasons (i.e., storms), but was not fixed due to a lack of maintenance as a result of low community participation, which in turn caused water pollution.

Design and implementation

- Leaks in cistern or structural problems;
- Collapsing or broken lids;
- Damaged gutters or downspouts;
- Broken pipes or valves;
- Pumps or equipment linked to warranty.

Maintenance

- Clogged filters or issues with purification system;
- Damaged drinking water spouts;
- Poor water quality in cistern;
- Lack of cleaning and emptying of settling tanks;
- Dirty, contaminated rooftops (that in turn can cause clogging of gutters or pipes).

External factors

- Safety issues, vandalism, and so on.
- Failure in electric supply (that could damage pumps or electric equipment);
- Strong winds or storms (causing overflows or ripping apart of membranes/ protective coverings, for instance).

Theft was not seen to be a significant problem in any of the sites, although some vandalism was reported on rare occasions, mainly affecting the cistern structure or water quality (i.e., hurling of rocks or debris towards cisterns).

As all systems were tested and delivered in fully functional conditions, it is assumed that there were no pre-existing flaws due to poor installation and that any issues observed later on were either due to a lack of maintenance, or eventual failing of equipment due to external reasons. Most equipment, as well as the installation itself, was covered by warranty for the first year, in which inspections were frequent and many minor problems were fixed.

6.5 COMMUNITY PARTICIPATION AND LEADERSHIP

The performance of the systems was seen to be related to the degree of participation and leadership during the period of evaluation (Table 6.2). Leadership was provided by a range of actors, including official figures (such as head teachers) or informal community leaders who decided to 'champion' the project. Participation refers to the on-going maintenance and involvement of the community with the system, not the installation. Water quality indicators were used to assess the performance of the systems. This was based on the reduction in chemical oxygen demand (COD) and coliform bacteria from the entry point (settling tank) to the drinking water tap (see Adler *et al.* 2011, 2013). The assessment of the overall condition of the system was based on the observed state of the system, including aspects such as cleanliness, leaks, condition of pipes and cistern lids, and other general maintenance issues. In order to have a standard basis for comparison, the following scale was utilised for all the parameters listed in Table 6.2:

+++ Excellent
++ Satisfactory
+ Average
– Missing or lacking altogether (i.e., very poor conditions; system not working)

In the Leadership column, a negative score (–) refers to a complete lack of clear leadership and a '+' (average score) to communities that had a high rotation of head teachers, for instance, where there may not have been an adequate transfer of skills and responsibilities.

Table 6.2 Comparison of system performance and community participation.

ID	Community/ sampling site	Overall system condition	System efficiency	Community participation	Leadership
1	Rancho Nuevo Guadalupe	++	+++	+++	++
2	San Antonio de La Joya	+	++	++	++
3	Don Juan	++	+++	++	+
4	La Aurora	++	–	+	–
5	S. Miguel Viejo – Classroom	+++	++	+++	+++
6	S. Miguel Viejo – Kitchen	+++	+++	+++	+++
7	Augustin Gonzalez – Clinic	+	+	+	–
8	Augustin Gonzalez – School	+	+++	++	+
9	El Salitre	+	++	++	+
10	Montecillo de Nieto	+	+++	+	+++
11	Boca de la Cañada	++	++	++	++

Source: Adler (2014).

Communities with greater participation and leadership tended to demonstrate better water quality and system performance. However, strong leadership in and of itself did not necessarily guarantee satisfactory results (Site 10). There were also instances where there was a strong community involvement without clear leadership (Sites 8 and 9), but even if efficiency was high, the overall condition and maintenance of the system tended to suffer as a consequence.

6.5.1 Training and succession

A phenomenon commonly observed in projects that are handed down to communities by NGOs or public programmes, is the lack of follow-up and stakeholder succession, particularly once the project has stepped out of the limelight and any political objectives have been achieved. For example, in the community of Montecillo de Nieto (Site 10) an ambitious dry toilet installation for the entire school had been abandoned for a number of years. Taking up valuable space and attracting flies and odours for a long period, it created a problem instead of a solution for the beneficiaries. When questioned about it, the head teacher vaguely mentioned a 'foreign NGO' that had donated the equipment, no doubt with the best of intentions, but with no follow-up on behalf of the community or the organisation.

The context in which such failure occurs was observed repeatedly with the RWH systems of the present study. The budget for all the installations included a training programme and the provision of an illustrated manual, so that users could know exactly what maintenance was required, where to purchase the necessary supplies and who to address in case of problems. These training sessions were all duly completed, with signed commitments to maintain the systems. Some of the most pro-active communities even implemented 'water committees' to follow up and pass the knowledge on to the future generations of parents in the schools, or staff in clinics. At the start, many of the projects received great ceremony and attention from the local press. In the more prominent sites, the city Mayor came in person, along with state officers and other leading figures, to attend a formal inauguration ceremony, with lofty speeches and offers to continue expanding the RWH agenda, as well as the promise of supporting the communities with some funding for yearly maintenance. There was never any formal commitment for maintenance funding, but the very promise of it created a sense of expectation in the communities, with the unintended effect of undermining local responsibility for the care of the systems.

After one or two academic years had passed, groups of children left the primary schools and along with them the parents who had been involved with the projects from the beginning. On a few occasions staff and head teachers changed too, creating a widening gap that resulted inevitably in poor maintenance and lack of understanding as to the operation of the systems. During the 2012 sampling round, for instance, it was noted in some communities that user manuals had been lost, or keys misplaced, barring access to the filters and pumps. In the case of the clinic, this became even more complicated, as doctors and nurses rotate regularly in rural Mexican health centres. Some training sessions were repeated on request for the new generations, on a *pro bono* basis, but without any funding this became increasingly difficult.

The issue of high rotation of government staff and local leaders is hard to resolve, particularly in schools and clinics. This can be partially circumvented by identifying early on influential people in the community who are not necessarily linked to the more transient roles of power or authority (such as head teachers or government officers). This was the case in one such community (Site 1), where despite a very high rotation of the local school's head teachers, the system delivered excellent outcomes (Table 6.2), thanks to the participation of local residents and parents, who were perceived as proactive 'leaders' by the community, even if they did not have any formal role. The high level of organisation and involvement of such individuals from the start guaranteed the continued success of that particular programme.

6.5.2 Technical complexities

Another lesson learned was that of technological know-how and the form by which it was transmitted to beneficiaries. Although most community members were familiar with the idea of a filter, a basic water pump or a cistern, the silver ion unit was an unknown component that was not immediately familiar to many. For the sake of simplicity, they were instructed, both in training sessions and in the user manuals, to inspect the cells only once a year and report any anomaly with the device, such as the indicator lights being off or malfunctioning. While this was not complicated in and of itself, the fact that not enough effort was put into explaining the mechanism of how it worked, created a certain distancing and apathy, which could be interpreted as a fear of tampering with the unit. This lack of familiarity resulted in the silver ion units never being inspected or replaced by the community members, even in those sites where other maintenance activities were dutifully carried out, such as emptying setting tanks and cleaning cisterns and roofs. Their perception of these somewhat sophisticated units as a mysterious 'black box' generated issues with their upkeep and necessitated external technicians conducting simple maintenance tasks that could otherwise have been dealt with locally.

In schools, which represent the majority of sites studied, children were also not involved enough in general. Rather, it was left to the teachers to decide how best to involve or inform them of the significance of the system. More effort in this direction would have greatly enhanced the project's capability and social participation. Greater involvement of students would have been of direct educational value and would have helped to support the continuity of the system itself and across the longer-term. In the case of Santuario de Atotonilco, for example (one of the abandoned systems described previously), the incidence of vandalism may have been reduced with greater student involvement.

6.6 CONCLUSIONS

Installing new technical systems, such as RWH, into local community buildings and facilities will have an impact, however minor, on the structure and social dynamic of a community. In contrast to centralised municipal water systems where a water utility is responsible for maintenance and operations, decentralised systems require active participation by community leaders, volunteers, residents and beneficiaries. New technical systems often require that local organisational structures adapt to the new infrastructure, with its associated operation and maintenance demands. Since drinking water is such a delicate vital issue, a great deal of emotional and even political charge can be expected when planning and operating such a project. The very success of the systems, in the long term, relies heavily on local politics and the involvement of local actors and stakeholders (Chauhan & Bihua, 1983).

From the research presented in this chapter, the following key recommendations for community based water projects have been proposed:

- Ensure the system meets a genuinely perceived need of the community;
- Identify leadership and follow-up responsibilities from all stakeholders at the outset of the project;
- Ensure a robust design, which requires the minimum (or the simplest) maintenance possible;
- Involve beneficiaries (e.g., school children) as much as possible, explaining system design, operation and maintenance and value to the community;
- Ensure that contact details for problems and emergencies are clearly posted in accessible locations;
- Ensure site visits and inspections occur on a regular and predictable basis;
- Avoid making promises or commitments that may be hard to follow-up (e.g., additional funding);
- Choose technology that is easy to fix with consumables that are locally available, whenever possible.

The cases evaluated in this chapter demonstrate that the success of alternative water systems requires the development of alternative social and institutional structures within communities. Social factors need to be incorporated at the conceptual stage of decentralised water system design and before feasibility studies are undertaken. Long term engagement between municipal authorities, technical experts and local communities is vital to maintain not only the technical system but also the social systems that support it.

ACKNOWLEDGEMENTS

The authors would like to thank Rachel Smith (IRRI-Mexico) for her help in collecting valuable site data, as well as CONACYT (Mexican Council of Science and Technology), the main sponsor of the lead authors's PhD. They are also highly indebted to the Municipality of San Miguel de Allende for their continued support. for their continued support, as well as Engineers Without Borders (EWB-UCL) and CATIS-Mexico.

REFERENCES

Adler I. (2011). Domestic water demand management: implications for Mexico City. *International Journal of Urban Sustainable Development*, **3**(1), 93–105.

Adler I. (2014). Application of a system based on filtration and silver ions to purify rainwater for potable applications in rural Mexican communities. Doctoral dissertation, University College London.

Adler I., Hudson-Edwards K. A. and Campos L. C. (2011). Converting rain into drinking water: quality issues and technological advances. *Water Science & Technology: Water Supply*, **11**(6), 659–667.

Adler I., Hudson-Edwards K. A. and Campos L. C. (2013). Evaluation of a silver-ion based purification system for rainwater harvesting at a small-scale community level. *Journal of Water Supply: Research and Technology—AQUA*, **62**(8), 545–551.

Chauhan S. K. and Bihua Z. (1983). Who Puts The Water in The Taps. Earthscan paperback, London, p. 92.

Dixon D., Murray J. and Gellat J. (2006). America's Emigrants: US Retirement Migration to Mexico and Panama. Migration Information Source, the online journal of the Migration Policy Institute. http://www.migrationinformation.org/feature/display.cfm?ID=416 (accessed 23 September 2006).

Ecosystem Sciences Foundation [ESF] (2006). Well Water Quality in San Miguel de Allende: Results and Conclusions. Boise, ID. www.ecosystemsciences.com (accessed 28 March 2011).

Ecosystem Sciences Foundation [ESF] (2008). Annual Report 2007 (p. 1). Boise, ID. www.ecosystemsciences.com (accessed 28 March 2011).

Garcia y Garcia E. (2006). El Agua en San Miguel de Allende (*Water in San Miguel de Allende*). San Miguel de Allende, Mexico, PTF S.C.

Hemson D., Kulindwa K., Lein H. and Mascarenhas A. (2008). Poverty and Water Explorations of Reciprocal Relationship. Zed Books, London, p. 210.

INEGI (2011). National Institute for Statistics and Geography. www.inegi.org.mx (accessed 20 March 2013).

Servicio Meteorologico Nacional [SMN] (2010). Normales Climatogicas 1951–2010. http://smn.cna.gob.mx/climatologia/Normales5110/NORMAL11093.TXT (accessed 20 March 2013).

Stoddart H. (2009). Water World: Why the Global Climate Challenge is a Global Water Challenge. Report prepared for the COP15 meeting in Copenhagen, December 2009.

Székely M., López-Calva L. F., Melendez A., Rodríguez-Chamussy L. and Rascón E. (2007). Poniendo a la pobreza de ingresos y a la desigualdad en el mapa de México. *Economia Mexicana*, **16**(2). http://www.economiamexicana.cide.edu/num_anteriores/XVI-2/03_SZEKELY.pdf (accessed 28 Mach 2013).

Xie Y. (2004). Disinfection Byproducts in Drinking Water: Formation, Analysis, and Control, CRC Press, Boca Raton, Florida.

Chapter 7

Assessing domestic rainwater harvesting storage cost and geographic availability in Uganda's Rakai District

Jonathan Thayil-Blanchard and James R. Mihelcic

7.1 INTRODUCTION

7.1.1 Self supply

Self supply is a promising policy framework, which seeks to supplement conventional methods of supplying water by encouraging and enabling users to make small investments in incremental, easily replicable improvements to their own supply (Sutton, 2008). In its most rudimentary form, self supply is the ability of a household to access water using their own resources. It has been defined as the improvement to household or community water supply through user investment in water treatment, supply construction and up-grading, and rainwater harvesting (RWH) (Sutton, 2008). It is based on incremental improvements in steps that are easily replicable, with technologies affordable to users. As such, it has been standard practice for millennia, especially to those populations considered unserved by improved water sources. Only in recent years has an effort been made to develop a framework of self supply that brings it into the mainstream of water supply planning. Differing from other frameworks, self supply is an approach to supply water that concentrates intervention and management at the lowest level (RWSN, 2003).

7.1.2 Domestic rainwater harvesting

Domestic rainwater harvesting (DRWH) is a form of self supply that refers to the practice of utilising water that falls as rain on a hard roof. This roof runoff is then directed to a storage device for purposes such as drinking, cooking, cleaning, hygiene and sanitation (Martinson & Thomas, 2003). DRWH is a core component of the self supply effort, encompassing a broad range of practices. These include informal efforts such as placing pots under eaves during a rainstorm or investment by households in elaborate systems with large built-in-place tanks that may serve as the sole water source all year round (Danert & Sutton, 2010).

The proximity of rainwater harvesting (RWH) sources to households can offer a high level of service and a consequent improvement in health. For example, a regional analysis in West Africa estimated that during the rainy season, a storage device as small as 200 litres could be optimal for enhancing the water supply of many urban households with small, simple roofs (Cowden *et al.* 2008). Furthermore, water storage from DRWH of as little as 400 litres was estimated to reduce the diarrheal disease burden (measured as disability adjusted life years, DALYs) by as much as 25% (Fry *et al.* 2010).

7.1.3 The Ugandan context

Eighty-five percent of Uganda's population is classified as rural (UBOS, 2010), two-thirds of the land area experiences more than 1200 mm of rain per year and over two-thirds of the roofs it falls on are constructed from galvanised iron (Danert & Motts, 2009). This suggests that most rural Ugandan households already have the basic climatic and catchment requirements for a basic DRWH system (Danert & Motts, 2009). However, while rural access to improved water sources has increased significantly from around 20% in 1990, it has stagnated at around 60% since 2001 (Danert & Motts, 2009).

Currently, DRWH constitutes the most popular method of private investment in water supply. Approximately 28% of the 15,000 or so tanks with a capacity greater than 6000 litres in Uganda have been privately financed (MWE, 2010), thus fitting the definition of self supply. Ugandan DRWH storage devices broadly fit into three categories: (1) traditional/informal methods (for which formal markets may not exist, but which have been practiced for a long time); (2) manufactured products (centrally produced tanks in a wide range of sizes, available for sale in nearly any town large enough to have a hardware store); and (3) built-in-place tanks constructed by trained artisans. These general categories of storage devices provide the foundation for the focus of this chapter.

7.1.4 Motivation and objectives

During his two years serving as a water/sanitation engineer with the U.S. Peace Corps in Uganda, this chapter's lead author had significant experience with people and institutions using DRWH as a water source. It was observed that while some manufactured products (especially small plastic tanks) were widely and consistently available, many other DRWH storage techniques were disparate and scattered. Additionally, knowledge regarding alternatives for implementing the approach was fairly limited from location to location. Furthermore, there was no collective knowledge resource of DRWH methods in practice to store rainwater in Uganda.

This observation is reinforced in the self supply litreature (Cruddas, 2007; Danert & Sutton, 2010). The materials that are used to construct roofs and gutters are fairly standard, but there are a number of creative methods for water storage spread throughout the country, generally limited in geographic scope and availability to at most a few sub-counties. Reproduction of existing storage methods and learning from the success and failure of previous efforts, two core values of self supply intended to foster the independent spread and uptake of effective water resource utilisation, are impeded by this lack of readily available, centralised information. The work described in this chapter was conceived in an effort to fill part of this knowledge gap. Accordingly, the objectives of the research presented here were to: (1) present a comprehensive collection of well-established and diverse rainwater storage options in Uganda that also includes a cost analysis; and (2) demonstrate the geographic disparities in the distribution of household water storage options within Uganda's Rakai District. Though rainwater has been assessed as safer than water from unimproved water supplies (Dean & Hunter, 2012), this chapter does not address how tank material impacts the water quality of stored water. A recent publication has addressed this important issue (Schafer & Mihelcic, 2012) and readers are referred to this reference for further reading.

7.2 DOMESTIC RAINWATER HARVESTING IN UGANDA

DRWH is a well-researched phenomenon in Uganda, even when examined separately from self supply. For example, a benefit/cost ratio analysis of several rainwater usage schemes, in combination with supplementary sources, had previously concluded that sole supply from DRWH is probably an inappropriate objective. This is when taking into account both the finance of the investment required and the realities of how rural households use water (Thomas & Rees, 1999). A later study concluded that the pursuit of sole source water provision with DRWH requires tanks 10–50 times larger than otherwise required. This leads to DRWH being overpriced, which has hampered enthusiasm for its adoption (Martinson & Thomas, 2003). Furthermore, it has been suggested that between the financially optimal tank size and the size required for sole source use lies a range of medium performance DRWH that can be just as convenient and reliable as many conventional point water sources. However, in order for a community or household to make an informed decision among a diverse set of available technologies, they will require information about how different size systems behave, as well as the costs and trade-offs involved in different designs (Martinson & Thomas, 2003). This is a crucial cornerstone of the self supply approach. In a more recent visit to Uganda (Thomas, 2011), it was noted that subsidies have a tendency to destroy private initiative. That is, if there is even the slightest possibility of a future subsidy, potential customers will not invest in a DRWH system on their own.

The most comprehensive overview of RWH policy in Uganda was conducted in 2004 by the Uganda Rainwater Harvesting Association (URWA) (URWA, 2004). In a survey of several districts that also included a broad look at the country as a whole, URWA found a generally immature market for RWH. Parts and supplies were generally unavailable and a good commercial structure and supply chain were only identified in approximately 15 of Uganda's hundreds of sub-counties. Moreover, URWA concluded there was limited awareness by consumers of the diverse range of available technologies or even where to obtain most of them (URWA, 2004). Most importantly in relation to this chapter, the URWA report introduced the concept of a RWH ladder (Figure 7.1). Each of the six rungs is described in Figure 7.1, demonstrating how incremental investment in DRWH by a household slowly increases their infrastructure while at the same time bolstering their dependence on rainwater.

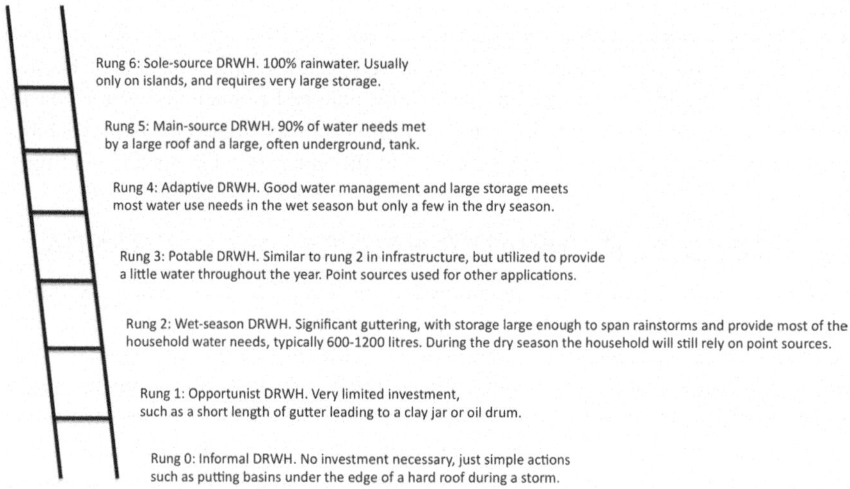

Figure 7.1 The RWH ladder shows how increasing household investment impacts the type of storage infrastructure and the importance of rainwater in meeting a household's total supply.

7.3 METHOD

Field work was conducted in Kalisizo, a town located in Rakai District, whilst the lead author was undertaking the Master's International Program in Civil and Environmental Engineering (Mihelcic *et al.* 2006; Mihelcic, 2010). Figure 7.2a shows where the Rakai District is located in the south of Uganda. The District abuts Tanzania to the south and Lake Victoria to the east. The most recent census in 2002 placed the population of Rakai District at 405,631 (UBOS, 2002). Figure

7.2b shows that Rakai District is sub-divided into three counties (Kooki, Kakuuto and Kyotera) and 20 sub-counties.

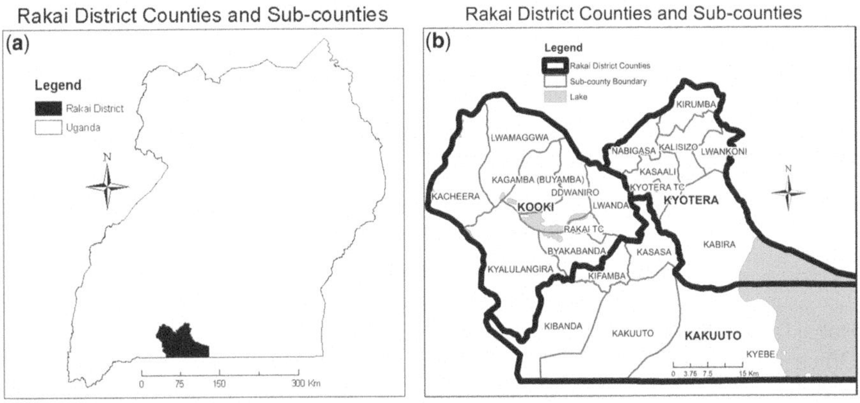

Figure 7.2 (a) Location of Rakai District within Uganda (b) Administrative boundaries of Rakai District.

It was observed that most of the residents of larger towns had access to a piped water system, while in rural areas boreholes and springs were the major sources of water. RWH was observed, but was, overall, a minority option for most residents. Two documents that have touched on these topics without a comprehensive review (Danert & Motts, 2009; Thomas, 2010) formed the starting point for the research methodology. Both documents describe the many available storage options for DRWH without a detailed examination of locations, costs and programmes by which specific storage technologies had been implemented, or how wide commercial adoption had spread. In particular, Danert and Motts (2009) observed that small manufactured products were available on a wide commercial basis, but larger manufactured storage options were available only in very large cities. As for built-in-place constructed tanks, Danert and Motts observed that they were only available where they had been promoted by a specific non-governmental organisation (NGO) and even in those areas the technology was not widely available. In addition, due to subsidies and a lack of focus on private sector uptake, many programmes that promoted the construction of built-in-place storage tanks failed to produce continuing businesses.

Building on this division of rainwater storage into manufactured and built-in-place constructed options, the current study was undertaken in three phases. The first phase was a series of upper-level meetings during the months of July, August and September 2011 with organisations having an advisory or oversight role in RWH. These included the Ministry of Water and Environment (MWE) at the national, regional and district levels; the Appropriate Technology Center (ATC) and the Uganda Rainwater Association (URWA). All of the stakeholders that participated in this study are described in Table 7.1. In-depth descriptions of the stakeholders are provided elsewhere (Blanchard, 2012).

Table 7.1 Summary of the stakeholders with knowledge of DRWH interviewed for this study.

Organisation	Description
MWE	Arm of national government responsible for national water policy and implementation
TSU 7	Regional advising office of the MWE for Rakai and surrounding districts
Rakai District Water Office	Rakai District office of MWE
URWA	Ugandan NGO promoting, studying and improving RWH across Uganda; has implemented ferrocement tanks and mortar jars
ATC	National center advising in appropriate water and sanitation technologies including RWH
World Vision	International NGO working in Rakai; has encouraged tarpaulin tanks
SNV	International NGO working in Rakai and elsewhere in Uganda
ACORD	International NGO working in Rakai; has implemented tarpaulin, ferrocement and partial underground tanks
CIDI	Ugandan NGO working in Rakai; has implemented tarpaulin tanks
COWESER	Local Rakai NGO; has implemented ferrocement tanks
Brick by Brick	Local Rakai business constructing Interlocking Stabilised Soil Brick tanks

A total of six meetings were held: two with the MWE at the national level and one each with the other levels of government and the organisations. Each stakeholder confirmed the central hypothesis of this research: that a centralised documentation of rainwater storage options is lacking and would be useful. Subsequently, each meeting had two outputs: (i) to understand the stakeholder's perspective on all the commonly available storage technologies available in Uganda; and (i) to gather knowledge regarding those organisations that had been involved in implementing each technology in the Rakai District.

The second phase of the study consisted of a series of meetings with the implementing organisations identified during Phase 1. These organisations were the Agency for Cooperation and Research in Development (ACORD), Community Welfare Services (COWESER), Netherland Development Organisation (SNV) and URWA. URWA does some implementation activities as well as advising and oversight activities. These organisations were also asked to confirm that some form of centralised documentation of rainwater storage options would encourage and enable uptake. They were then asked about the volumes, prices and locations of the various programmes implementing each kind of water storage tank. The stakeholders were also asked to provide documentation that would substantiate this information. Finally, they advised whether they had implemented other types of storage tanks or if they were aware of other organisations implementing the same or other types of tanks. Only ACORD identified an additional stakeholder (here referred to as CIDI), who had also built tarpaulin tanks in Rakai District. ACORD also suggested adding the partially underground tank technology, which they were promoting in the south-west region.

The first two phases, described above, provided data on constructed tanks. The third phase collected data on manufactured tanks. In order to obtain data on the availability and pricing of manufactured products (plastic tanks, corrugated metal tanks and oil drums), a survey of Rakai District and the closest large town (Masaka, including its suburb, Kyabakuza) was conducted. It was determined, in consultation with a local resident familiar with the district, that there were only 10 or so trading centres in the district large enough to have hardware stores selling these smaller manufactured tanks.

To obtain costs indicative of those at which Ugandans could actually purchase these products, a Ugandan resident of the Rakai District visited all of these stores between 17 August and 6 September, 2011. The resident was instructed to examine, as an interested consumer, the types of manufactured storage products available at every commercial source and inquire as to their purchase price. The material (plastic, metal), brand (where relevant) and size of available manufactured tanks were noted and respective costs solicited and recorded. In addition to the survey of Rakai stores, a major national supplier of plastic tanks ranging from 60–24,000 litres supplied their price list. Costs are reported in U.S. dollars ($) using the exchange rate from September 2011 of 2425 Uganda shillings (UGX) per dollar.

7.4 RESULTS

7.4.1 Traditional/informal storage methods

Three distinct informal storage technologies were identified, meaning they are not being actively promoted by any institution and consequently data were generally unavailable. The first informal technology for rainwater storage was handmade clay pots, which have been phased out in favour of cheaper and relatively durable products (such as the 20-litre 'jerry can'). The second method was to simply arrange pots and basins underneath the edge of a roof during a rainstorm. This rudimentary approach does not require guttering and while storage capacity is low, it can provide at least a day's worth of water for cooking, drinking, washing and possibly bathing. Moreover, the marginal cost for RWH with this method is negligible, since it uses existing cooking vessels and plastic basins. Finally, there were an abundance of brick masonry tanks utilising standard burned clay bricks and concrete mortar. From first-hand observations, it would appear that most of these tanks were old and a high percentage of them were inoperative.

7.4.2 Manufactured products

Manufactured products represent the most widely and readily available method of rainwater storage. This was because they were available for purchase in many locations and could generally be easily transported. Transportation was fairly well organised: the larger manufacturers offered to deliver anywhere in the country, while the informal transport sector was well developed. The towns and the number of stores in the study area where manufactured tanks could be purchased are listed in Table 7.2. The town of Masaka was included, even though it is not technically within Rakai District, because it is the nearest large town and it is common for Rakai District residents to source from Masaka goods that are unavailable locally. Ten towns were identified that had a local commercial entity that sold manufactured tanks and 8 of the 10 towns had only one to three stores that provided such a service. In addition, two central manufacturers of large plastic tanks could provide products to all of Rakai District.

Table 7.2 Number and availability of manufactured storage tank vendors serving Rakai District (Uganda) by sub-county.

Town	Number of stores selling tanks	55-gallon drums	Corrugated metal tanks	Small plastic tanks (≤1000)
Masaka	11	×		×
Kyabakuza	3		×	
Kalisizo	1	×		×
Kyotera	3	×		×
Lwamaggwa	1			×
Mutukula	3			×
Ssanje	2			×
Kibaale	5	×		×
Rakai	1	×		×
Kasensero	2			×

Figure 7.3 shows the locations of towns that had a commercial store that served the Rakai District. Residents could purchase manufactured tanks from 32

different stores in 10 towns widely spread throughout the district, in addition to large plastic tanks that could be acquired from either of the national distributors. The remainder of this section describes the available manufactured products. Greater detail regarding price and volume of specific manufactured products identified in this study is available elsewhere (Blanchard, 2012). Figure 7.3 and Table 7.2 demonstrate that Rakai district is served by 10 commercial centres with at least one vendor of manufactured products (in addition to the national distributors) and that small plastic tanks are the storage mechanism most readily available in 9 out of 10 commercial centres. There were other self-sustaining options widely available, which provide a positive environment for the policy of self supply to thrive in as residents had multiple choices from which to make informed decisions, as well as the opportunity to emulate the success of their neighbours.

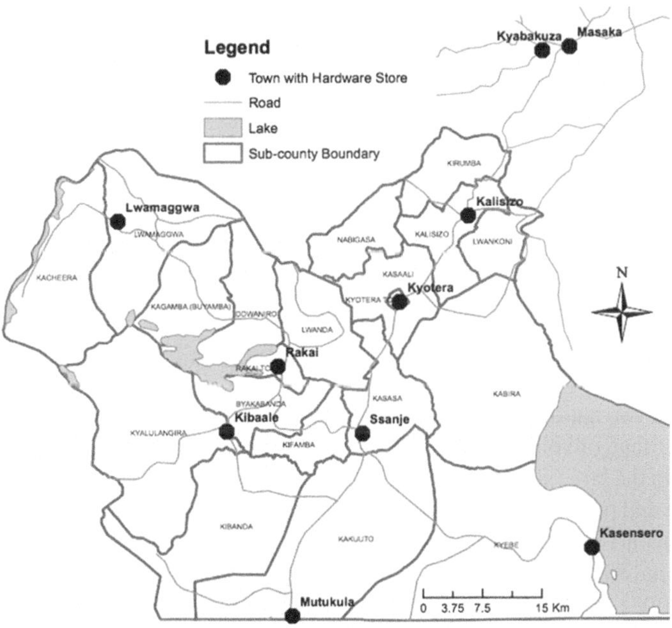

Figure 7.3 Locations (dark circles) of commercial suppliers of manufactured rainwater storage tanks serving Rakai District (Uganda).

7.4.2.1 Fifty-five gallon metal drums

Reclaimed metal drums with a volume of 55 gallons (208 litres) and emptied of oil or other original product, were commonly used for rainwater storage. It was not uncommon to see even the smallest house direct rainfall from a metre of guttering into a reused metal drum. These drums could be categorised as traditional, since they seem to have been used in Uganda longer than the other manufactured products. However, because they were available for purchase in many hardware stores and in many applications are used exclusively for RWH, they were considered a manufactured product. The survey identified six stores where these drums were sold at consistent prices of either \$31 or \$33. There was one store located in each of Rakai Town, Kibaale, Masaka and Kalisizo and two stores in Kyotera (refer to Figure 7.3 for geographic locations).

7.4.2.2 Corrugated iron tanks

Storage tanks constructed of curved, corrugated iron sheets welded into cylindrical tanks were a common sight on Ugandan roadsides. These were generally not sold at hardware stores, but at specialised metalworks. None of the metalworks operated within the Rakai District, but there were three in the town of Kyabakuza, just outside of Masaka (see Figure 7.3). Each metalworking facility used 24 or 26 gage iron sheeting, in similar volumetric configurations with comparable prices. The pricing was fairly consistent from the three metalworking facilities and the gage was a major contributor to cost. The thicker 24-gage tanks were more expensive than 26-gage tanks, were generally larger sized tanks (≥8000 litres) and tended to last longer. The thinner gage material was typically used on tanks ≤4000 litres. Tank size ranged from 2000 to 15,000 litres. A 24-gage metal corrugated tank that could store 8000 litres cost \$370, while a 15,000 litre tank cost \$620. A 26-gage metal corrugated tank of 2000 litres cost \$120 with a 4000 litre tank costing \$230.

7.4.2.3 Plastic tanks

There were two national, centralised manufacturers and distributors of plastic tanks in a wide range of volumes (100 litres–24,000 litres), though they did most of their business in the large range (>1000 litres). One of these manufacturers supplied their catalogue and price list, while the other was unresponsive to inquiries, though from discussions it is believed they were similar in price and quality. This conclusion is reinforced by a previous study that was able to compare the two manufacturers (Rowe, 2007). The price was \$830 for an 8000 litre tank and \$2800 for a 24,000 litre tank. A 100 litre tank cost \$24, a 1000 litre tank \$130 and a 4000 litre tank \$460. As both of these distributors would deliver to any location in Uganda, all of Rakai District is considered to have access to these large plastic tanks.

Other manufacturers were identified that focused on smaller tanks (identified here as ≤1000 litres). The selection of a small plastic storage tank could be broken into three categories: (1) One overwhelmingly dominant brand; (2) brands available at more than one location, but not widely competing with the dominant brand; and (3) tanks available at only one location. The dominant supplier was available at 28 separate stores throughout the district and at least one store in every commercial trading center. The second and third category of tanks were not included in this analysis due to their relative scarcity and general consistency with the dominant brand, both in regard to volumetric configuration and price.

Prices for the 65-, 120- and 220-litre tanks of the dominant brand were found to be quite consistent between different stores with an average price of \$7.50, \$11 and \$19, respectively. The 120- and 220-litre tanks were determined to have the highest availability, with eight and nine towns having them available, respectively. In contrast, the 65 litre plastic tank was only sold at four locations. The analysis consistently showed that larger plastic tanks (≥1000 litres) were a more expensive option than other available technologies (data in Blanchard, 2012).

7.4.3 Built-in-place products

It is difficult to discern exactly where private sector capacity for trained artisans constructing built-in-place tanks exists. However, no comprehensive compilation of RWH interventions exists for Rakai District. Three reports from major training programmes in the district were available and in conjunction with information obtained from interviews, it is believed that the results presented here provide a fair representation of what exists.

7.4.3.1 Mortar jars

Mortar jars were an inexpensive option for storing moderate volumes of water at households. Sizes ranged from several hundred to several thousand litres. The jars were constructed by pouring a circular concrete base, into which the tap was embedded. A wooden mould, approximating the interior shape of the jar, was erected on the base and thin layer of mud was applied to the exterior, in order to provide a smooth surface for plastering. The exterior was plastered with a 10–12 mm-thick layer of cement and allowed to cure for at least 48 hours. After the wooden mould was removed, the mud was scraped from the inside before an additional 1–2 mm thick waterproofing layer of cement was applied to the interior. The jars were transported to households in a handcart or by vehicle if properly protected.

The URWA conducted training of rural masons in this technology in seven sub-counties of Rakai District in 2006. Three masons were trained per sub-county, as well as a total of 12 apprentices. Costs for these tanks are reported to be \$63 for a 420 litre jar, \$130 for a 2000 litre jar and \$210 for a 3000 litre jar. The seven sub-counties of the Rakai District where the subsidised mortar jars were constructed by the Uganda Rainwater Association are shown in Figure 7.4a. These sub-counties were situated in the central region of the district: Byakabanda, Dwaniro, Lwanda, Lwamagwa, Kakuuto, Kifamba and Nabigasa. In all, 426 mortar jars were installed for rainwater storage, with the number of jars identified in each of these seven sub-regions ranging from 40 to 71.

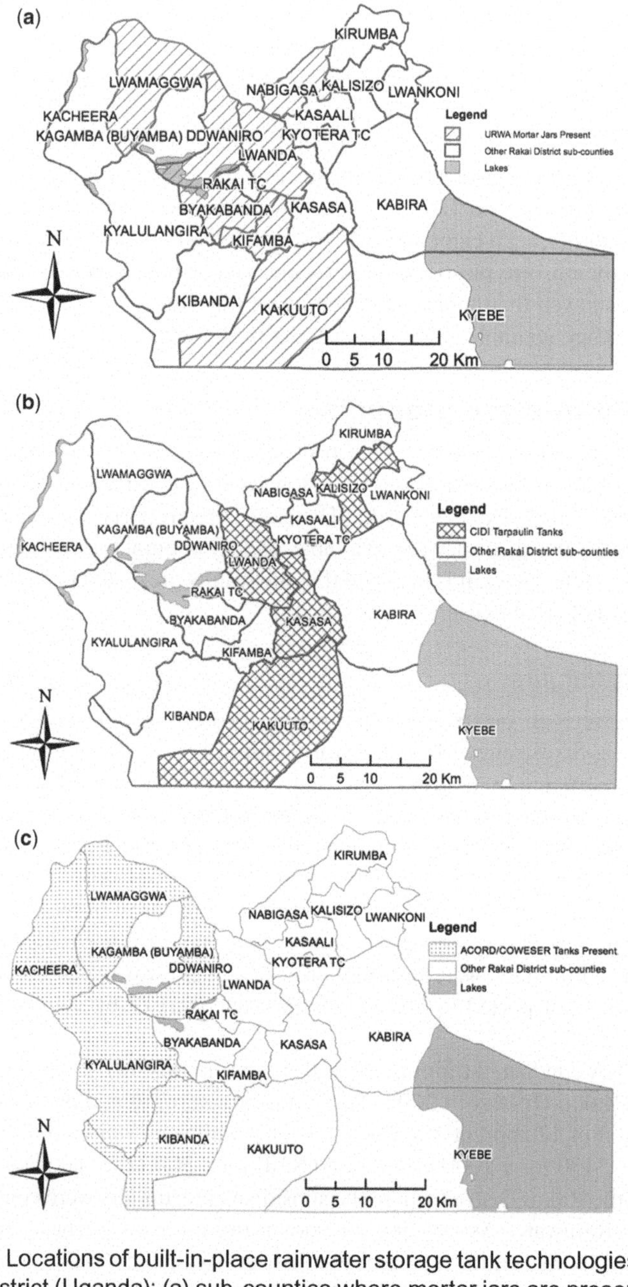

Figure 7.4 Locations of built-in-place rainwater storage tank technologies identified in Rakai District (Uganda): (a) sub-counties where mortar jars are present; (b) sub-counties that have implemented the use of tarpaulin tanks; and (c) sub-counties with ferrocement tanks present.

7.4.3.2 *Tarpaulin tanks*

Tarpaulin tanks were another low-cost option for rainwater storage. Typically, a hole was excavated by hand and covered by a structure that consisted of a small brick wall, wooden beams and a roof made of iron sheets. The pit was then lined with a locally available plastic tarpaulin. Tank volumes could range from 8000 to 25,000 litres. The cost of such structures was identified to be $140 for the 8000 litre tank, $220 for a 15,200 litre tank and $480 for a 25,000 litre tank. Figure 7.4b shows the four sub-counties where tarpaulin storage tanks were installed in the Rakai District.

7.4.3.3 *Ferrocement tanks*

The ferrocement tank construction method has become popular in recent years. It consisted of a wire mesh framework around which a tarpaulin is wrapped. Cement mortar is packed against the tarpaulin and around the reinforcement from the interior. Once the inside had dried (usually 2 or 3 layers), the tarpaulin was removed and the process repeated on the outside (see Mihelcic *et al.* 2009 for additional detail on the method). Tank volumes in the study location ranged from 5000 to 50,000 litres, though two of the organisations did not typically construct tanks above 20,000 litres.

Both ACORD and COWESER had built ferrocement tanks extensively throughout certain sub-counties of Rakai District (Figure 7.4c). URWA had not held any training or constructed any ferrocement tanks specifically in Rakai District, but they were actively promoting the technology nationally and their cost estimations for the method were relevant for the central region of Uganda in general.

In 2010, ACORD implemented a project for the building of ferrocement tanks in the Rakai sub-counties of Kachera, Lwamagwa, Kyalulangira and Ddwaniro. They trained 68 masons (51 female, 17 male), who subsequently built 170 tanks across the four sub-counties in 2010. From 2006 to 2008, COWESER implemented, on behalf of the Network for Water and Sanitation in Uganda (NETWAS (U)), the construction of 233 household and institutional ferrocement tanks in Kibanda and Kyalulangira sub-counties. This project, titled the 'Roof Catchment Rainwater Harvesting and Management Pilot Project', was funded by the African Development Bank and also included similar efforts in Bugiri and Kamwenge districts.

Prices for ferrocement storage tanks based on size and constructed by the three different organisations are provided in Figure 7.5. Prices were fairly comparable between the three organisations implementing ferrocement tanks at sizes ≤10,000 litres. For example, the price range for the 6000 litre tank differed by only $100, or less than 20% of the lowest priced tank for that volume. The prices diverge as size increases, and it appears ACORD was significantly more efficient at building larger tanks. The organisation claims to be able to construct a 30,000 litre tank for $1100, which is less than either URWA or COWESER, which could build a storage tank of 20,000 litres.

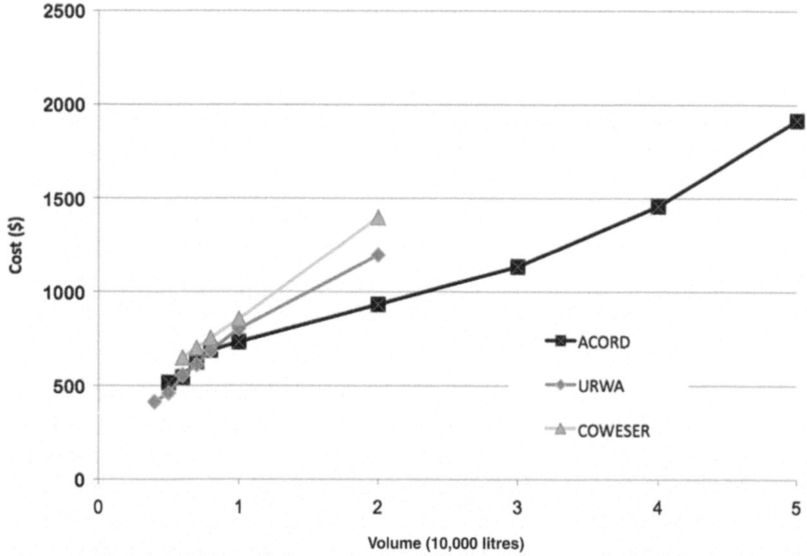

Figure 7.5 Ferrocement storage tanks constructed by three organisations in the study area are shown to diverge in cost after 10,000 litres.

7.4.3.4 Partially below-ground ferrocement tanks (PBG)

ACORD was encouraging and promoting the use of partially below-ground ferrocement (PBG) tanks. This type of tank was similar in form to above-ground ferrocement tanks, but the below-ground feature offered opportunities for material savings. The excavated pit offered external resistance to water pressure, which meant that the steel reinforcement – a major source of expense in above ground tanks – could be reduced. A small dome covering the tank and with an access point or tap for the pump, was all that was visible above ground. These tanks are also low cost, though uptake seemed slower than for the ferrocement and mortar jar options. This may have had something to do with the perceived prestige conferred on a homeowner by having a tank visible above ground.

No specific sites were identified where this technology was being implemented in Rakai District, but the ACORD office in Mbarara was actively promoting it in the south-west region of Uganda. For the purposes of this study it was considered a proven technology with the potential for application elsewhere. ACORD reported that a 6000 litre tank would cost $200.

7.4.3.5 Interlocking stabilised soil brick (ISSB)

The most recent contributor to rainwater storage facility construction in Rakai District is 'Brick by Brick'; a business constructing rainwater tanks out of

Interlocking Stabilised Soil Bricks (ISSBs). ISSB's were bricks formed from a moistened mixture of Ugandan sub-soil and 5–10% cement. They were subsequently compressed using a manual steel press to create an interlocking brick, with tongue and groove on opposite ends, as well as the top and bottom of the brick. Straight bricks could be made for standard building applications or a separate curved brick press could create curved bricks for use in rainwater tanks.

When rainwater tanks were constructed with this technology, cement mortar was used between every horizontal and vertical joint between bricks. The walls were then plastered both inside and out. The roof could be made of iron sheets spread over wooden beams or a concrete roof could also be integrated into the design. Dr. Musaazi of Makerere University was involved with fostering and propagating the use of this technology throughout Uganda for most of the last 20 years, though it is believed Brick by Brick is the most ambitious commercial application. Brick by Brick is based in Kalisizo, but is prepared to work throughout the district and beyond because of the portability of the press. Brick by Brick's standard volumes for tanks and respective prices are shown in Table 7.3 and, based on the previous discussion, were a competitive alternative for rainwater storage compared to other manufactured or built-in-place options.

Table 7.3 Volumes and associated prices for interlocking stabilised soil brick (ISSB) constructed rainwater storage tanks in Uganda.

Tank volumes (litres)	Cost ($)
10,000	820
15,000	1100
20,000	1300
25,000	1400

7.5 DISCUSSION

7.5.1 Technologies

The study presented in this chapter identified 11 distinct rainwater storage technologies, ranging in storage volume from as little as 5 litres to as much as 50,000 litres and ranging in cost from zero to over $3300. Uganda thus has access to a diverse selection of rainwater storage methods encompassing a wide range of volumes and costs.

7.5.2 Access

It was concluded that households in the Rakai District had access to a wide and consistent variety of manufactured rainwater storage options. Residents could purchase small plastic tanks from several dozen different hardware stores in 10 towns widely spread throughout the district. These stores had very similar prices, indicating a competitive and well-developed private sector for manufactured products. Alternatively, households could acquire larger plastic tanks from the centralised distributors, who would arrange for delivery anywhere in Uganda. Residents could also purchase 55-gallon metal drums from six different stores in five different towns – not as widely spread as the small plastic tanks (see Table 7.2 for comparison), but still available to anyone who wanted to acquire one. Finally, Rakai residents could choose the corrugated iron tanks available from three metalworks that were only located in one town. These results indicate a competitive and developing private sector for manufactured products.

In contrast to the manufactured sector, a household's access to built-in-place technologies for water storage was much more limited. This is shown graphically in Figure 7.4 for three of the built-in-place technologies. ISSB's are not considered further, as while Brick by Brick is an active and ongoing enterprise, willing and able to travel, it has not yet achieved the market penetration necessary to facilitate the claim that all of Rakai District has access to its service. Having access within a sub-county to at least one, but preferably several types of rainwater storage device, is important for advancing self supply as households are better able to make informed choices and imitate what works for their neighbours.

Table 7.4 summarises the sub-counties that had a choice of built-in-place technologies. The table shows that eight sub-counties had zero access to built-in-place technologies and a further seven had access to only one. Only five sub-counties had a choice between two different built-in-place technologies and none were able to choose between all four. This demonstrated that access to artisan-constructed storage options was limited, with significant gaps between areas where there was sufficient private sector capacity for implementation of the various methods. These data also demonstrate the large available opportunities for built-in-place artisans to expand existing, or set up new, businesses.

7.5.3 Cost

Figure 7.6 (not to scale) represents the financial cost associated with specific steps a household could take towards increasing their rainwater storage capacity. Shading indicates the storage technology that offers the lowest cost per volume of storage within a given volume range. The price and volume points indicate a step up where the next storage technology is available and offers more storage per unit cost than the previous step. Large plastic tanks were not included in this analysis, as they were associated with the largest cost on a unit volume basis for tank sizes ≥1000 litres. The grey transitions in Figure 7.6 represent the cost and volume where the next technology offers a lower cost per litre than the previous technology. These points were determined by a linear interpolation between specific detailed tank sizes and costs (available in Blanchard, 2012).

Table 7.4 Built-in-place tank choices available to Rakai District households by sub-county.

Sub-county	Mortar jar	Tarpaulin tank	Ferrocement tank	Interlocking Stabilised Soil Brick (ISSB)
Kyebe				
Kabira				
Lwankoni				
Kirumba				
Kasaali				
Kyotera TC				
Rakai TC				
Kagamba (Buyamba)				
Kalisizo		×		×
Kasasa		×		
Kifamba	×			
Byakabanda	×			
Nabigasa	×			
Kibanda			×	
Kyalulangira			×	
Kachera			×	
Kakuuto	×	×		
Lwanda	×	×		
Ddwaniro	×		×	
Lwmaggwa	×		×	

Figure 7.6 demonstrates that in this location, if a household had less than \$89 to spend on DRWH storage, they could purchase one or more small plastic tanks that could provide a storage capacity of several hundred litres. However, if they could spend \$89, they might want to invest in a mortar jar, which would increase their storage to 1000 litres. Mortar jars were determined to be the most financially viable in terms of volume per unit cost up to \$120 investment, at that point a household could purchase 2000 litres of storage volume by investing in a corrugated iron tank. However, for an additional \$20 investment, a household could purchase 6000 litres of storage by investing in a tarpaulin or a below-ground ferrocement tank. For households that had the ability to invest \$1000 to

$1900, they could increase their storage capacity to ≥25,000 litres and the most financially viable technologies would be ferrocement and interlocking stabilised soil brick storage tanks.

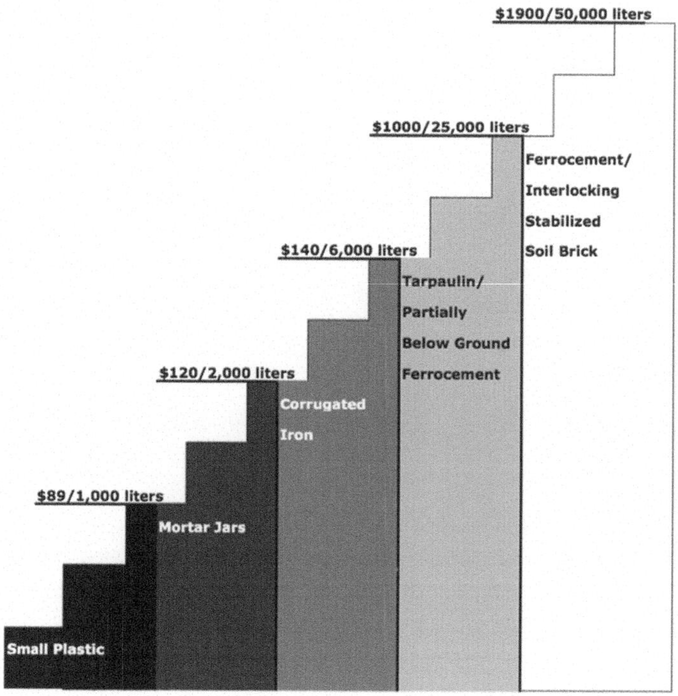

Figure 7.6 Incremental steps households could take to increase rainwater storage.

It should be noted that some of these increments were only marginally beneficial. For example, a user should only invest in a corrugated iron tank if they wanted to spend more than $120 but no more than $140. Outside this narrow band of costs, a user would achieve a larger storage volume for their money with mortar jars or tarpaulin/partially below-ground ferrocement tanks on the lower and upper bounds, respectively. Tarpaulin and partially below-ground ferrocement storage tanks, as well as ferrocement and ISSB tanks, were grouped together because they are priced similarly enough that users might want to choose either.

Figure 7.6 can be related to the RWH ladder presented previously (Figure 7.1). The lower steps represent informal or opportunistic RWH situations, while the upper levels represent main source, or in rare cases, sole-source utilisation of RWH. The middle ranges, depending on how water is collected and the individual needs of the household, represent wet-season, potable or adaptive RWH. While Figure

7.6 is based only on cost per volume of storage, it demonstrates that this is clearly not the only factor in a household's decision to invest in storage. For example, 55-gallon oil drums do not appear in the ranking because they are roughly twice as expensive as the plastic tanks of equal size. Nonetheless, they are a popular storage mechanism. It is postulated that the increased durability and possible theft deterrence-potential posed by the greater weight of the oil drums, may increase their attractiveness despite their higher cost. A more complete analysis of tank costs appears elsewhere (Blanchard, 2012).

Figure 7.6 also excludes the expected life of the tank, which is a dimension of understanding with regard to RWH that could be the basis for future study. Furthermore, it is known that water quality degrades with increasing time of storage and water temperature (Schafer & Mihelcic, 2012). Schafer and Mihelcic observed there was a statistical difference in the microbial water quality between polyethylene, fibreglass and cement water storage tanks as measured by *E. coli* counts. This could increase the health risk posed to household residents associated with possible microbial growth in the stored water. This understanding is not reflected in Figure 7.6, but also needs to be studied further. For example, PBG tanks may lessen the health risk because they would maintain a cooler water temperature, as they are constructed partially below ground.

7.6 CONCLUSIONS

The objectives of the study presented in this chapter were: (1) to present a comprehensive collection of rainwater storage options in Uganda; and (2) demonstrate the geographical disparities in the distribution and cost of those options. This was in order to assist the self supply concept in providing households with reliable, safe and sustainable water supplies. With regard to the first objective, the information was presented in a graphical hierarchy (Figure 7.6), which was organised by cost and storage volume. This graph is useful for water users in making informed decisions regarding selection from the variety of water storage mechanisms available to them. This type of centralised hierarchy had not previously existed for RWH in Uganda and should be helpful in expanding the impact of self supply. Information gathered and presented in this chapter to achieve the second objective, should aid those promoting the self supply concept in targeting its efforts at NGOs, communities, government agencies and businesses, in order to more effectively aid the acquisition of safe and reliable water supplies.

It was observed that a wide variety of domestic rainwater storage techniques were available to users in the Rakai District. The study identified a number of useful observations. Firstly, it was identified that 11 distinct technologies were in use: clay pots, pots and basins, brick masonry tanks, plastic tanks, 55-gallon metal drums, corrugated iron tanks, mortar jars, tarpaulin tanks, ferrocement tanks, partially below-ground ferrocement tanks and interlocking stabilised soil brick (ISSB) tanks. Secondly, the availability of manufactured rainwater storage products was documented as they were well distributed and marketed by many commercial entities. Thirty-two hardware stores were also identified that were selling manufactured tanks spread across 10 towns in Rakai District, as well as two national distributors of larger plastic tanks.

Thirdly, it was identified that in addition to the widely prevalent small plastic tanks (≤1000 litres), 55-gallon metal drums were reclaimed and available for purchase and corrugated iron tanks were actively manufactured and distributed in one location. In contrast, built-in-place tanks were not as well distributed on a geographical basis. This resulted in a major gap between areas, where households had a real choice of many rainwater storage options and other locations where built-in-place options were not as accessible. It was determined that of the five types of viable built-in-place tanks identified, eight sub-counties had no access to any, eight had access to one and only four could choose between two of the five technologies. Accordingly, access to artisan-constructed storage options was limited, with significant gaps between locations where there was sufficient private sector capacity for the implementation of the various technologies.

Finally, with regard to cost, it was identified that for tanks with storage volume less than 1000 litres, costs ranged from $0.075 to $0.30 per litre of storage. For volumes between 1000 and 10,000 litres, costs ranged between $0.017 and $0.14 per litre of storage. Above 10,000 litres of storage, tanks ranged from $0.014 to $0.14 per litre of storage. Figure 7.6 shows the incremental steps a user could take to increase their storage: up to an $89 investment, small plastic tanks offered the lowest cost per litre of storage. Ferrocement and ISSB tanks were found to occupy the high end of the storage range on a cost per storage volume basis. In terms of modularity, it was determined that the ISSB tanks were the most modular, because the bricks were interlocking. Consequently, additional layers could be added at a later time, if the roof was removable. In contrast, tarpaulin and below-ground ferrocement tanks were determined to be the least modular, because installation would require a new tank that would require a new excavation.

Three reasons were proposed for the lack of access to certain types of DRWH, for those households who wished to implement DRWH as a method to advance self supply. The first was the heavy use of subsidies when implementing DRWH programmes. Previous studies had identified that the possibility of subsidy made private investment unlikely in this location. All of the programmes that implemented built-in-place technologies in Rakai District funded the tanks through a grant of some kind, which subsidised construction in some way. This was with the exception of Brick by Brick (providers of ISSBs), which was a completely private enterprise.

Secondly, there appears to be a disconnect between the goals of those promoting DRWH storage technologies and a successful self supply approach. That is, the goals of several stakeholders were focused on solely building a specific number of tanks within a budget – but not to create an environment with the proper technical knowledge for private initiative to continue in a self supply scheme. Apart from Brick by Brick, only one stakeholder identified in this study (the Uganda Rainwater

Association) had this second goal with their promotion of mortar jars, intending to create fully functional businesses with supply chains continuing to operate. Unfortunately, it was believed that high subsidies subdued the initiative once URWA's involvement ended. Similarly, there appears to be a disconnect between the goals of self supply, as described by the Rural Water Supply Network (RWSN) and the government of Uganda. A basic component of the self supply philosophy is that users should be assisted to take incremental steps toward sufficient water quantity and quality, encouraging private investment by seeing even a small step as a good one. The Government's approach, however, in advancing water provision appears to be one of 'all or nothing'. This is because the minimum volume necessary for a tank to provide sole-source access for an average household throughout the dry season is calculated as being 6000 litres. This large storage requirement does not support any size smaller than that which as shown in this case study are widely available (and affordable).

The final reason for lack of access to certain types of DRWH storage is the failure to truly understand in which products investment should be made. URWA is actively promoting mortar jars, which would appear to be of good value, as they are certainly one of the least expensive built-in-place tank types. However, due to reasons such as their small size, poor performance during the dry season and their vulnerability to damage due to high sun exposure, the Rakai office of the Ministry of Water and Environment does not widely support their use. Further research is required to assess the true limitations of such tanks, as well as to determine the impact of the differing tank types on overall rainwater collection and usage.

ACKNOWLEDGEMENTS

The authors wish to thank the many Ugandan stakeholders who generously shared their expertise with us. Kirsten Danert of RWSN provided input in the selection of the topic studied here. This work is based upon work supported by the National Science Foundation under Grant No. 0965743. Any opinions, findings, conclusions or recommendations expressed in this report are those of the authors and do not necessarily reflect the views of the National Science Foundation.

REFERENCES

Blanchard J. (2012). Rainwater Harvesting Storage Methods and Self Supply in Uganda. Unpublished M.S. Thesis, University of South Florida, Florida, U.S.A.

Cowden J. R., Watkins D. W. Jr. and Mihelcic J. R. (2008). Stochastic rainfall modeling in West Africa: Parsimonious approaches for domestic rainwater harvesting assessment. *Journal of Hydrology*, **361**, 64–77.

Cruddas P. (2007). An Investigation into the Potential to Reduce the Cost of Constructed Rainwater Harvesting Tanks in Uganda. Unpublished M.S. Thesis, Cranfield University, U.K.

Danert K. and Motts N. (2009). Uganda Water Sector and Domestic Rainwater Harvesting Sub-Sector Analysis. http://www.rmportal.net/library/content/translinks/translinks-2009/enterprise-works-vita-relief-international/paper_ugandadomesticrainwater harvesting.pdf (accessed 30 May 2013).

Danert K. and Sutton S. (2010). Accelerating Self Supply: A Case Study from Uganda 2010. Report RWSN 2010–4, St. Gallen, Switzerland.

Dean J. and Hunter P. R. (2012). Risk of gastrointestinal illness associated with the consumption of rainwater: a systematic review. *Environmental Science & Technology*, **46**, 2501–2507.

Fry L. M., Cowden J. R., Watkins D. W., Jr. Clasen T. and Mihelcic J. R. (2010). Quantifying health improvements from water quantity enhancement: An engineering perspective applied to rainwater harvesting in West Africa. *Environmental Science and Technology*, **44**(24), 9535–9541.

Martinson D. B. and Thomas T. (2003). The roofwater harvesting ladder. *Proceedings of the 11th IRCSA Conference*, Texcoco, Mexico.

Mihelcic J. R. (2010). The right thing to do: graduate education and research in a global and human context. In: What is Global Engineering Education For? The Making of International and Global Engineering Educators, Part I and II: Synthesis Lectures on Global Engineering, G. L. Downey and K. Beddoes (eds), Morgan and Claypool Publishing, San Francisco, pp. 235–250.

Mihelcic J. R., Myre E. A., Fry L. M., Phillips L. D. and Barkdoll B. D. (2009). Field Guide in Environmental Engineering for Development Workers: Water, Sanitation, Indoor Air. American Society of Civil Engineers (ASCE) Press, Reston, VA.

Mihelcic J. R., Phillips L. D. and Watkins D. W. (2006). Integrating a global perspective into engineering education and research: engineering international sustainable development. *Environmental Engineering Science*, **23**(3), 426–438.

Rowe N. (2007). Factors Affecting The Cost of Prefabricated Water Storage Tanks for use with Domestic Roofwater Harvesting Systems in Low-Income Countries. Unpublished M.S. Thesis, Cranfield University, U.K.

RWSN (2003). Self Supply – Small Community and Household Water Supplies – Concept Note. Report RWSN, St. Gallen, Switzerland.

Schafer C. A. and Mihelcic J. R. (2012). Impact of storage tank material and maintenance on household water quality. *Journal American Water Works Association*, **104**(9), E521–E529.

Sutton S. (2008). An Introduction to Self Supply: Putting the User First. Incremental Improvements and Private Investment in Rural Water Supply. Report RWSN, St. Gallen, Switzerland.

Thomas T. H. (2011). Report on Tour of Roofwater Harvesting Systems in Southern Uganda – 22nd–24th February 2010. http://ewb-mit.wikispaces.com/file/view/Report+on+Tour+of+Roofwater+Harvesting+Systems+Feb+2010.pdf (accessed 12 May 2014).

Thomas T. and Rees D. (1999). Affordable Roofwater Harvesting in The Humid Tropics. http://www2.warwick.ac.uk/fac/sci/eng/research/civil/crg/dtuold/pubs/reviewed/rwh/dgr_tht_ircsa_1999/04_01.pdf (accessed 30 May 2012).

UBOS (2002). 2002 Uganda Housing and Population Census. UBOS Report, Kampala, Uganda.

URWA (2004). SEARNET Rainwater Harvesting Policy Study: Uganda Study. URWA Report, Kampala, Uganda.

Chapter 8

Incentivising and charging for rainwater harvesting – three international perspectives

Sarah Ward, Fernando Dornelles, Marcio H. Giacomo and Fayyaz Ali Memon

8.1 INTRODUCTION

Incentivising and charging for rainwater harvesting (RWH) is not a new topic. Cities and countries in which more recent technology-based proprietary systems have been implemented for significant periods of time have a wealth of experience in this area.

For example, in Australia, in response to frequent droughts and a growing urban population, incentive and RWH system subsidy schemes have been ongoing in a number of areas, although current potable water prices can make the situation challenging even there (Rahman *et al.* 2012).

Germany is another county about which information on RWH incentives and subsidies is widely known (Konig, 2001). In Germany, the government is keen to support environmental technologies, as well as alleviate surface water management and water quality issues related to rainwater entering sewers. Downpipe (downspout) disconnection schemes have been in action in the Rhine-Ruhr metropolitan area since the 1990s and charging is done either through the Emscher Association (in the case of direct dischargers) or through municipalities (in the case of households) (Herbke *et al.* 2006; Geretshauser & Wessels, 2007).

Furthermore, 'smart regulation' has recently been considered as a way to support the RWH implementer: smart regulation is regarded as the interaction of three financial instruments (water abstraction fees, water supply and effluent fees and subsidies), rather than their implementation in isolation. Additionally, it is recognised that actors at various levels ('change agents', 'blocking agents') need to be mobilised to establish the value of smart regulation (Partzsch, 2009).

In contrast to the information about these contexts, limited information is available for other contexts where RWH is only beginning to take off or where legislation has recently changed. This chapter aims to present perspectives on incentives and charging for three contexts, one of which has a long-established history of RWH implementation, but has experienced recent regulatory change and two of which do not have long histories of RWH implementation, but have also experienced recent regulatory change in relation to alternative water supply systems.

Consequently, this chapter intends to: (1) provide a brief overview of how the water sector operates in three international locations (UK, Brazil and Texas (USA)); (2) describe the current and potential future markets for RWH in these locations; (3) summarise the main legislation and policies that apply to RWH in these locations and (4) highlight the incentives and charging mechanisms (or lack thereof) in use for facilitating the appropriate adoption of RWH at a range of levels. The first two sections cover the UK, the second two Brazil and the final two the USA. A concluding section draws together the main points of the discussion raised in the chapter.

8.2 FIRST INTERNATIONAL PERSPECTIVE – UK

8.2.1 Legislation and emerging markets

In 2009, an estimated 9000 RWH systems were installed in the UK, not counting DIY installations, compared to the German market of ~60,000 new rainwater tanks in 2009 – the largest market in Europe (Ziegler, 2010). The UK market has been stimulated to a limited degree by the Code for Sustainable Homes (voluntary guidance on sustainable buildings for the new build construction sector), the introduction of several pertinent British Standards (BS 8515: 2009 on RWH; BS 8525: 2010 on greywater and BS 8595: 2013 on the selection of water reuse systems) and a recent focus on flood and water management, due to alternative flood and drought events over the last ten years. The latter has resulted in the introduction of two landmark pieces of legislation: the Flood and Water Management Act (2010) and the Water Act (2014), which promote the increased consideration of sustainable drainage systems and water reuse.

The non-residential market accounted for 65% of the RWH market by value in 2010, with limited growth in the residential sector, due to limitations of existing proprietary systems for the household scale. However, the residential sector is forecast to increase its share value by 5% to 40% and its share volume to 70% by 2014 (MTW, 2010). The introduction of innovative retrofit systems is likely to stimulate this market significantly, once their relative sustainability benefits have been better quantified (Melville-Shreeve *et al.* 2014). Taking into account certain residential dwelling features (being owner-occupied and metered, having suitable structural design, the nature of owner motivations and a suitable non-potable water demand profile), it is estimated that approximately 13.7 million properties in the

UK could be suitable for retrofit RWH (Ward, 2013), where appropriate, once socio-technical hurdles are overcome. Such hurdles include (Ward *et al.* 2012):

- The limited range of suitable systems currently available on the market;
- A lack of incentive schemes to encourage the early adoption and diffusion of such innovations into the built environment;
- Health and safety concerns over harvested rainwater quality.

Innovation is required to move away from the proprietary, off-the-shelf, high-volume (~2–5 m^3) RWH systems, as rainwater storage requirements for residential properties have and may continue to decrease. This is due to water efficient appliances now being widely promoted by a number of organisations, including water companies in England and Wales under their (current) statutory regulatory reporting requirements. Recent research indicates that only around 44% of people with outdoor areas that require watering, actually water them (Pullinger, 2013). This suggests that in the future the main application for low-volume residential RWH might be in toilet flushing. Therefore downpipe/roof-level take-off and storage of rainwater could become the most feasible type of RWH system for the retrofit market. An example of such a system is being developed in the UK at the University of Exeter, which is a low-volume RWH system that uses a low-energy pump to off-take rainwater from a downpipe to a roof-level storage tank. However, in order to prepare the residential market for the entry of such systems, innovation may first be required in the financial areas of incentives and charging for RWH, as well as in water company business models.

8.2.2 Incentives and charging mechanisms

Despite the emerging legislative and market drivers outlined in the previous section, central implementation of incentive schemes for RWH, to parallel those for renewable energy (feed-in-tariffs, renewable heat incentives) have not been forthcoming from the various governing bodies in the UK. The only existing long-running incentive scheme (to date) is the Enhanced Capital Allowance scheme HMRC (2014), which is a purchase tax reclaim mechanism. Any *business* customer purchasing eligible equipment from the Water Technology List (WTL – an online catalogue of water-efficient and water reuse products), can claim back the tax paid on the items. This scheme applies only to business customers however; therefore even if innovative retrofit RWH systems targeted at the residential market were to feature on the WTL, the ECA scheme would require significant modification for homeowners to benefit (Ward *et al.* 2012).

With regard to charging for harvested rainwater, recent research has investigated implications for water companies in England and Wales in relation to losses in revenue, to account for potential future increases in utilisation of harvested rainwater by residential customers (WRc, 2012). Where RWH systems are installed, effluent from a property remains at the same volume as if potable mains water was being used, but the volume of potable mains water supplied is less than the discharged volume. This leads to a discrepancy in charges, due to the effluent volume for residential properties usually being calculated based on 95% of the potable water supplied. Consequently, water companies may be treating the same volume of effluent, but receiving reduced revenue for this service. Arguably the rainwater would end up in the sewer eventually anyway, but its treatment would be counterbalanced by revenue from potable mains water usage (in a non-RWH scenario).

Options for reconciling the discrepancy in charging to maintain revenue include:

- Developing innovative metering regimes to facilitate sub-metering of RWH systems (and other alternative water systems, such as greywater reuse systems) and charging for such services;
- Developing innovative charging or tariff structures for implementation where RWH systems are prevalent (i.e., adjusting the 95% rule to a different percentage/ratio);
- Developing a RWH system maintenance service charging model that could incorporate a fixed cost element to cover a proportion of the lost revenue. Such a service might also ensure that systems were properly maintained and therefore potentially reduce health and safety concerns.

In relation to metering, the installation of a sub-meter is potentially not straight forward depending on how long after the RWH/greywater system was installed the meter is fitted (difficulty in accessing pipework), the nature of flow in the pipework (if gravity flow there will be a lack of head to force the water through a meter) or the water may contain a high level of particulate matter reducing the meter's effectiveness. The latter issue could easily be overcome by introducing 3-tiered filtration, which is recommended by BS 8515: 2009, before the meter installation point. The matter of maintenance of such filters would remain however, as RWH systems are not 'fit and forget'. The recent release of new metering technologies such as electromagnetic and ultrasonic meters means that issues with particulate matter can be overcome, although such meters are unlikely to initially be cost-effective. A further issue with the potential to complicate metering is that although the location and ownership of a potable water meter is covered by legislation, such legislation does not exist for non-potable metering. Therefore ownership, reading and maintenance liability disputes could potentially occur (WRc, 2012).

The other side of the charging argument is that although water companies may treat more effluent for less revenue, they may receive other financial benefits from residential RWH. For example, distributed storage tanks within a development may result in the localised retention of rainwater (HR Wallingford, 2011), further discussed in Chapter 4, resulting in spare capacity in sewers potentially alleviating the need for additional capacity addition to cope with rainwater in sewers (Hurley *et al.* 2008). Furthermore, less raw water may need to be abstracted, treated and distributed, which could result in water company operating and capital expenditure cost reductions, although the exact level of such gains would vary depending on the water company operating area (due to variations in tariffs and operating costs relating to topography (pumping energy costs) and other such parameters).

On balance and in line with a recent review on water charging mechanisms (Walker, 2009), it is acknowledged that excessive charges, which could potentially discourage the appropriate use of RWH systems, are not currently acceptable and should be avoided. However, should the market for residential retrofit RWH systems rapidly expand, this would likely be reviewed to adjust sewerage charges in order to account for revenue deficits or to develop another mechanism through which to retain a balance between non-potable water supplementation and revenue maintenance. It is estimated that such a review would be unlikely to occur within the next 50 years, based on the view of the financial regulator for water in England and Wales (Ofwat) (WRc, 2012).

To understand how these perspectives from the UK align with the international RWH arena, the next section focuses on the situation in Brazil.

8.3 SECOND INTERNATIONAL PERSPECTIVE – BRAZIL

8.3.1 Legislation and market

Since 1934, water in Brazil has been managed as a public property, decreed by the *Código das Águas* (Water Regulation) and the National Law number 9.433/1997. This law established the National Policy on Water Resources and created the National Water Resources Management policy, which further defines water as a limited resource, prioritising water supply for human and animal utilisation.

The first water supply systems in Brazil began to operate during the early decades of the 20th century due to federal government actions to attract private companies to provide services on sanitation (water supply and sewage collection), although few cities benefited from this service. This scenario continued until 1971 when the National Sanitation Plan (PLANASA) was introduced. Although implementation of the plan led to 86% of the population being served in 1991, this was by municipal companies rather than private companies. In 1994, momentum gathered around private company participation, although this is not widely accepted by the public due to risks and uncertainties about regulation in the water sector (Saiani *et al.* 2009).

Partly in response to this, RWH has become more widely accepted, although this is not consistent across all regions due to the low cost of mains water. RWH is also being cautiously viewed as a useful contributor to attenuate surface water and assist in the alleviation of urban flooding, with its main negative being the inability to guarantee the tank will be empty during a storm event (HR Wallingford, 2011 and Chapter 4). The market for RWH is being driven by its strong social and environmental appeal in both rural and urban areas (where mains supplies may be non-existent or unreliable or intermittent) and many cities in Brazil have initiatives in place to create legislation for RWH.

There are different types of legislation: some impose an obligation to include RWH in all new construction projects, others only impose such a requirement on construction projects with a specific roof area above a threshold value. Where such legislation applies, construction and commissioning documents can be withheld if RWH is not implemented. Often such legislation is developed in conjunction with instruments for rational water consumption such as metering, water efficiency and water loss (leakage) reduction. Usually legislation relating to RWH contains limited or no technical information on sizing, preliminary treatment or maintenance. However, technical information is provided by the Brazilian Association of Technical Standards (ABNT, 2007), which is valid for the entire country and restricts RWH utilisation to non-potable end uses, although potable use is indicated in rural areas without mains supplies.

The primary market for RWH is within rural communities where surface water resources are low during certain seasons. In such locations, RWH storage tanks tend to be constructed using slabs of cement and with a capacity of 16 m^3 – enough to supply a family for six to eight months. Usually the construction is undertaken by a homeowner with their neighbours, which can generate a collaborative interaction for the social and economic growth of the community (ASA, 2013).

Although demand is high for RWH in industrial buildings with large roof areas and large non-potable demand requirement (as large savings on the water bill can be achieved), in contrast and despite increasing water consumption in urban regions, RWH is not as popular due to resistance from water companies based on the potential for loss of revenue on sewerage charges. This is based on the same principle as explained in the previous section on the UK – where the proportion of mains water used by a property with RWH becomes erroneous for the sewage effluent charge calculated for a property. As in the UK, there is no standardised or approved legislation or technique to facilitate the measuring or recovery of such revenue. At present, the charge for sewage is obtained by similar calculation to that used in the UK, which, as mentioned previously, is based on mains water supplied measured by a meter. However, the Brazilian calculation is based on a proportion ranging from 0.6 to 1.0, which accommodates the sewage volume arising from utilisation of rainwater from the RWH system (Dornelles *et al.* 2012).

8.3.2 Incentives and charging mechanisms

The level of incentives for RWH in Brazil, similarly to that previously described for the UK, is low with the main incentive for the urban construction sector taking the form of tax exemption or reduction. However, in contrast, rural regions in the semiarid zone can take advantage of the recently introduced '*One Million of Cisterns*' programme of the Federal Government. To date a total of approximately 500,000 cisterns (RWH storage tanks) have been constructed for families earning less than half the minimum per capita income salary (U$ ~340) and who are without regular access to enough food. Preliminary results of the programme have been to encourage new programmes to sustain the semiarid population (~5 million) through guidelines on utilising rational use of harvested rainwater for human consumption, watering animals and irrigation of subsistence crops (ASA, 2013).

In relation to charging for RWH and in an attempt to resolve the sewerage charge deficit, Dornelles *et al.* (2012) proposed a methodology for estimating the volume of sewage generated using RWH, based on the characteristics of the RWH system and a percentage of the volume of metered potable water. They determined their method was viable for estimating the volume of sewage generated by the use of rainwater, which had immediate practical application as it required no investment in any additional equipment. At present such methods for incorporating a percentage factor for sewage generated using RWH have not been implemented in the UK, but as outlined in Section 8.2.2, such methods may be considered for implementation in the future (WRc, 2012).

For the urban areas, there are at present no differential charging schemes or incentives and the situation is unlikely to change in the short term. As previously mentioned and in line with the position in the UK, the increasing profile of RWH in Brazil is primarily for its environmental appeal, observed by the issuing of environmental certification that enhances a construction projects' value. However, as also in the UK, it will be necessary to address two aspects before RWH implementation in Brazil becomes more straight forward:

- Improve the estimation of sewage generated in properties utilising RWH, to reduce rejection of RWH by water/sewage companies;
- Develop the plumbing and construction professionals' understanding that the RWH storage tank cannot be guaranteed as a surface water runoff control measure, but that using techniques emerging from other contexts (e.g., the UK (HR Wallingford, 2011)) could lead to tank designs capable of meeting stormwater attenuation objectives. This may lead to more appropriately sized tanks more suited to the dual functionalities of water supply and stormwater attenuation.

Examining the situation in the Brazilian context has identified certain parallels between legislation, incentives and charging with the UK. The next section presents Texan perspectives from the USA, to identify if there are similar parallels to the two contexts presented above.

8.4 THIRD INTERNATIONAL PERSPECTIVE – USA

8.4.1 Legislation and market

The use of RWH systems as a complementary or alternative water supply source to centralized water supply systems has risen in recent years in the United States (Lye, 2002). Many drivers contribute to the rise in RWH usage, including the frequency of droughts, increasing demands caused by population growth and increasing costs of centralized systems and well drilling (Kalaswad & Arroyo, 2008). Nevertheless, as with the situation in the UK and Brazil, many barriers to expanding RWH remain as a result of technical, regulatory, financial and cultural issues. As the water sector in the USA is subject to state legislation, this section will focus on the state of Texas.

With regard to rainfall, patterns in Texas vary greatly; the eastern border with the state of Louisiana receives an average of 1422 mm of rainfall per year, but the western border with New Mexico receives only 152 mm of precipitation per year (Texas Rainwater Harvesting Evaluation Committee *et al.* 2006). Consequently, RWH is not an effective approach for all regions of the state. However, the majority of the population resides in large cities lying in the central-eastern region with the metropolitan areas of Houston, Dallas and San Antonio accounting for approximately 60% of Texas residents (U. S. Census Bureau, 2010). Average annual rainfall depths for these cities are 1264, 918 and 820 mm, respectively.

According to the 2012 Texas State Water Plan (TWDB, 2012), the daily per capita water consumption for Houston, Dallas and San Antonio are 601, 976 and 556 litres per capita per day, respectively. The estimation of daily per capita consumption is controversial due to a lack of a unified methodology for estimating the volumes of consumption and the populations, but these numbers indicate a great potential for water conservation; RWH can help decrease the burden on water supply systems. Further to this, demographic projections indicate that the Texas population will double in less than 50 years (TWDB, 2012), posing an incredible pressure on the ability of water systems to reliably supply increasing water demands. Prolonged periods of droughts are likely to exacerbate the gap between supplies and demands. In recognition of these stressors, the state of Texas has been investing in water conservation initiatives and RWH is one of the strategies that has been adopted by consumers and promoted by water utilities, generating a substantial market for these types of system.

With regard to legislation, the surface water sector in Texas operates under a blend of two dominant water doctrines in the USA: prior appropriation and riparian rights. These result from historic legal systems imported by Spanish and Anglo-American settlers, respectively (Sansom *et al.* 2008). The prior appropriation system is based on the principle of 'first come, first served'. Under this system, water is owned by the state of Texas and held in trust for users who apply for permits. In the event of a drought, senior water rights holders take priority of use over junior water rights holders. The riparian rights system gives landowners next to stream the right to use water that passes through their land. However, this flow is inadequate in regions subjected to low flow periods. In Texas, both systems have been combined, although the prior appropriation system dominates with the riparian rights represented by an exemption that grants riverside landowners a certain volume for domestic and livestock consumption (Sansom *et al.* 2008).

Subsequent to this legislation, during the last 20 years RWH has been promoted by the Texas State Legislature and implemented by the Texas Water Development Board (TWDB) and local water entities (Kalaswad & Arroyo, 2008). Table 8.1 lists some of the Texas Legislature regulations that have related to RWH over the last 20 years. The measures these regulations enact demonstrate a comparative deficit in action in the UK and Brazil, although the requirement for buildings with certain roof areas to have RWH is similar in Brazil. One of the more prominent measures enacted by legislation in 2005 was the creation of the Texas Rainwater Harvesting Evaluation Committee, which consists of members from the Texas Water Development Board (TWDB), the Texas Commission on Environmental Quality (TCEQ), the Texas Department of State Health Services and the Texas Section of the American Water Works Association Conservation and Reuse Division. This committee is directed to evaluate the potential for RWH in Texas and to recommend minimum water quality guidelines, standards and treatment methods for potable and non-potable indoor uses of rainwater, as well as developing promotion strategies for RWH. In 2006 the Committee undertook a review, which identified three key findings and ten recommendations (listed in Table 8.2), which resulted in the 2007 and 2011 legislation summarised in Table 8.1.

As a result of state foresight, implementation of legislation and the collaborative efforts of the organisations with responsibilities for RWH in Texas, incentives and charging mechanisms for systems are more advanced than those currently in action in the UK and Brazil. Such schemes and mechanisms are discussed in the following section.

8.4.2 Incentives and charging mechanisms

In order to incentivise the adoption of RWH, the TWDB published the Texas Manual on Rainwater Harvesting (TWDB, 2005) to provide commercial and residential owners information about RWH systems. The manual is a comprehensive review of the technical, financial and regulatory aspects concerning RWH systems. It describes the main system components, discusses considerations on water quality and treatment, presents a methodology for designing and estimating costs and lists financial and other incentives for undertaking RWH in Texas. It also contains descriptions of 13 case studies located in different cities around Texas.

Since 2007, the TWDB has sponsored the 'Texas Rain Catcher Award', which is a competition and recognition program designed to promote RWHS in Texas. The annual competition presents awards to RWH system installations in three categories: residential, commercial/industrial and educational/governmental. The projects are judged using the following criteria:

- Demonstration of how RWH helped conserve surface and/or groundwater and reduce dependency on conventional water supply systems;

Table 8.1 Texas Legislature regulations related to rainwater harvesting.

Document	Year	Measures
Proposition 2	1993	• Gives property tax relief to commercial and industrial facilities that implement rainwater harvesting
Senate Bill 2	2001	• Gives local taxing entities the authority to exempt all or part of the assessed value of property on which water conservation modifications, including RWH, are made • Provides sales-tax exemptions for rainwater harvesting equipment
House Bill 645	2003	• Prevents homeowner associations from implementing new covenants banning outdoor water conservation measures
House Bill 2430	2005	• Established the Texas Rainwater Harvesting Evaluation Committee
House Bill 4	2007	• Directs new state facilities with roof areas greater than 930 m^2 to incorporate RWH • Encourages Texas institutions of higher education and technical colleges to develop curricula and provide instructions about RWH • Exempts homes using RWH as their sole source of water supply from water quality regulations • Requires facilities using both public and RWH supplies to have safeguards for cross connections
House Bill 3391	2011	• Mandates RWH on new state facilities with a roof area of at least 4645 m^2 located in a region with average annual rainfall of at least 508 mm • Encourages municipalities to promote RWH through incentives such as subsidizing rain barrels or offering rebates for water storage facilities • Instructs TWDB to provide training on RWH for members of permitting staffs of municipalities and counties • Encourages school districts to implement RWH at local facilities

Source: Adapted from Kalaswad and Arroyo (2008).

Table 8.2 Texas Rainwater Harvesting (RWH) Evaluation Committee findings and recommendations.

Finding	Recommendations
There is significant untapped potential to generate additional water supplies in Texas through RWH, particularly in urban and suburban areas	• Direct State facilities with a roof area greater than 4645 m^2 to incorporate RWH • Develop incentive programs to encourage the adoption of RWH by residential, commercial and industrial facilities • Consider a biennial appropriation of US$500,000 to the TWDB to help match grants provided for RWH demonstration projects
With the application of appropriate water quality standards, treatment methods, and cross-connection safeguards, RWH systems can be used in conjunction with public water systems	• Direct TCEQ and other state agencies to continue to exempt homes with RWH as their sole source of water supply from water quality regulations that may be required for public water systems • Direct TCEQ and other state agencies to require those facilities with both public and RWH for indoor purposes to have appropriate cross-connection safeguards and use the RWH for non-potable indoor purposes • Appropriate funds to the Texas Department of State Health Services to conduct a public health epidemiologic field and laboratory study to assess pre- and post-treatment water quality from different types of RWH systems • Direct Texas cities to enact ordinances requiring their permitting staff and building inspectors to become more knowledgeable about RWH
There is a need to develop training and educational materials on RWH to help design appropriate systems	• Direct a cooperative effort by TCEQ and the Texas State Board of Plumbing Examiners to develop a certification program for RWH installers • Direct Texas Cooperative Extension to expand their training and information dissemination programs to include RWH for indoor uses • Encourage Texas institutions of higher education and technical colleges to develop curricula and provide instruction on rainwater harvesting technology

Source: adapted from Texas Rainwater Harvesting Evaluation Committee (2006).

- General benefits to the environment, including reduction of runoff generations;
- Demonstration of how much money was saved;
- Originality and innovation; and
- The uniqueness of the system.

The financial incentives for implementing RWH are provided through state tax breaks or municipal/local incentives (TWDB, 2005). At the state or county level, property tax exemptions can be granted to properties that adopt pollution control equipment, which includes water conservation equipment. In addition, the state of Texas provides sales tax exemptions for RWH equipment and supplies. Additionally, State legislation has been enacted to support RWH system adoption, such as promoting educational initiatives and training, as well as providing financial incentives; but there is still much more that can be done. Further education and training opportunities are required for RWH licensers and developers. More financial incentives may engage municipalities that so far have not considered RWH a viable option to help decrease the pressure on water supply systems. In addition, further research is required to help address financial concerns, including the cost of rainwater treatment and also lowering the capital cost of implementing RWH as far as possible. This echoes the situation previously described in the UK, where suitable residential retrofit systems are urgently required to reduce installation and maintenance costs (Ward *et al.* 2012b). Ultimately, public awareness and education is vital because it teaches local decision makers and consumers to recognise that they have the power to make water supply systems more sustainable in the long term and that RWH can help achieve that goal.

At the municipal level, some cities encourage residents and businesses to adopt water conservation measures, including RWH, in the form of rebates and discounts. For example, the city of Austin provide rebates of US$0.50 per gallon for non-pressurized systems and US$1.00 per gallon for pressurized systems, to a maximum amount of US$5,000, which should not exceed 50% of the system's cost. The rebate can be used to pay for labour and materials, including the tank, mounting pad, screens, filters, first-flush devices and pipes. However, it cannot be used to pay for gutters, irrigation system components or backflow preventers.

According to the TWDB (TWDB, 2005), more than 6,000 rain barrels (water butts) have been installed through the incentive program. The city of Fort Worth sponsors rain barrel sales by subsidising approximately 50% of the retail price City of Fort Worth (2014). In the city of San Antonio, a large-scale retrofit rebate program exists for commercial, industrial and institutional water users served by San Antonio Water Systems (SAWS). The value of the rebate, up to 50% of the installation costs, is determined by the amount of savings and the life of the equipment. According to the terms of the rebate program, the project and equipment must remain in use for at least 10 years or the life of the equipment, whichever is less. Additionally, water volume data must be collected before and after the retrofit and must be reported, the project must show clear potential for water consumption reduction and an annual report must be submitted for five years following the implementation. This is in contrast to the ECA scheme in the UK, which does not require such monitoring. This has the advantage of keeping the scheme relatively simple, but means a lot of possibly useful data on RWH installations is not collected or analysed, which limits updates to the scheme based on the possible findings of such research.

The efficiency of RWH systems depends on the climatic characteristics of a region and the size of the system, which can incur in high capital costs. Therefore many municipalities are investing in other water conservation initiatives, such as demand reduction equipment (high efficiency toilets and shower heads, xeriscaping). For instance, since 1994, SAWS has replaced 240,000 toilets in commercial and resident facilities with low-flow models and has recently introduced a rebate program for residential irrigation design that offers a maximum of US$800 for projects that redesign landscapes adopting low-water-use irrigation systems. This implies that RWH is only part of the water management toolkit, but is a valuable tool nonetheless and greater effort is required to facilitate its appropriate implementation across the international arena.

8.5 CONCLUSIONS

Population pressures, climate change impacts and increasing urbanisation are resulting in serious consideration of alternative sources of water to complement centralised water supply and drainage systems. Consequently, the market for RWH in the UK and Brazil is increasing in size and future population projections for Texas, USA indicate that the market there, which is already substantial, is also likely to grow significantly over the next 50 years.

In all three contexts, different pieces of legislation have or will be enacted, which aim to promote RWH and increase its appropriate uptake. Texas has a long history (two decades) of developing promotion strategies, education programmes and incentive schemes, the more recent of which have arisen as a direct result of the creation, under law, of the Texas Rainwater Harvesting Evaluation Committee. The comprehensive Texas Manual on RWH has also been developed, providing comprehensive guidance on the technical, financial and regulatory aspects of RWH. Although the UK and Brazil have produced guidance on aspects of RWH over the last five years, they could learn a number of lessons from the schemes that have been developed and implemented in Texas.

With respect to building-level regulations for RWH, Brazil appears to be ahead of the UK, as it has rules in place (similar to those in Texas) that make RWH compulsory in buildings with roof areas over a certain area. Similarly all three locations have tax-related incentive schemes, although those in operation in the UK and Brazil appear to be limited in comparison to those in place in Texas (with the exception of rural areas in Brazil, which receive higher incentives based on per capita income). Brazil and the UK are also developing and considering the implementation of differential charging schemes for sewage generated using rainwater from RWH systems, in order to protect the revenue of water companies.

The issues examined in this chapter have highlighted that the legislation, markets, incentives and charging schemes relating to RWH are similar across the international locations represented, despite different overall water management regimes (private, public-private and public, respectively for the UK, Brazil and Texas), although they are implemented to differing degrees. Policy-makers and water managers in each location would certainly benefit from greater consideration of the strategies operating across the international arena. For example, Texas could perhaps learn from the UK and Brazil's development of RWH charging schemes and the UK and Brazil could learn from Texas' dedicated approach to facilitating collaboration on RWH strategies.

REFERENCES

ABNT (2007). NBR 15527: *Água de Chuva-Aproveitamento de áreas urbanas para fins não potáveis – Requisitos.* (NBR 15527: Rain water harvesting for urban non-potable uses – requirements). Brazilian Association of Technical Standards, Rio de Janeiro, October 2007, p. 8.

ANA (2010). *Atlas Brasil: abastecimento urbano de água – Panorama Nacional* (Atlas Brazil: urban water supply – National panorama). Prepared by the Engecoprs/Cobrape Consortium for the National Water Agency, V.1.

ASA (2013). Semiarid Brazil. http://www.asabrasil.org.br/Portal/Informacoes.asp?COD_MENU=105 (accessed 22 April 2013).

City of Fort Worth (2014). Rain barrel and compost bin sale. http://www.rainbarrelprogram.org/cityoffortworth (accessed 2 June 2014).

Dornelles F., Goldenfum J. A. and Tassi R. (2012). *Metodologia para ajuste do fator esgoto/água para aproveitamento de água de chuva* (Methodology of sewage/water factor adjustment for rain water harvesting). Revista Brasileira de Recursos Hídricos (RBRH) (Brazilian Water Resources Magazine) Jan/Mar 2012, **17**(1), 111–121.

Geretshauser G. and Wessels K. (2007). The stormwater management information system – a GIS portal for the close-to-nature management of stormwater in the Emscher region. *Proceedings of the 6th International Conference on Sustainable Techniques and Strategies in Urban Water Management (Novatech)*, 24–28 June 2007, Lyon, France.

Herbke N., Pielen B., Ward J. and Kraemer R. A. (2007). Urban Water Management, Case Study II: Emscher Region. PowerPoint Presentation. http://ecologic.eu/download/vortrag/2006/herbke_milan_emscher.pdf (accessed 03 April 2009).

HMRC (2014) First year allowances for water efficient technologies. http://www.hmrc.gov.uk/capital-allowances/fya/water.htm (accessed 2 June 2014).

HR Wallingford (2011) Stormwater Management Using Rainwater Harvesting: Testing the Kellagher/Gerolin Methodology on a Pilot Study. HR Wallingford Ltd, Report SR 736.

Hurley L., Mounce S. R., Ashley R. M. and Blanksby J. R. (2008). No Rainwater in Sewers (NORIS): Assessing the Relative Sustainability of Different Retrofit Solutions. *Proceedings of the 11th International Conference on Urban Drainage*, Edinburgh, Scotland, UK.

Kalaswad S., and Arroyo J. (2008). Rainwater Harvesting in the State of Texas. *Water Quality Magazine*, **13**(10), 14–15.

Konig K. W. (2001). The Rainwater Technology Handbook: Rainharvesting in Building. Wilo-Bran, Dortmund, ISBN 3-00-008368-5.

Lye D. (2002). Health risks associated with consumption of untreated water from household roof catchment systems. *Journal of the American Water Resources Association*, **38**(5), 1301–1306.

MTW (2010). Rainwater Harvesting Market Research and Analysis UK 2010. http://www.marketresearchreports.co.uk/Rainwater-Harvesting-Market/Rainwater-Harvesting-Market-Size.htm (accessed 22 April 2013).

Melville-Shreeve P., Ward S. and Butler D. (2014). A preliminary sustainability assessment of innovative rainwater harvesting for residential properties in the UK. *Journal of Southeast University(EnglishEdition)*,**30**(2),135–142.DOI:10.3969/j.issn.1003-7985.2014.02.001. http://ddxbywb.paperonce.org/oa/pdfdow.aspx?Type=pdf&id=201402001 (accessed 31 July 2014).

Nobrega R. L. B, Galvão C. O., Palmier L. R. and Cebalhos B. S. O. (2012). *Aspectos políticos-instituicionais do aproveitamento de água de chuva em áreas rurais do semiárido brasileiro* (Institutional and political aspects of the rainwater use in rural semiarid Brazil). Revista Brasileira de Recursos Hídricos (RBRH) (Brazilian Water Resources Magazine), Oct/Dec 2012, **17**(4), 109–124.

Partzsch L. (2009). Smart regulation for water innovation – the case of decentralized rainwater technology. *Journal of Cleaner Production*, **17**, 985–991.

Pullinger M., Browne A., Anderson B., and Medd W. (2013). Patterns of Water – The Water Related Practices of Households in Southern England, and Their Influence on Water Consumption and Demand Management. https://www.escholar.manchester.ac.uk/api/datastream?publicationPid=uk-ac-man-scw:187780&datastreamId=FULL-TEXT.PDF (accessed 15 April 2013).

Rahman A., Dbais J., Islam S. M., Eroksuz E. and Haddad K. (2012). Rainwater harvesting in large residential buildings in Australia. In: Urban Development, S. Polyzos (ed.), InTech, Croatia, pp. 159–178, ISBN 978-953-51-0442-1.

Saiani C. C. S., Menezes R. T. and Toneto Junior R. (2009). *Desestatização do abastecimento de água no Brasil: efeitos sobre o acesso e a desigualdade de acesso* (Water supply privatization in Brazil: effect on access and inequality of access). *Proceedings of the 37th Brazilian Economics Meeting*, Brazil.

Sansom A., Armitano E. R. and Wassenich T. (2008). Water in Texas: An Introduction. University of Texas Press, Austin.

SNIS (2012). *Diagnóstico dos serviços de água e esgoto – Brasil 2010* (Diagnosis of water services and sewage – Brazil 2010). Sistema Nacional de Informações sobre Saneamento (National Information System on Sanitation), Brasilia, Brazil.

Texas Rainwater Harvesting Evaluation Committee, Texas Commission on Environmental Quality, Texas Department of State Health Services and Texas Section of the American Water Works Association Conservation and Reuse Division (2006)

Rainwater Harvesting Potential and Guidelines for Texas. Texas Water Development Board, Austin, TX, 45.

TWDB (2005). The Texas Manual on Rainwater Harvesting. Texas Water Development Board, Austin, TX. https://www.twdb.texas.gov/publications/brochures/conservation/doc/RainwaterHarvestingManual_3rdedition.pdf (accessed 1 September 2012).

TWDB (2012). Water for Texas 2012 State Water Plan. Texas Water Development Board, Austin, TX. http://www.twdb.state.tx.us/waterplanning/swp/2012/ (accessed 1 September 2012).

U.S. Census Bureau (2010). Annual Estimates of the Population of Metropolitan and Micropolitan Statistical Areas: April 1, 2010 to July 1, 2012. http://www.census.gov/popest/data/metro/totals/2012/ (accessed 21 January 2014).

Walker A. (2009). The Independent Review of Charging for Household Water and Sewerage Services (Walker review). Department for Environment, Food and Rural Affairs, London, UK.

Ward S., Barr S., Butler D. and Memon F. A. (2012). Rainwater harvesting in the UK – socio-technical theory and practice. *Technological Forecasting and Social Change*, **79**(7), 1354–1361.

Ward S., Barr S., Memon F. A. and Butler D. (2012b). Rainwater harvesting in the UK: exploring water-user perceptions. *Urban Water Journal*, **10**(2), 112–126.

Ward S. (2013). The Market Potential for Residential Retrofit Rainwater Harvesting Systems. Internal Report, University of Exeter.

WRc (2012). Metering and Charging for Non-Potable Water Supplies. Report P9209, Personal Communication, 20th May 2013.

Ziegler H. (2010). Raindrops Keep Falling in My Reservoir – Harvesting Rain in Germany and The EU. http://sustainablecitiescollective.com/helmuthziegler/16781/raindrops-falling-my-reservoir-harvesting-rain-germany-and-eu (accessed 16 March 2013).

Chapter 9

Air conditioning condensate recovery and reuse for non-potable applications

Pacia Diaz, Jennifer Isenbeck and Daniel H. Yeh

9.1 INTRODUCTION

Condensate is the resulting waste product from air conditioning. The air conditioning (AC) process requires humidity removal from the air in order to provide thermal comfort to building occupants. As humid air blows past the cooling coils, the moisture in the air condenses and is routed away from buildings and disposed of as waste (Figure 9.1). This 'nuisance' water (commonly referred to as 'clear waste' by mechanical engineers) is now being seen in a new light as a sustainable strategy that contributes toward sustainable buildings as well as increased resilience in urban areas.

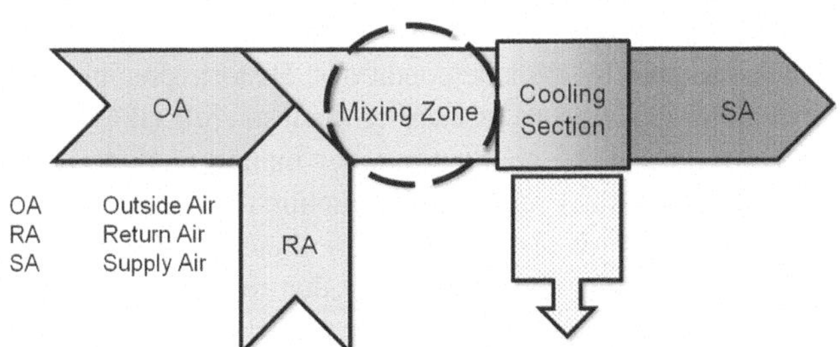

Figure 9.1 Air conditioning condensate production.

Global and regional average temperatures will continue to rise as a result of changing climate; temperatures could rise by as much as 11.5°F by the end of the century, depending on different emissions scenarios (US Global Change

Research Program, 2009; The World Bank, 2012). Higher temperatures will increase evaporation levels, causing for more water to stay in the air (humidity). Air conditioning may become the norm and humidity removal from buildings may become more prevalent as climate changes. Therefore, capitalizing on the waste stream from this compulsory and energy intensive investment in buildings is compelling, given our contemporary climatic and anthropogenic issues.

As AC condensate recovery becomes a more acceptable alternative source to satisfy water demands, it is important to understand the applicability and potential treatment necessary for the collected condensate. In some cases, condensate can go virtually untreated; in others, treatment methods should be considered based on efficacy, cost, safety and long term maintenance. The case studies included here provide useful examples of implementation, considerations and lessons learned on how to implement strategies related to condensate collection.

9.2 MOTIVATION

9.2.1 A solution to urban water supply issues

Issues related to water scarcity in urban areas are a large motivator to look to alternative water supply. Even in metropolitan areas where drought planning and redundant infrastructure has been constructed, reliable water supply efforts sometimes fail. Water scarcity can be caused by anthropogenic impacts, such as the high concentration of people at a single location. This condition applies pressure on resources, such as water and energy. Population increases will affect US cities directly, with an anticipated increase of water use of 50 percent by 2025 (Kumar, 2011). Over 81 percent of the population of the United States is found in urban environments (Scruton, 2010). A large portion of that population – 52 percent – lives in coastal areas and is expected to grow significantly, which translates into higher demands for water, higher potential for exploitation of sensitive coastal aquifers, and increased pressure on already stressed infrastructure (Barlow & Reichard, 2010; NOAA, 2011; Konikow, 2013). Water scarcity is also exacerbated by climatic conditions. Historic precipitation trends are changing by extending dry seasons or causing them to shift (IPCC, 2007; NOAA, 2012). These changes affect the ability for water managers to anticipate seasonal water supplies or to design water related infrastructure (Milly *et al.* 2008; Galloway, 2011). Finally, water scarcity is also impacted by the economy. Approximately six billion gallons of treated water is lost each day due to leaky pipes in the US, the equivalent of 14 percent of the nation's water use (ASCE, 2010). A lack of funds has caused water infrastructure to fall into disrepair; an estimated one trillion dollars is needed to repair and replace the existing drinking water system (Lien-Mager, 2012).

9.2.2 A water-energy infrastructure synergy

Improving water supply systems by implementing on-site strategies does more than just provide alternative water supply; it also conserves energy. The water-energy nexus – the interdependency that water and energy have on each other – makes this possible; consider that water is essential to generate power and that power is necessary to treat and move water (Sandia National Laboratories, 2005). In order to better manage dwindling water supplies, some regions of the United States 'may need to reassess the value of energy and water resources and consider new technologies and approaches to optimize economic growth.' According to the Department of Energy's report to Congress, one way to address this challenge is to seek water-energy infrastructure synergies (US Department of Energy, 2006). AC condensate recovery is one such example. In 2006, energy consumption in residential and commercial buildings was dominated by air-conditioning; 39 and 32 percent, respectively (Kelso, 2008); the end products are cooled air and water. AC condensate recovery is a synergistic opportunity which needs to be harnessed as a potential alternative water source, making the best out of an energy intensive process. Given the above issues with the reliability and efficiency of water supply in the urban context, the installation of a condensate recovery system in air conditioned buildings can improve the resilience of water supply within buildings and urban communities.

9.3 QUANTITY: VOLUME POTENTIAL

Before implementing a condensate recovery and reuse programme, it is important to quantify the volume of condensate that can be generated/recovered from a building with an air-conditioning system. This section briefly discusses factors influencing condensate volume recovery and available methods for its quantification. Geographic location plays a significant role in how much condensate can be obtained. Ideal locations are those where warm and humid climate is prevalent as well as where AC systems have a high number of cooling days. For instance, the American Society of Heating, Refrigerating and Air Conditioning Engineers (ASHRAE) climate maps and charts depict this climate criteria being prevalent in zones 1, 2 and 3 of the Central and Eastern United States, as shown in Figure 9.2. The numbers on Figure 9.2 coincide with climate zones within the United States and represent the portion of heating and cooling degree days defined by thermal and hygric criteria used by mechanical engineers. The letters designate if an area is humid, dry or marine (Briggs *et al.* 2003).

Collecting condensate from buildings is relatively simple, since air conditioning systems are already designed to remove moisture from the air. The potential for collecting condensate, however, can vary significantly. There are several factors that must be considered when calculating the condensate volume potential. These factors vary between building use and geographical area.

- *Dehumidification for buildings varies*: One important factor to consider is not only the size, but the type of building. All air conditioning processes are not created equal, especially when it comes to dehumidification (Doty, 2009).

For example, air conditioning systems for industrial buildings may require less dehumidification than those for commercial or residential buildings.

- *Airstream composition*: Collection potential also depends significantly on the airstream's composition. Generally, the air stream is made up of dry air and water vapour. The air that passes through the air conditioning process (in large buildings) will generally consist of a ratio of return air (recycled air from within the building) and outdoor air. Each of these can be of different temperature and humidity levels. Therefore, evaluating each source separately and estimating the thermodynamic state of the mixed air is a necessary first step for the estimation of condensate volume.
- *Ton hours per season*: Air conditioning units do not run constantly, 24 hours a day, 365 days a year. Units cut on and off throughout the day. Settings can vary between seasons. With the current green movement coupled with the economic climate, there are those that may choose natural ventilation during select times of the year to conserve energy, especially in residential applications. These factors should be taken into consideration and adjustments made for accurate monthly/yearly condensate production rates.

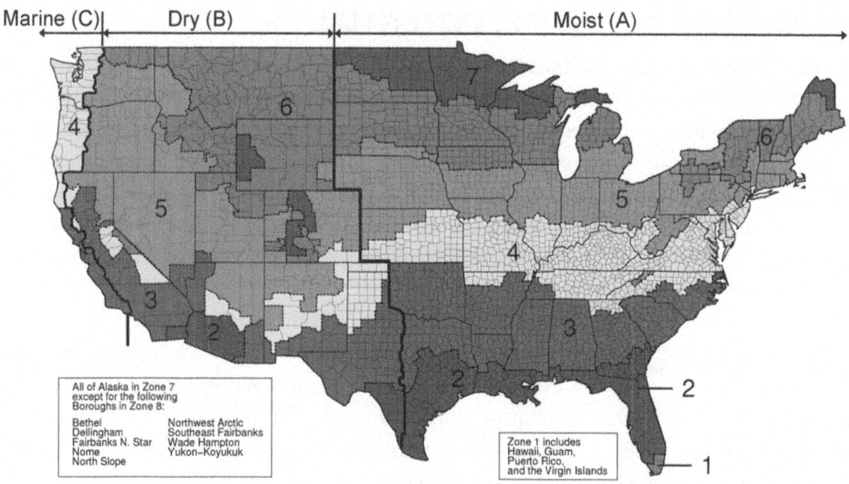

Figure 9.2 US climate zone map (Briggs *et al.* 2003).

At the moment, there is no commercially available software that accurately calculates the condensate volume potential. However, depending on the level of accuracy needed, there are a number of methods in practice for estimating the volume potential. One of these is a technical method, traditionally used by mechanical engineers, and involves the use of the psychrometric (humidity) chart. The method involves using properties of the ambient and outdoor air which the system is treating as well as the volumetric flow rate of the air. By knowing any two independent properties of the air, the thermodynamic state can be determined (Cengel & Boles, 2008). Additionally, the following limitations should be taken into consideration when estimating the condensate volume potential:

- Calculations using the psychrometric chart would reflect the estimated production rate at 'peak' conditions. Actual production rates would most likely be less. Production rates would be similar to that of a juicer – an initial high flow rate when air conditioning system turns on and a dwindling rate as the system continues to run and removes humidity from the air.
- Not all the water vapour would be condensed out of the air. Some water could potentially be reabsorbed as vapour into the air as it blows past the cooling coil and drip pan.

Another method for estimating condensate potential is through the use of online calculators, like the one developed by the San Antonio Water system (BuildingGreen, 2011). These calculators do not account for fluctuating conditions like those mentioned above. However, these calculators can serve to provide a range of production for very general estimates when needed.

The most desirable method for estimating production (or recovery) volume would be to have measured (metered) production rates for various building types in the various climate zones in different geographical locations. Unfortunately, literature provides few documented case studies of metered condensate production. Those that have been published only provide calculated estimates. The EPA's *Laboratories for the 21st Century: Best Practices, Water Efficiency Guide for Laboratories* cites findings on condensate recovery rates based on the load factor and cooling equipment tonnage. Depending on the rate of ambient humidity, the conclusion was that 'from 0.1 to 0.3 gallons of condensate could be collected for every ton-hour of operation' (Wilcut & Lillibridge, 2009; Guz, 2005; Carlisle, 2005). Another study bases its estimates on predicted weather data and calculated production rates, rather than measured results (Lawrence *et al.* 2010).

9.4 QUALITY: FIT-FOR-PURPOSE

9.4.1 Microbial concerns

Water that collects on the evaporator coils of cooling systems can *initially* be considered as fairly high quality water; it is practically the same as distilled water. However, once condensate comes in contact with air conditioning equipment surfaces (including drain pans and coils), dust, mildew and other elements can potentially contaminate the water being collected. Quality diminishes due to the organic content drawn in through the air conditioning system during the formation of condensate. Left chemically untreated, stagnant, warm-water sources provide an ideal environment for *Legionella* (OSHA, 1999). Although condensate leaves the cooling coil at a temperature well below that which *Legionella* typically grows, it can still survive at low temperatures (US EPA, 1999). Therefore, it is important to keep in mind the potential health risks once it is collected and stored.

The exposure pathways for *Legionella* can be through either aerosol effect or direct contact. In the case of aerosol, the inhalation of aerosol droplets, usually less than one meter's distance from the source can be sufficient for exposure. Although there are no documented cases of *Legionella* exposure caused by aerosols generated by toilet flushing or air streams from air handling equipment, precautions should be taken in treating condensate that is being used for this application (Barker & Jones, 2005; Lye, 2010).

Since an individual disinfection method has not always been successful against *Legionella*, a combination of disinfection treatments is recommended. For example, in addition to a chlorine residual, a supplemental disinfection method, such as ultraviolet light sterilization, may be used in order to effectively prevent outbreaks (Lye, 2010; US EPA, 1999).

Choosing the appropriate type of disinfection or treatment is dependent on the intended use of condensate. For example, if recovered condensate is intended for irrigation purposes, it may require minimal treatment, especially if used with drip irrigation systems. When utilised for plant process water, water treatment procedures are already in place for anti-microbial and anti-scaling. Therefore, the condensate becomes integrated with the treatment process. Operators and maintenance personnel working with potential exposure to cooling tower overspray should wear appropriate personal protection equipment.

9.4.2 Metals

An additional health risk that could result from the use of recovered condensate can come from the presence of heavy metals brought on by contact with the cooling coils and other parts of the air conditioning equipment. Some literature also mentions the 'slight risk of lead contamination (from solder joints in the evaporative coils) building up to dangerous levels in soil continually irrigated with the water' (Alliance for Water Efficiency, 2010). These issues should be kept in mind when deciding on which use to apply recovered condensate.

9.4.3 Other issues

The commingling of condensate should be considered with treatment quality in mind. For instance, in some cases, some may choose to collect condensate in a stormwater vault for reuse elsewhere. Where the condensate is similar in quality to distilled water, stormwater runoff from roads, parking lots and roofs will most likely contain oils, debris and other materials that could be detrimental to the reuse system. As a precaution, filters and first flush diverters have been used in some cases to eliminate debris prior to the reuse of stormwater. In other cases, careful consideration should be given to include sand filters or other catchment devices to protect the equipment utilising the commingled condensate and stormwater.

When mixing with potable water, it is extremely important the prevent backflow of condensate to potable water supply. All connections should be indirect or protected by a backflow preventer that is maintained on a regular basis. Prior to considering this type of application, local municipal codes and standards should be reviewed and available guidance followed.

9.5 USES AND BENEFITS

For recovered condensate to be used effectively it should be matched to its intended use; the term coined for this is fit-for-purpose. An example is the ineffective use of potable water in outdoor applications. Most outdoor uses of water and a significant portion of indoor water use are often non-potable in nature. Yet, most non-potable applications are satisfied using high-quality, treated potable water. Recovered condensate could certainly substitute potable water in non-potable applications, but the most desirable strategy for recovered condensate should be its reuse with minimal additional treatment. The following can be considered as appropriate applications for using condensate as an alternative water source (Building Green, 2011):

- Toilet flushing
- Irrigation
- Water cooled equipment
- Decorative fountains and water features
- Evaporative coolers
- Rinse water for washing vehicles and equipment
- Water for laundry operations
- Steam boiler make-up water
- Closed loop cooling/heating systems

Mechanical engineers agree that the best use for AC condensate is as cooling tower make-up water for three reasons:

(1) It is of an ideal quality for the intended use. Almost no dissolved solids (calcium, magnesium, chloride and silica) are present in recovered condensate, so blowdown – the water drained to remove mineral build-up – is reduced, yielding more efficient water use (US Department of Energy, 2014).
(2) Its neutral properties enable the reduced use of chemicals, reducing the overall chemical treatment costs.
(3) There is no need for storage. Condensate can be applied directly to the cooling tower, keeping the implementation costs down.

A benefit that applies to all applications is the reduction of sewage costs. In some locations, it is common practice to route condensate discharge from systems into the building's sewer connection. However, most municipalities request condensate be routed to storm or other locations rather than sanitary sewer (Boston Water and Sewer Commission, 1998; City of Phoenix, 1999). If buildings are charged for quantities of wastewater needing treatment, they would be charged for the volume of condensate as well. In the case of open loop systems such as cooling towers, municipalities will allow credit meters to allow for the water that is evaporated (City and County of Honolulu, 2012; US Department of Energy, 2014; Portland Water Bureau, 2014). When using condensate in this manner, it is important to tie in the condensate separately after the credit meter with an appropriate backflow preventer or check valve as required by most municipalities and water service providers (International Code Council, 2007).

The recovered condensate is *not* suitable for potable applications. Condensate 'should never be used for human consumption' due to the quality issues mentioned previously, particularly those pertaining to metals (Alliance for Water Efficiency, 2010).

9.6 CASE STUDIES

This section contains case studies located in ASHRAE climate Zones 2 and 3: two university campuses in Tampa, Florida; one student project at another university campus in Macon, Georgia. These zones are ideal for condensate recovery, as climate is generally hot and humid and most buildings are equipped with air conditioning that includes dehumidification. The section mainly provides examples of buildings where condensate recovery and its subsequent reuse have been implemented.

9.6.1 Case study: University of Tampa

Project 1: Alternative water for irrigation and landscape features

Background: The University of Tampa is a metropolitan university with approximately 6900 students. The signature building, Plant Hall, was built in 1894 as a hotel serving a nearby railroad station. Since then, it has been transformed into a building that serves multiple purposes for the University, including classrooms, administration offices, ballrooms, as well as a museum. The demand for water in this building has increased considerably. The original water infrastructure is now aging and under-designed to meet present demands. In addition, the City of Tampa's stormwater main serving the majority of the campus is restricted in capacity and is overtaxed with campus growth over the past 10 years. Further, many buildings are served by City potable water lines which are over 50 years old. Therefore, there was an obvious need to take action: either to conserve potable water or build new infrastructure that could compensate for these challenges. Finally, with multiple sports fields (baseball, softball, soccer, and lacrosse), grounds maintenance and general campus maintenance (such as pressure washing sidewalks, buildings), demands for non-potable uses adds to the burden on the potable supply system.

Condensate recovery and reuse system implementation: Most of campus irrigation is supplied by one of the nine campus wells. During a gymnasium renovation, it became viable to incorporate condensate recovery within the scope of the project. Two 5.7 m^3 (1500 US gallons) above ground cisterns temporarily store (Figure 9.3) gravity fed condensate, which is later pumped for distribution by the existing irrigation system prior to drawing groundwater from a nearby well. The volume of recovered condensate offsets approximately 1135 m^3 (300,000 US gallons) of groundwater annually.

Figure 9.3 Condensate Storage at Bob Martinez Athletic Centre.

Additionally, in an effort to augment the capacity of stormwater loads during intense rain storms, the campus has built a series of underground storage vaults, which provided the opportunity for the commingling of recovered condensate from building condensate collection systems to be stored together with the stormwater. Commingled water is then pumped from the vaults and used to satisfy irrigation demands. Although condensate makes up a small fraction of the water collected, it is currently estimated that approximately 2130 m^3 (562,400 US gallons) of condensate replaces groundwater used for irrigation annually.

Benefits: For the University of Tampa, sustainability was of primary consideration rather than the actual measured savings. When considering the University's tiered rate structure with the City, potable water is billed at approximately $\$1.44/m^3$ (\$4.07 per 100 cubic feet), this amounts to nearly \$1500 US dollars for saving approximately 1000 m^3. If the project were to be judged purely by a Life cycle cost analysis for the infrastructure, water meters and sensor technology, it would easily exceed 10 years. However, the motivation for implementation was not attributed to financial gains or quick payback. Rather, recent trends toward integrated sustainable design are beginning to take hold in multi-disciplinary programmed projects, allowing the water energy nexus to become a value driven proposition. The opportunity here was in allowing for the collection, treatment and monitoring of AC condensate as a practical and quantifiable alternative water source at a site.

Project 2: Utility plant process make-up water

Background: Process water used on large campus facilities can amount to 20 to 40 percent of the total water demand. It is used for a variety of purposes in utility plants; cooling tower make-up and steam boiler make-up are the most common. Ideally, treated water from a municipality should not be used, but in urban or confined site settings, options may be limited. Groundwater abstraction wells, depending on water availability and quality, could be used, but will require permitting, energy and maintenance. Steam boiler systems are the most sensitive to water quality further resulting in an analysis and potential treatment of make-up water systems which should be implemented. The higher temperatures of water will allow more dissolved solids, which end up being caustic to the piping and the equipment. Softening of water is typical for steam and condenser water systems. In both heating and cooling process water systems, there is a continual need for use of make-up water that offsets the evaporative effect of the heat.

The University began a series of master planning exercises to better utilise campus facilities, green space and determine a utility corridor for current and future utilities in order to promote sustainable growth. Overhead aged electrical lines were replaced with underground switchable loops and outdated crumbling potable water lines were replaced with larger centralised looped lines (individually metered at the building). In the Southeast climate, air conditioning can amount to nearly 70% of the electric bill. When the air conditioned space of buildings on a campus (regardless of use) exceeds 18,500 m^2 (200,000 square feet), it can be more cost effective over the long term to centralise the air conditioning operations of a facility, like using a centrifugal chilled water plant; such was the case with the university's campus (Ceden Engineering, 2014).

Condensate recovery and reuse system implementation: In 2011, the University built its first large chiller plant (currently 2200 tons of cooling capacity with the ability to expand to 6000 tons), serving over 81,000 m^2 (approximately 875,000 square feet) of conditioned space (variety of spaces consisting of classroom buildings, residential halls and athletic complex). The related infrastructure improvements include large chilled water piping extending over 2400 meters (approximately 8000 linear feet) supplying approximately 7°C (44°F) water to various buildings. The project included a multi-building condensate collection piping network which feeds back to the plant as a make-up water source for the cooling towers. While the potential efficiency improvement is difficult to document, the temperature of the cooler water from AC condensate return line to the cooling tower reduces the energy required to cool the condenser water, from 35°C (95°F) to 29°C (85°F) as well as requiring less fan energy.

As long as the plant is operating, there will be a requirement for make-up process water. The campus air conditioning load and ambient conditions will dictate the evaporation rate and necessity for make-up. The condensate collected is stored in a 12 m^3 (3100 US gallons) underground storage tank, but a system of floats and variable speed pumps dependent on plant water pressure requirements keep a continuous flow of condensate to the system.

Benefits: Operating since April of 2013, the collected condensate has offset nearly 160 m^3 (42000 US gallons) while operating at only partial capacity. Upon completion, it is expected to collect and reuse over 7600 m^3 (2.0 million US gallons) annually. A simple payback for the installed systems is estimated to be just over 11 years. However, in areas where potable or even reclaimed water is more expensive, the savings and payback would make the system even more practical from a financial perspective.

9.6.2 Case study: University of South Florida

Background: The University of South Florida (USF) serves over 47,000 students through three separately accredited campuses in the Tampa Bay area (University of South Florida, 2013). USF is dedicated to taking steps towards sustainability, as is evidenced by its numerous initiatives (USF Office of Undergraduate Affairs, 2009; USF Office of the Provost and Executive Vice President, 2011; USF Magazine, 2012).

Condensate recovery and reuse systems implementation: The water source for the Tampa campus is mainly through its groundwater wells located onsite, although some buildings are connected to the City of Tampa's municipal potable water distribution system. In recent years, USF has taken steps to implement alternative water sources whenever possible in several projects, some of which have implemented condensate. Table 9.1 lists information about each of these.

Patel Centre for Global Solutions – The centre, established in 2005, promotes research to 'creating real solutions that deliver a sustainable quality of life for all people' (Prieto *et al.* 2008). Constructed in 2011, the building was certified by LEED (Leadership in Energy and Environmental Design), an internationally recognised green building certification system, at a Gold level. One of its distinctive aspects is strategically aligned with one of the principle areas of research: potable water and sanitation issues. The building is the first one on campus to capture and use Rainwater PLUS (i.e., harvested rainwater supplemented by AC condensate) for toilet flushing within the building and irrigation of the site landscaping (Patel

College of Global Sustainability, 2013). Rainwater and condensate is collected and stored in a 113.6 m^3 (30,000 US gallon) underground cistern (Figure 9.4), which is treated using ultraviolet light for disinfection (Cline, 2011).

Table 9.1 Condensate recovery at USF Tampa campus.

Building/ Feature	Building type	Use	Storage requirement		Treatment
			gallons	m^3	
Patel Centre for Global Solutions	Offices (education)	Flushing toilets, irrigation	30,000	114	Ultra-Violet Light
Marshall Student Centre	Offices, meeting rooms, dining areas	Decorative water feature	None		Chlorine
Leroy Collins Welcome Fountains	Library	Decorative water feature	None		Chlorine

Figure 9.4 Patel Centre cistern installation.

Marshall Student Centre – The USF Marshall Centre is a 29,700 m^2 (320,000 square feet) building constructed in 2009, replacing the original student union building. The new building contains a food court, ball room, an auditorium, sports grille, 30 event and functions spaces, office space for student organisations, two computer labs, an art gallery and several other facilities which serve the Tampa Campus students. The Running of the Bulls Fountain, located at the building's entrance, uses condensate recovered from air handling units which serve the building's auditorium. The condensate is treated with chlorine and used as a make-up water for this iconic decorative water feature (Figure 9.5).

Leroy Collins Welcome Fountains – Leroy Collins Boulevard is the main entrance to the Tampa Campus. As part of a traffic improvement project, decorative fountains were installed together with turn lanes to adjacent parking lots and bus stops. Condensate is recovered from air handling units at the library (shown in the background) and used as make-up water for the fountains (Figure 9.6) (USF News, 2011).

Figure 9.5 Marshall Student Centre decorative fountain using recovered condensate.

Figure 9.6 Leroy Collins welcome fountains using recovered condensate.

Benefits: As was the case for the University of Tampa, USF's commitment to sustainable strategies related to water and energy is its main driver in implementing condensate recovery rather than the seeking the economic justification prior to implementation. Additionally, USF looks for ways to incorporate alternative water sources in order to preserve natural water sources and reduce potable water consumption.

9.6.3 Case study: Mercer University

Project: Senior design project

Background: The Science and Engineering Building is a two-storey 4830 m^2 (52,000 square feet) structure containing faculty offices, student work rooms, laboratories, and classrooms. The building envelope consists of brick, building wrap and 2 inch fiberglass board over a metal frame. Construction was completed in June 2007. The air conditioning system consists of two air handlers with a design flow rate of 983 m^3 per minute (34,700 $feet^3$ per minute) each. The systems cooling setpoint is 70°F with a relative humidity of 55 percent.

Condensate recovery and reuse system implementation: In the summer of 2011, a senior engineering design team installed a condensate recovery system on one of the air handlers. The recovered condensate was to be used to irrigate an area adjacent to the Science and Engineering Building. This system (Figures 9.7 and 9.8) had two main components: a condensate capture system and a system to deliver the captured condensate to a drip irrigation system. The condensate capture system consisted of a catch pan installed underneath the condensate drain line. Condensate was removed from this catch pan and delivered to the irrigation system via a pump controlled by a pair of float switches. The catch pan also had an overflow fitting installed above the highest level float switch that would route flow to the floor drain in the event of a pump failure. A pump cycle counter was used to monitor condensate production. From August to October of 2011, 72,000 litres of condensate were recovered. Daily condensate capture data is presented in Figure 9.9. The zero condensate production indicated on August 26 was the result of the pump impeller failing while the zero productions indicated from October 2 to 11 were due to the collection pan becoming misaligned and not level. The misalignment of the pan was causing the captured condensate to flow over the side and into the floor drain before the level required to trigger the pump circuit was reached. While this system functioned well, it was not part of the original construction of the building and could not be operated without daily monitoring. The system was therefore shutdown and removed.

9.6.4 Additional condensate recovery and reuse examples

In addition to the cases highlighted above, the Florida Aquarium, also located in Tampa, recovers AC condensate for use as cooling tower make-up water and irrigation. Located on Tampa Bay and with high visitor traffic, considerable condensate volume is generated. Many other locations, particularly throughout Texas, collect condensate for reuse within many libraries, malls and research facilities (Carlisle, 2005; US EPA, 2005; American Institute of Architects, 2008; Alliance for Water Efficiency, 2009; US EPA, 2009; BuildingGreen, 2011). The

University of Texas at Austin is one specific case, where water sources (AC condensate, once through cooling process water used for lab equipment, cooling and swimming pool overflow) are comingled and collected through tunnelled pipelines that are pumped back to the cooling towers at the main power plant as make-up water. It should be noted that the use of pool overflow water is acceptable since cooling tower water is ultimately treated with biocides. Not all reuse systems include pre-treated water that consist of chlorine and corrosion inhibitors found in closed and open loop piped systems. Water collection and reuse should be evaluated on a case-by-case basis. Over the past three years, condensate collection at the University of Texas at Austin is averaging nearly to nearly 190,000 m^3 (50 million gallons) annually. The US EPA has also implemented AC condensate recovery and has estimated a savings of 14,000 m^3 (3.8 million gallons) of water per year (US EPA, 2012).

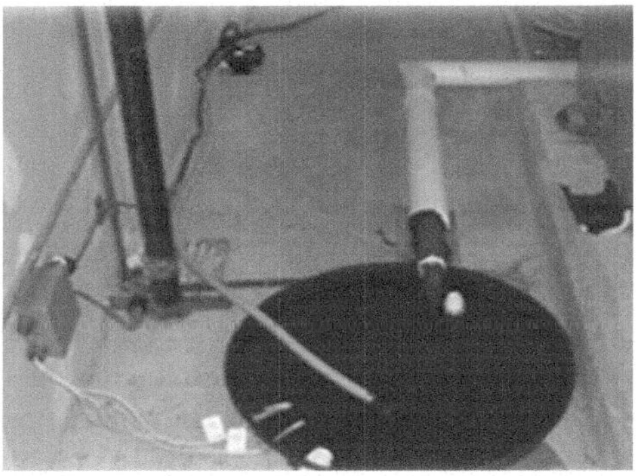

Figure 9.7 Condensate capture pan, pump, and pump control system.

Figure 9.8 Condensate holding tanks and irrigation pump.

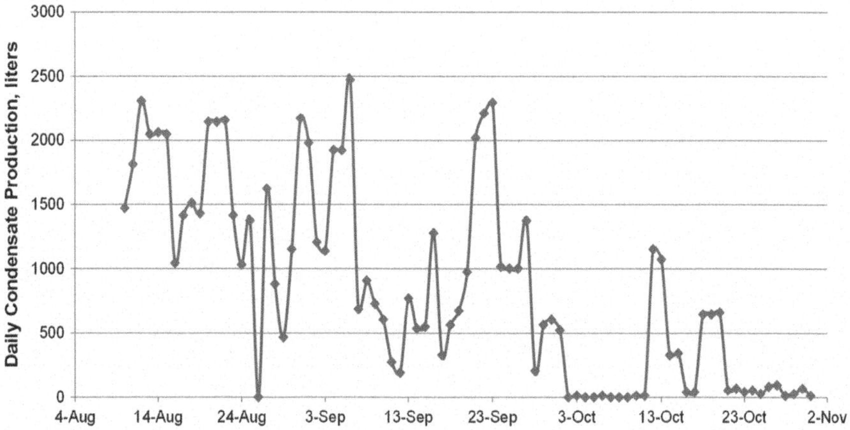

Figure 9.9 Condensate production from one of the two air handlers during August, September, and October, 2011.

9.7 LESSONS LEARNT AND DISCUSSION

Where condensate is used for irrigation or water features, the collected water has not been metered for verification. There are two primary issues, the first being a meter that can accurately read and report flow. A paddle or mutating disk type meter may read a flow, but since the flow is typically by gravity, the pipe is not flowing at full capacity, and therefore does not allow for accurate flow quantification. Ideally, a meter will read accurately when placed in the horizontal run, pressurized flow. For existing systems, the condensate system typically has been gravity run from a variety of air handling units down to a storm drain, roof drain or approved ground location (French drain). Locating this meter in a maintainable location where accurate readings can be taken and flow monitoring integration into the building's energy management system has been a challenge.

In the case of the Patel Centre, where a cistern was used for storage of commingled condensate and rainwater for use for flushing water closets and urinals, the float sensors failed to trigger the potable water back-up supply. Therefore the tank ran dry during the winter season when precipitation levels are lowest. The first few building occupants to use the restroom had the unfortunate experience of not being able to flush the water closet. This was resolved with a redundant potable water back-up supply connected with a backflow preventer at the main plumbing supply line.

While pure condensate is considered clear water waste, when stored onsite in underground vaults, considerations should still be made regarding minimising dirt, debris, lawn clippings, and so on from entering the vault. At the University of Tampa, the underground condensate collection vault was located in an active construction site. Once in operation, the pumps often got clogged with dirt and debris. Unfortunately, the dirty water made its way back to the chiller plant.

A dual filtration system was retrofitted into the existing make-up water system to prevent potential damage of plant equipment. The potential damage was initially anticipated to be at the water softener, but it could have easily been detrimental to the cooling towers and the chiller condenser tubes as well.

It should be noted that due to its high quality, condensate does not need to be softened as is often recommended for most plant equipment (municipal potable water and well water systems may have high mineral content). Allowing the condensate to by-pass the softening loop will save on cost. As always, the condensate system should be separated from any potable water supply systems with a backflow prevention device. Newly installed condensate lines should be flushed and cleaned prior to tying into plant systems to prevent damage to equipment. Newly installed piping systems are currently required to be flushed to remove slag, slues and debris related to pipe installation. Currently there is not an industry or code standard for flushing and pressure testing of condensate lines. However, the following recommendations should be followed:

- *Flushing*: A minimum of 0.6 meter/second of flushing with clean water for two hours would be recommended.
- *Pressure testing*: Pressure testing should be applied for pressurised lines, at least 1.5 times the operating pressure. Gravity condensate lines should be able to withstand a stack test (air or hydrostatic).
- *Separation from potable systems*: Cleanouts should be installed in all condensate piping at all turns similar to other gravity piping installation. The cleanouts will facilitate access for maintenance if foreign materials are accidentally caught in the lines.

In Florida, especially when campus cooling is essential, the feasibility of implementing condensate recovery systems becomes not only sustainable, but provides financial incentives when incorporated into design and operation of systems. This is especially prevalent when the infrastructure needs to be installed in the first place. Looking at campus type environments which include everything from educational facilities, large corporations, plant processes, healthcare facilities and even centralised municipal utility plants, condensate will be available as free water from air conditioning systems and can be utilised to offset irrigation and plant process needs. Therefore, the cost of installation, collection, treatment and reuse can provide an overall payback as municipalities increase costs for potable supply and sanitary collection. Further, in some cases, municipalities are charging substantial impact fees for large campuses and institutions to help offset the growing needs for infrastructure.

9.8 FUTURE RESEARCH

Lord Kelvin once said, 'To measure is to know. If you cannot measure it you cannot improve it.' Without data to back theoretical calculations, it is difficult to take the leap of faith of implementing condensate recovery. Next steps that should be taken begin with the integration of metering into condensate collection systems in order to verify volume potential with the theoretical calculations and trends. Meters could be read manually (daily, monthly, annually), but ideally these meters should be tied into the buildings' energy management system (BMS or EMS) for better real-time tracking. If properly trended simultaneously with significant variables such as weather properties, dry bulb, wet bulb, building air conditioning load (tonnage), correlations could then be developed for more accurate condensate production rates. Much can be learned about seasonal production rates and the relevant links to geographical locations and building types.

9.9 CONCLUSION

As our contemporary climatic and anthropogenic issues continue to impact urban water supply, viable solutions must be implemented. Solutions which consist of water-energy synergies are most desirable. Although AC condensate is not fit as an adequate source of drinking water, it certainly provides sufficient flows for it to be considered as a reliable alternative source for non-potable uses. It improves the reliability of non-potable water supply within buildings, with minimal infrastructure investment, water treatment and energy input if applied as cooling tower make-up water or for other non-potable applications. It ultimately conserves the use of potable water and improves community resilience by offsetting its use. However, although a good alternative source of water for non-potable applications, it is important to keep in mind water quality issues and system design parameters when considering the intended use for the recovered condensate.

Many buildings throughout the hot and humid regions of the United States (mainly the southeast) are implementing AC condensate for supplementing onsite and building water demands. The uptake of condensate recovery systems can be further improved through the provision of targeted training of professionals and the development and enforcement of enabling policies and legislation.

ACKNOWLEDGEMENTS

The authors would like to thank Chris Cuthbert (The Florida Aquarium, Tampa), Suchi Daniels (USF), Steve Doty (Colorado Springs Utilities), Dennis Lye (US Environmental Protection Agency), Philip T. McCreanor (Mercer University), Mischa Morgera (USF), Gary Osborne (The Florida Aquarium, Tampa) and Kyle Tejashri (University of Texas, Austin) for providing access to technical information on the case studies presented in this chapter.

The development of this chapter is based on work supported by grants from the Alfred P. Sloan Foundation and the National Science Foundation under Grant Number 0965743. Any opinions, findings, and conclusions or recommendations expressed in this material are those of the author(s) and do not necessarily reflect the views of the above mentioned organisations.

REFERENCES

Alliance for Water Efficiency (2009). Watersmart Guidebook - Alternate On-Site Water Sources. Alliance for Water Efficiency, http://www.allianceforwaterefficiency.org/uploadedFiles/Resource_Center/Library/non_residential/EBMUD/EBMUD_WaterSmart_Guide_Alternate_On-site_Water_Sources.pdf (accessed 21 August 2009).

Alliance for Water Efficiency (2010). Condensate Water Introduction. http://www.allianceforwaterefficiency.org/Condensate_Water_Introduction.aspx? terms=condensate (accessed 23 March 2012).

American Institute of Architects (2008). AIA/EOTE Top Ten Green Projects 2008. http://www2.aiatopten.org/hpb/#2008 (accessed on 21 August 2009).

ASCE (2010). America's Report Card. http://www.infrastructurereportcard.org/ (accessed 27 October 2011).

Barker J. and Jones M. V. (2005). The potential spread of infection caused by aerosol contamination of surfaces after flushing a domestic toilet. *Journal of Applied Microbiology*, **99**(2), 339–347.

Barlow P. M. and Reichard E. G. (2010). Saltwater intrusion in coastal regions of North America. *Hydrogeology Journal*, **18**(1), 247–260.

Boston Water and Sewer Commission (1998). Regulations Governbing the Use of Sanitary and Combined Sewers and Storm Drains. http://www.bwsc.org/REGULATIONS/SewerRegulations.pdf (accessed 12 May 2014).

Briggs R. S., Lucas R. G. and Taylor Z. T. (2003). Climate classification for building energy codes and standards: Part 1 – Development Process. *Transactions – American Society of Heating Refrigerating and Air Conditioning Engineers*, **109**(1), 109–121.

BuildingGreen (2011). Alternative Water Sources: Supply-Side for Green Buildings. http://www.buildinggreen.com/auth/image.cfm?imageName=images/1705/UsesCA_chart.gif&fileName=170501a.xml (accessed 13 September 2011).

Carlisle N. (2005). Water Efficiency Guide for Laboratories. http://www.nrel.gov/docs/fy05osti/36743.pdf (accessed 21 August 2009).

Ceden Engineering (2014). Centralised Vs Decentralised Air Conditioning Systems. http://www.seedengr.com/Cent%20Vs%20Decent%20AC%20Systems.pdf (accessed 12 May 2014).

Cengel Y. A. and Boles M. A. (2008). Thermodynamics: An Engineering Approach, Sixth Edition. McGraw-Hill Companies, Inc, New York, NY.

City and County of Honolulu (2012). Revised Cooling Tower Sewer Credit Policy Letter. http://www1.honolulu.gov/env/wwm/policyrevisedcoolingtowersewercreditpolicylettertocustomer31912.pdf (accessed 12 May 2014).

City of Phoenix (1999). Policy 61: Storm, Surface, Cooling & Uncontaminated Process Waters. http://phoenix.gov/webcms/groups/internet/@inter/@dept/@wsd/@esd/documents/web_content/091999.pdf (accessed on 12 May 2014).

Cline K. (2011). Rainwater Harvesting: Achieving sustainability with reuse. *Storm Water Solutions*, **36**(3), 10–11.

Doty S. (2009). Energy Engineer, Colorado Springs Utilities. Personal communication.

Galloway G. E. (2011). If stationarity is dead, what do we do now? *Journal of American Water Resources Association*, **47**(3), 563–570.

Guz K. (2005). Condensate Water Recovery. *ASHRAE Journal*, **47**(6), 54–56.

International Code Council (2007). International Mechanical and Publmbing Codes, General Requirements Condensate Disposal. http://publicecodes.cyberregs.com/icod/ (accessed 12 May 2014).

IPCC (2007). Climate Change 2007: Synthesis Report: An Assessment of the Intergovernmental Panel on Climate Change. Wageningen Institute for Environment and Climate Research, IPCC, Geneva, Switzerland. ISBN 9789291691227.

Kelso J. D. (2008). Buildings Energy Data Book. US Department of Energy, Silver Spring, MD.

Konikow L. F. (2013). Groundwater Depletion in the United States (1900–2008): USGS Scientific Investiations Report 2013–5079. USGS, Reston.

Kumar P. (2011). A crowded world's population hits 7 billion. *Reuters*, http://www.reuters.com/article/2011/10/31/uk-population-baby-india-idUSLNE79U04N20111031.

Lawrence T., Perry J. and Dempsey P. (2010). Capturing Condensate by Retrofitting AHUs. *ASHRAE Journal*, **52**(1), 48–55.

Lien-Mager L. (2012). New Report Says $1 Trillion Needed over 25 Years for Drinking Water Systems. http://www.acwa.com/news/water-supply-challenges/new-report-says-1-trillion-needed-over-25-years-drinking-water-systems (accessed 1 April 2012).

Lye D. J. (2010). Research Microbiologist, EPA, Personal communication.

Milly P., Betancourt J., Falkenmark M., Hirsch R. M., Kundzewicz Z. W., Lettenmaier D. P. and Stouffer R. J. (2008). Stationarity Is Dead: Whither Water Management? *Science*, **319**, 573–574.

NOAA (2011). The US Population Living in Coastal Counties. http://stateofthecoast.noaa.gov/population/welcome.html (accessed 4 November 2011).

NOAA (2012). US Seasonal Drought Outlook. http://theworldisurban.com (accessed 9 March 2012).

OSHA (1999). OSHA Technical Manual. Section 3, Chapter 7, http://www.osha.gov/dts/osta/otm/otm_iii/otm_iii_7.html (accessed 7 January 2010).

Patel College of Global Sustainability (2013). Green Building. http://psgs.usf.edu/usf-office-of-sustainability/initiatives/green-building/ (accessed 1 July 2013).

Portland Water Bureau (2014). Commercial Water Efficiency Heating & Cooling Fact Sheet. http://www.portlandoregon.gov/water/article/372318 (accessed 12 May 2014).

Prieto A. L., Hernandez P., Birk E., Lewis C., Mills C. and Upadhyay N. (2008). AASHE Poster LEED Certification Assessment Class Project, Dr. Kiran C. Patel Center for Global Solutions. http://www2.aashe.org/conf2008/uploads/dl.php?f=69 (accessed 1 July 2013).

Sandia National Laboratories (2005). The Energy/Water Nexus: A Strategy for Energy and Water Security. http://www.sandia.gov/energy-water/docs/NEXUS_v4.pdf (accessed 20 September 2013).

Scruton P. (2010). The New Urban World. http://theworldisurban.com/wp-content/uploads/2010/12/URBAN_WORLD_28061.jpg (accessed 9 March 2012).

The World Bank (2012). Turn Down the Heat: Why a 4 degree C Warmer World Must be Avoided. The World Bank, Washington, DC.

University of South Florida (2013). Overview. http://www.usf.edu/about-usf/index.aspx (accessed 1 July 2013).

US Department of Energy (2006). Energy Demands on Water Resources: Report to Congress on the Interdependency of Energy and Water. US Department of Energy, Washington, DC.

US Department of Energy (2014). Best Management Practice: Cooling Tower Management. http://energy.gov/eere/femp/best-management-practice-cooling-tower-management (accessed 12 May 2014).
US EPA (1999). Legionella: Human Heath Criteria Document EPA-822-R-9-001. EPA, Washington, DC.
US EPA (2005). US EPA Region 6 Environmental Services Branch. http://www.epa.gov/oaintrnt/documents/houston_wmp_508.pdf (accessed 21 August 2009).
US EPA (2009). Greening EPA. http://www.epa.gov/greeningepa/facilities/houston.htm (accessed 20 August 2009).
US EPA (2012). Air Handler Condensate Recovery at EPA. http://www.epa.gov/oaintrnt/water/airhandler.htm (accessed 8 July 2013).
US Global Change Research Program (2009). Global Climate Change Impacts in the United States Report. Cambridge University Press, New York, NY.
USF Magazine (2012). Sustainable Leader. http://magazine.usf.edu/2012-summer/features/sustainable-leader.aspx (accessed 16 June 2013).
USF News (2011). A Grand Entrance. http://news.usf.edu/article/templates/?a=3624 (accessed 1 July 2013).
USF Office of the Provost and Executive Vice President (2011). Strategic Initiatives. http://www.acad.usf.edu/Office/Strategic-Initiatives/Sustain-A-Bull-USF.htm (accessed 16 June 2013).
USF Office of Undergraduate Affairs (2009). USF's Mission, Goals, Values and Vision. http://www.ugs.usf.edu/catalogs/0809/mission.htm (accessed 16 June 2013).
Wilcut E. and Lillibridge B. (2009). Condensate 101-Calculation & Applications. http://www.techstreet.com/cgi-bin/detail?product_id=1146833 (accessed 21 August 2009).

Section II

Greywater Recycling Systems

Chapter 10

Greywater reuse: Risk identification, quantification and management

Eran Friedler and Amit Gross

10.1 INTRODUCTION

Onsite decentralised treatment of greywater is becoming a popular alternative source of water for non-potable uses such as landscape irrigation, toilet flushing as well as wash-water for various purposes. Greywater reuse can significantly decrease domestic water consumption, while alleviating stress from existing water resources and contributing to sustainable water use. However, inappropriate reuse of greywater might negatively affect the environment and human health, as it often contains a range of pathogens (bacteria and viruses) as well as substances with the potential to induce environmental consequences such as soil hydrophobicity (repelling water), accumulation of salts and damage to plants. Treatment is therefore needed for safe greywater reuse. Unlike large treatment systems, maintenance of small onsite greywater systems (e.g., a single-family home) is usually performed by the home owners themselves with limited (if at all) professional intervention and/or support. Therefore, unless these onsite systems are reliable, environmental and public health might be compromised.

The aims of this chapter are to quantitatively identify and discuss the major risks associated with greywater reuse, then to portray design and management means to mitigate these concerns during the design, installation and operation of the various onsite treatment systems. The chapter starts with characterisation of greywater followed by a short review of existing treatment technologies. Then major risks associated with greywater reuse are reviewed and their potential impacts are discussed. This is followed by quantification of the health risk associated with greywater reuse performed by employing a QMRA (Quantitative Microbial Risk Assessment) methodology. Similarly, current knowledge on environmental risk assessment is reviewed. Methods to minimise the health risks associated with greywater reuse during the design stage of greywater treatment systems are discussed and management practices to avoid malfunctions are demonstrated.

These include designing according to the fault tree analysis (FTA) approach, where potential risks are identified during the design process and measures to mitigate them are taken in the reuse system production/construction stage. Finally, the reliability of real-world full-scale single-family greywater treatment and reuse systems, designed and constructed in accordance with the FTA, is analysed and its implications on the systems maintenance programme discussed.

10.2 GREYWATER CHARACTERISATION AND MAJOR RISKS ASSOCIATED WITH ITS REUSE

Domestic in-house specific water demand in industrialised countries approximates 100–150 l/c/d (litre/capita/day), of which 60–70% is transformed into greywater, while most of the rest is consumed for toilet flushing and released as blackwater (Friedler *et al.* 2013). Greywater typically includes the liquid waste streams generated from bathroom sinks, baths and showers, and may also include the stream discharged from laundry (i.e., washing machines and hand washing of laundry). Some definitions include liquid waste streams from kitchen sinks and dishwashers (termed 'dark' greywater in some places), although there is no consensus on this (Queensland, 2003; Friedler, 2004). Greywater reuse for toilet flushing can reduce the in-house net water consumption by 40–60 l/c/d, leading to a potential reduction of 10–20% in urban water consumption, which is significant especially under water scarcity situation (Friedler & Hadari, 2006). An additional reduction of 40% or more can be achieved by reusing greywater for garden irrigation (Gross *et al.* 2007), which is a considerable water consumer in some semi-arid regions (Australia, California, Israel). Moreover, water saving from greywater reuse is expected to have an effect on a national scale. For example, Adel *et al.* (2012) predicted that under a moderate penetration ratio of greywater reuse systems of 20–30% (proportion of houses having greywater reuse units installed) in Israel, water savings of over 150 million m^3/y could be achieved in 2050. This potential water saving accounts for about 10% of the projected urban water consumption for 2050 and equals the capacity of a medium size seawater desalination plant.

Although conceived to be 'clean', greywater is polluted (Table 10.1) and exhibits high variability in the concentrations of various pollutants. COD concentrations can range from 7 to more than 2500 mg/l, faecal coliforms of about 10^2–10^8 cfu/100 ml and significant concentrations of detergents, salts (boron, sodium and chlorides) and so on (Friedler, 2004; Gross *et al.* 2008). Therefore if greywater is used without proper treatment, it may pose health risks and exhibit negative environmental and aesthetic effects, especially in warm climates where higher ambient temperatures increase organic matter degradation and enhance pathogen regrowth. As a result of the above, it is important to adequately characterise the quantity and quality of domestic greywater in each regional and cultural setting. This, in turn, will help to develop appropriate system designs and guidance for their operation in different contexts.

Table 10.1 Greywater quality characteristics.

Parameter	Mean	Range	SD	Parameter	Mean	Range	SD	%0[1]
pH	7.24	6.4–10	0.37	SO_4^{-2} (mg-SO_4/l)	157	0.5–72	146	–
TSS (mg/l)	52	2–1070	28	TAN (mg-N/l)	3.46	1–75	3.27	–
VSS (mg/l)	45	6–413	22	NO_3^- (mg-N/l)	1.21	0.1–17	1.48	–
Turbidity (NTU)	28	20–279	19	NO_2^- (mg-N/l)	4.9	0.04–0.4	7.2	–
COD (mg-O_2/l)	174	7–2570	30	TN (mg/l)	10.5	0.1–128	7.5	–
TOC (mg-C/l)	27	73–93	7.7	Faecal coliforms (cfu/100 ml)	3.0×10^5	2×10^2–6×10^6	3.8×10^5	0
Cationic surfactants (mg/l)	0.64	NA	0.30	Heterotrophic plate count (cfu/ml)	8.8×10^6	8×10^6–3×10^7	7.9×10^6	0
Anionic surfactants (mg_{MBAS}/l)	2.87	1.4–56	2.20	*P.aeruginosa* (cfu/100 ml)	3.0×10^4	3×10^3–3×10^4	3.9×10^4	33
PO_4^{-3} (mg-PO_4/l)	1.9	0.1–49	1.0	*S. aureus* (cfu/100 ml)	1.2×10^4	2×10^3–1×10^4	1.8×0^4	67

[1]%0 – Proportion of samples where concentrations were below the detection limit, SD – Standard Deviation.
Source: Based on summary from Alfiya *et al.* (2012).

10.3 SHORT REVIEW OF EXISTING TREATMENT TECHNOLOGIES

Numerous technologies have been suggested for greywater reuse ranging from diversion systems with virtually very little treatment and maintenance to intensive membrane technologies (Gross *et al.* 2012). Yet, typically, many of the small scale onsite systems being proposed for greywater treatment are low-tech, low-cost technologies and often fall into one of two categories: filtration systems providing minimal treatment or down-scaled wastewater-treatment systems. Until recently, most of these systems were not designed to handle the differences in both flow and composition between greywater and wastewater, which resulted in insufficient treatment ability and unsatisfactory treated greywater quality as demonstrated by Gross *et al.* (2008) who tested the efficiency of six typical greywater treatment systems (Figure 10.1). Better understanding of the unique nature of greywater followed by establishment of new regulations has resulted in increasing research and development of greywater treatment systems that can meet stringent water quality regulations as demonstrated by three examples in Figure 10.2. As depicted in the figure, all three systems produced treated greywater that complies with the quality requirements for water reuse. After disinfection (either chlorination or UV irradiation) the concentrations of faecal coliforms were below the detection limit (1 cfu/100 ml; Figure 10.3 top left). Details on the systems or reasons for their successful or unsuccessful application is beyond the scope of this chapter and can be found elsewhere (Gross *et al.* 2007; Aizenchtadt *et al.* 2008; Zapater *et al.* 2011; Dekel Oz, 2011; Friedler *et al.* 2011). Complementary to proper treatment, introduction of management practices such as night-time or subsurface irrigation may decrease direct exposure of the population to the treated greywater and consequently reduce potential risks.

In summary, educated development/adjustment of appropriate technologies, application of barriers (such as subsurface irrigation), formulation of appropriate regulations and guidelines that are based on quantitative approaches such as the QMRA (explained below), as well as public education and communication, are necessary elements required to bring the potential risks to a negligible minimum.

10.4 QUANTITATIVE MICROBIAL RISK ASSESSMENT (QMRA)

As suggested above, reuse of greywater can compromise human and environmental health. Pathogens in greywater may cause diseases through direct contact, as well as through the consumption of contaminated plants (Shuval *et al.* 1997; Mara *et al.* 2007), and/or through peripheral vectors like mosquitoes (Morel & Diener, 2006). Additionally, greywater can contain elevated levels of surfactants, oils, boron and salts, which may alter soil characteristics, damage vegetation and pollute groundwater (Gross *et al.* 2005; Wiel-Shafran *et al.* 2006; Travis *et al.* 2008).

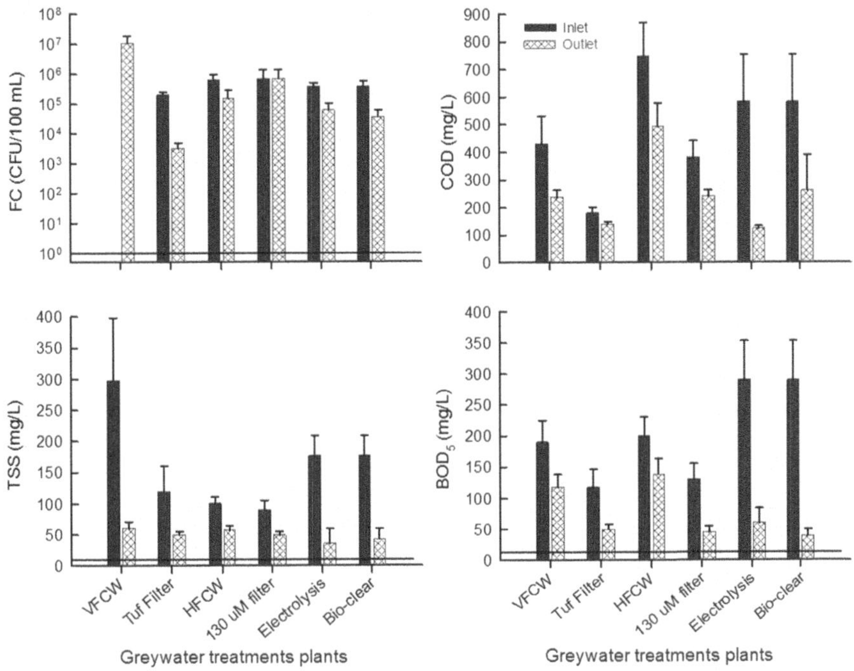

Figure 10.1 Example of six underperforming onsite greywater treatment systems: vertical-flow constructed wetland (VFCW), tuff filter; horizontal-flow constructed wetland (HFCW), 130 μ net filtration, electrolysis and an off-the-shelf proprietary system. Based on Gross *et al.* (2008). FC – faecal coliforms; Horizontal black lines – Upper concentration limit in the Israeli regulations for water reuse (Halperin & Aloni, 2003; Inbar, 2007).

Both the associated challenges and opportunities should be taken into account when considering greywater reuse policy. For greywater to be more accessible, reuse schemes must be relatively simple and economically feasible to the user, encouraging wider use, thereby maximising the quantity of water saved (Friedler & Hadari, 2006). At the same time, greywater reuse must be environmentally sound and avoid any public health compromise. Indeed, several jurisdictions have established standards/regulations for greywater reuse. However, the variation between policies in different countries (and in some cases between different regions of the same country) is significant in many cases. Often regulatory policies do not differentiate between black- and grey- waters, or even fail to formulate specific regulation for greywater reuse. On the other hand, several states in the U.S., states in Australia, some EU member states (e.g., the UK, Germany and Spain) and several other countries (such as Canada, Japan and Taiwan), do recognise the benefit of onsite reuse of greywater and have created highly detailed normative frameworks (Radcliffe, 2004; NRMMC, 2006; Rosner *et al.* 2006; WHO, 2006; National Water Commission, 2008. More details about greywater reuse regulations can be found in Chapter 7). Some of the above countries/states even offer various incentives to encourage people to adopt onsite greywater reuse practice.

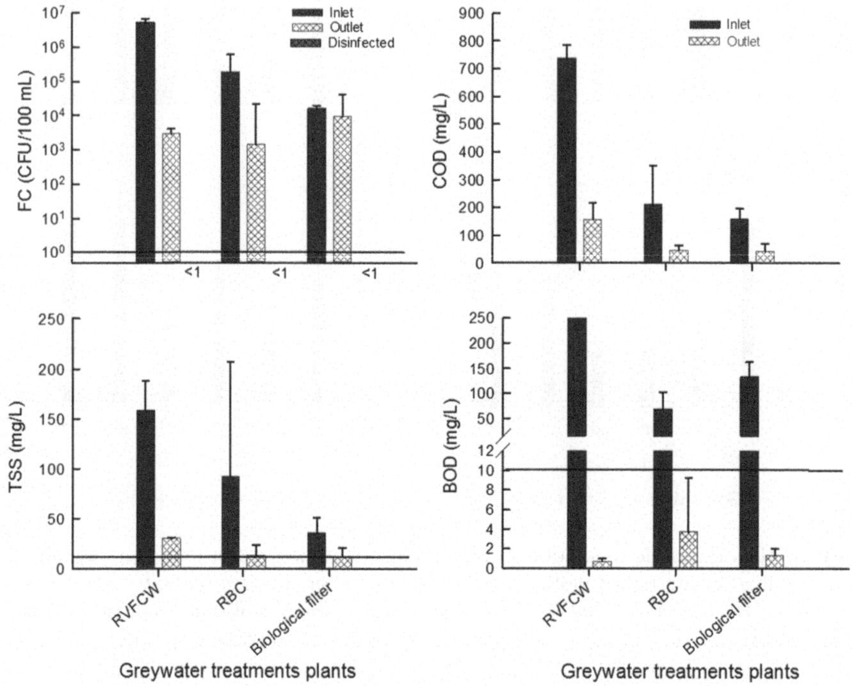

Figure 10.2 Examples of the successful performance of three onsite greywater treatment systems: recirculating vertical flow constructed wetland (RVFCW), rotating biological contactor (RBC) and biological filter. FC – faecal coliforms; <1 – lower than the detection limit (1 cfu/100 ml); Horizontal black lines – Upper concentration limit in the Israeli regulations for water reuse (Halperin & Aloni, 2003; Inbar, 2007).

In this section, risk assessment tools are demonstrated to form a baseline for a standardised evaluation of existing regulations and measures that should be taken to protect public and environmental health. The reason for choosing the QMRA methodology lies in its relative simplicity and evidence that its predictions compare well with those obtained by parallel epidemiological field studies (Mara *et al.* 2007). The QMRA methodology comprises four steps that lead to the fifth, managing the risk, where public health related guidelines should be elaborated or examined (Figure 10.3). These steps are described in detail in the following paragraphs.

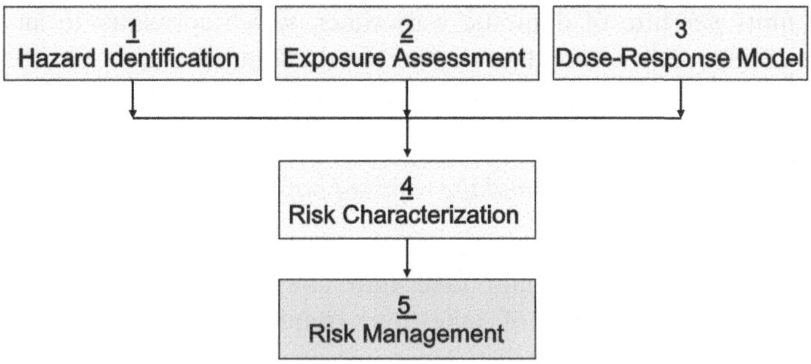

Figure 10.3 Schematic representation of the methodological steps of QMRA.

Step 1: Hazard identification – Defining the hazards, or finding index hazard agents that present the most prominent risks and assessing their prevalence in the relevant environment.

Since it is practically impossible to identify and account for all pathogens, indicator organisms are often used in risk assessments. For instance, traditionally, faecal contamination is a central parameter in wastewater quality monitoring and faecal coliforms is the most common indicator of the possible presence of other faecal pathogens. Moreover, they are considered as efficient indicators for measuring removal of bacterial pathogens (Mara, 2003). Although many reports demonstrate elevated concentration of faecal coliforms in greywater (Christova Boal *et al.* 1996; Eriksson *et al.* 2002), its relevance as an indicator for the microbial quality in greywater is disputed (Dixon *et al.* 1999; Ottoson & Stenstrom, 2003). It should be noted though that some faecal contamination does exist in greywater and may pose unacceptable health risks (Ottoson & Stenstrom, 2003).

Viruses constitute a key component of such faecal pathogens because of their high excretion rate from infected persons, low dose needed for potential infection and their high survival rate in the environment (Gerba *et al.* 1996; Ottoson & Stenstrom, 2003; WHO, 2006). Rotaviruses are a common cause of gastroenteritis in humans (Gerba *et al.* 1996), for which a dose-response model has been established. In a risk assessment conducted by Ottoson and Stenstrom (2003), Rotavirus was found to pose the most significant risk to human health from greywater. Therefore, the QMRA approach is demonstrated on Rotavirus within this chapter (Maimon *et al.* 2010).

Methods for quantifying rotavirus concentrations in greywater are not as straightforward and simple as those used for quantifying faecal coliforms or *E. coli*, which is a major group of faecal coliform bacteria and is often measured as a representative indication for faecal contamination (Mara, 2003). Various studies have correlated rotavirus loads with faecal indicators such as *E. coli*. The WHO guidelines (2006) suggest that there are between 0.1 to 1 rotaviruses for every 10^5 *E. coli* in 100 ml of domestic wastewater. The Australian national guidelines for water recycling (NRMMC, 2006) suggest an average concentration of 8000 rotavirus units *per* litre of domestic wastewater, which correlates to an average density of 10^7 *E. coli* per 100 ml, or in other words, 8 rotavirus units per 10^5 *E. coli*, which is roughly in the same order of the higher range reported by the WHO.

Step 2: Exposure assessment – Assessing the routes, frequency and duration of exposure to the hazard and the exposed populations.

Exposure rates are a key factor in determining the probability of an infection. An exposure assessment should take into consideration possible exposure pathways such as all forms of ingestion, frequency and the magnitude of exposure (e.g., the quantity ingested *per* one exposure event). The Australian guidelines (NRMMC, 2006) offer examples of estimated exposures based on the volume used in gardens irrigated with wastewater (Table 10.2). Other exposure routes, such as those associated with the ingestion of contaminated soil, crops or groundwater, can be adapted from risk assessments employed in agricultural wastewater irrigation studies. For example, it was estimated that a quantity of 10–100 mg *per* person *per* day of soil saturated with wastewater could be ingested by people working or playing in wastewater irrigated soils (WHO, 2006). Shuval *et al.* (1997) estimated the volume of irrigation water clinging onto 100 g of cucumber and 100 g of lettuce at 0.36 and 10.8 ml, respectively. If eaten unwashed, microorganisms in the greywater that were deposited on the crops during irrigation can be ingested. The Australian guidelines (NRMMC, 2006) adapted data from Shuval *et al.* (1997) to estimate the potential exposure to greywater following consumption of home-grown and commercially produced vegetables. An attempt to standardise and summarise the risks presented in exposure assessments is presented in Table 10.2.

Table 10.2 Possible exposure scenarios for greywater applications.

Frequency (events/year)	Quantity	Exposure scenario
1	100 ml	Accidental ingestion of greywater
90	1 ml	Routine indirect ingestion from touching plants and lawns
90	0.1 ml	Ingestion of greywater sprays from irrigation
According to the number of working days in the garden	10–100 mg	Ingestion of soil contaminated with greywater
7 for lettuce; 50 for other produce	0.36–10.8 mL/100 g; 5 mL per serve of lettuce; 1 mL for other produce	Eating a home-grown plant that was exposed to greywater

Exposure scenarios are based on: NRMMC (2006); Haas *et al.* (1999) and Shuval *et al.* (1997).

Step 3: Dose-response characterisation – Defining the quantitative connection between the rate of exposure to the probability of becoming infected and expressing it mathematically

The probability of infection due to exposure is driven by available dose-response models. The Haas's beta-poisson dose-response model for rotavirus is used as an example of this within a QMRA and is presented in Equations 10.1 and 10.2 (Haas *et al.* 1999):

$$P_i(d) = 1 - \left[1 + \frac{d}{N_{50}} \cdot \left(2^{\frac{1}{\alpha}} - 1\right)\right]^{-\alpha} \quad (10.1)$$

(For the rotavirus model $\alpha = 0.253$, $N_{50} = 6.17$)

$$P_{i(A)}(d) = 1 - [1 - P_i(d)]^n \quad (10.2)$$

where d is the dose of the pathogen; $P_i(d)$ is the probability of individual infection or the proportion of infected people in a community as a result of each of its members exposure to a single dose 'd' of a pathogen; N_{50} is the dose at which half of the population will be infected; α is the infectivity constant of the pathogen; $P_{i(A)}(d)$ is the annual risk of infection; and n is the number of exposure events per year.

Step 4: Risk characterisation – Integrating data from the previous steps, estimating the magnitude of risk in comparison to existing health targets, or to risks deemed 'acceptable'.

Utilisation of wastewater or greywater involves risk. Accordingly, there is a need to set a maximum acceptable risk level. Such thresholds involve ethical decisions and are a function of societal benefit-cost equations, balancing the benefits of saving water versus the costs of infectious disease. The DALY (Disability Adjusted Life Year) concept calculates both the number of years of life lost due to death (YLL) and the years lived with disability (YLD) and it is used to measure the healthiness of a society (Homedes, 1996). DALY is commonly used by the WHO and some countries (such as Australia) as an important tool to assess maximum tolerable risks by which health targets and public health management are decided. The WHO (2008) has set 10^{-6} DALYs *per* person-year as the maximum tolerable risk for waterborne diseases. In other words, a risk is deemed tolerable if one year of healthy life is lost due to waterborne diseases in a population of million people. The tolerable infection risk for rotavirus was calculated according to the 10^{-6} target and severity of the diseases it causes and was set as 1.4×10^{-3} infections *per* person-year (WHO, 2006). Consequently, looking at the entire population it is tolerable for about one person out of a thousand to become infected with a rotavirus, once a year. The DALY index details are beyond the scope of this chapter, for further details the reader is directed to the WHO (2006, 2008) and other publications on the subject (Homedes, 1996; NRMMC, 2006).

In order to find the 'safe' dose (d), it is possible to use an inverse solution to the dose-response model (Eq. 10.1) by introducing the tolerable infection risk (e.g., 1.4×10^{-3}) as the probability of infection ($P_i(d)$) as outlined in Equation 10.3

$$1.4 \times 10^{-3} = 1 - \left[1 + \frac{d}{6.17} \cdot \left(2^{\left(\frac{1}{0.253}\right)} - 1\right)\right]^{-0.253} \tag{10.3}$$

The safe dose, d, is therefore 2.4×10^{-3} rotavirus units, which means that if the population is exposed to a dose lower or equal to this dose the infection risk will be tolerable. Dividing the safe dose, d, by the estimated rotavirus densities in greywater as outlined in the Hazard Identification step (see step 1 above), would yield the maximum allowable volume of greywater that can be 'safely' ingested in a single occurrence (Table 10.3).

Table 10.3 Maximum greywater dose (ml) that can be 'safely' ingested by a person, assuming that the greywater contains between 0.01 to 0.8 rotavirus units/ml which is correlated to a count of 10^6 *E. coli*/100 ml, as estimated by three different sources.

Source	Rotavirus (organisms/ml)	Max dose (ml)
WHO (2006)	0.01–0.1	0.24–0.024
Ottoson and Stenstrom (2003)	0.17	0.014
NRMMC (2006)	0.8	0.003

The same rationale can be used to address multiple exposures using Equation 10.2. For example, the following hypothetical data is used in the following analysis: a routine ingestion scenario of 90 exposures to 1 ml per year (Table 10.2). This exposure was chosen as an example as it represents a high routine exposure in a scenario that is hard to avoid (routine indirect ingestion from touching plants and lawns). The probability of infection $P_i(d)$ followed by the safe dose (d) can be calculated as follows:

$1.4 \times 10^{-3} = 1 - [1 - P_i(d)\ ]^{90}$; the $P_i(d)$ is therefore 1.6×10^{-5}

The $P_i(d)$, is then used in Eq. 10.1 to determine the safe dose (d): $1.6 \times 10^{-5} = 1 - [1 + (d/6.17)(2^{1/0.253} - 1)]^{-0.253}$. The safe dose, d, is therefore 1.4×10^{-4} rotavirus units/ml.

Transforming the above figure to an *E. coli* concentration, based on ratios suggested by the WHO (2006) and the Australian guidelines (NRMMC, 2006), generates a safe *E. coli* concentration ranging between 10^2–10^4 (in the case of 90 events of 1 ml ingestion annually). These results suggest that the maximum tolerable concentrations of *E. coli* may lie between 10^2 and 10^4 cfu/100 ml. This considerably wide range may explain the differences between various regulatory guidelines. For example, the WHO wastewater irrigation guidelines limit *E. coli* concentrations to 10^3 cfu/100 ml (WHO, 2006), while the Israeli regulations require levels two orders of magnitude lower at 10^1 cfu/100 ml for *E. coli* (Inbar, 2007). Interestingly, the Australian guidelines suggest using log reductions (by treatment) rather than specifying a maximum *E. coli* concentration (NRMMC, 2006). It should be noted that in most risk assessments, computer simulations, such as the Monte Carlo method, with multiple trials are used to calculate risk levels (Ottoson & Stenstrom, 2003; WHO, 2006; Mara *et al.* 2007) rather than one exposure scenario as demonstrated above.

Such low infective doses demonstrate that the use of untreated greywater may be unsafe. However, as noticed by Dixon *et al.* (1999) and the Australian guidelines (NRMMC, 2006), the smaller the reuse cycle, the lower the pathogen risk. In other words, reusing greywater from a single house system is much safer than reusing greywater on a neighbourhood-scale system. Indeed, many states have separate regulations for single households and multi household systems, for example Arizona (Arizona Department of Environmental Quality, 2001), Utah, Nevada (Rosner *et al.* 2006) in the USA and Victoria (Victoria Environmental Protection Agency, 2006, 2008), South Australia (South Australia Department of Health, 2006), Northern Territory (Northern Territory Department of Health, 2007), New South Wales (New South Wales Department of Energy, 2007) in Australia. This distinction can often be attributed to historical reasons rather than a conscious strategy for lowering the associated risks.

Most regulatory programs allow restricted use of untreated greywater within the context of a single household property. Excluding kitchen effluents, enteric pathogens appear in greywater mainly if one of the people contributing to the system is a carrier. If there is one infected person in a household, others living at the same property may become infected by the pathogen through multiple pathways other than *via* greywater. Following this logic, any additional household connected to a system increases the risk of morbidity. Yet, even at the single household scale, issues such as pathogen survival or re-growth in greywater conveyance systems (Ottoson & Stenstrom, 2003) may pose unnecessary risk to the direct user of greywater. There is therefore a need to promote suitable treatment, such as the introduction of basic disinfection. It should be noted that several greywater treatment systems were found to reduce *E. coli* concentrations to low and even undetectable levels after the introduction of a disinfection unit (Friedler *et al.* 2006; Gross *et al.* 2007; and Figure 10.2 above). It should be noted that *E. coli* is not necessarily a sufficient indicator of bacteria and may even be less appropriate for viruses, protozoa and helminth (Mara, 2003). Another complimentary approach can be the establishment of barriers to minimise human contact with potentially hazardous bacteria (Dixon *et al.* 1999).

Currently, most of the relevant regulations rely on approaches that utilise such barriers. These can take the form of restrictions on the products and processes allowed to go into a recycling scheme, the level of treatment required or the reduction of exposure rates (NRMMC, 2006). Normative barriers can reduce the 'maximum risk' measured in a risk assessment to a negligible 'residual risk' following their adoption (NRMMC, 2006). The first barrier imposed is usually placed on the source of water allowed into the reuse scheme. For example, in California (CA) the reuse of water from the kitchen is completely prohibited (State of California, 2009), while in New South Wales (NSW, Australia) water from kitchen streams can only be allowed if an appropriate treatment device is in use. Similar to this approach is the restriction on the use of water from the washing of soiled diapers in Arizona (AZ) and CA (State of California, 2009) and the use of untreated water from that source in NSW. NSW has also recommended not using greywater when a person in the house has gastroenteritis. Other barriers focus on required treatment levels, the permitted uses of the water and other technical barriers. Some programs have established a tiered approach, in which there are different requirements for different types of reuse schemes. For example, AZ's and CA's tiered approach classifies utilisation according to the size of the system where uses and system requirements are based on greywater volume. Victoria's and NSW's tiered approach is driven by treatment levels. As a rule of thumb, barriers (other than the treatment itself) are lowered and additional uses of the treated greywater are allowed as treatment level increases.

In summary, most regulatory programs use multiple barriers to reduce exposure rates in order to eliminate health risks. The links between different exposure scenarios and recommended technical barriers suggested for their prevention are summarised in Table 10.4.

To date, no epidemiological survey supports claims that greywater usage at a single household scale is associated with higher morbidity (O'Toole *et al.* 2012). While the precautionary principle mandates a conservative approach to standard setting, the particularly widespread usage of greywater in Australia (55% of households; Australian Bureau of Statistics, 2007) may suggest that greywater use does not constitute an acute public health risk. However, attention should be paid to issues such as the under-reporting of gastrointestinal illness and other confounding factors that serve to mask associations between greywater use and disease. The dearth of empirical case studies and epidemiological surveys on the matter is regrettable as they would contribute to a higher quality of risk assessment. Despite the lower level of health risks typically associated with single household reuse (as compared to multiple sources, that is, multiple family systems), suitable treatment and disinfection are recommended prior to all greywater reuse, irrespective of scope. Regulations should also consider and weigh the added benefits provided by additional water, as a *resource*, against any *risk* associated with its utilisation. Most regulations provide better measures for protection of public health yet, other potential environmental risks, such as soil degradation and the pollution of ground- and surface-water, are often overlooked and still need to be studied.

Table 10.4 Exposure scenarios and related common barriers.

Exposure type	Exposure scenario	Summary of suggested barriers by different authorities
Direct	Accidental ingestion of greywater	Wearing protection when maintaining the system
		Marking the pipes as non-drinkable water
	Ingestion of greywater from the irrigation system	Human contact is avoided
		Restricted spray irrigation
		Water should not pond
		Marking the pipes as non-drinkable water
	Ingestion of soil contaminated with greywater	Applied as subsurface irrigation
		Overflow to sewer system
	Inhalation of aerosols from spray irrigation system	Restricted spray irrigation
	Eating fresh vegetables that were irrigated with greywater	Restricted food crop irrigation
Indirect	Groundwater pollution	Setback distance from groundwater level
	Surface water pollution	Water should not flow outside property boundaries
		Location outside drainage or flood zones
		Water should not get into open water bodies
		Overflow drains to sewer system
	Pathogen transmit through vectors such as mosquitoes	Water should not pond
		Overflow drains to sewer system

When the risk associated with greywater reuse is assessed and regulations are derived, there is a need to design reliable treatment systems that can meet desired treated greywater quality most of the time in order to minimise health risk. The following section demonstrates an approach to designing for reliability.

10.5 DESIGN FOR RELIABILITY AND RELIABILITY ANALYSIS

Appropriate design of greywater systems followed by undertaking a reliability analysis can further reduce potential risks associated with greywater reuse. The aim of this section is to demonstrate this approach by: (a) identification of potential causes for failures in biological greywater treatment and reuse schemes *via* the establishment of a fault tree analysis, which ranks failures in terms of the degree of possible impact on public health and the environment; (b) demonstration of how the design of a system can be executed in a way that reduces these faults, which can lead to self-containment of most faults within the system; and (c) demonstration of the applicability of this method through a case study in which the reliability of full-scale onsite treatment units (designed following the above principles) was tested in twenty single-family homes where the treated greywater effluent was used for landscape irrigation (Alfiya *et al.* 2013).

10.5.1 Using a fault tree analysis to identify system failures

Greywater treatment systems consist of various components that can be classified into three categories: structural components (tanks, pipes and media), electro-mechanical equipment (pumps and valves) and elements related to the biological process (biomass). The characteristics of failures that can occur in relation to these categories are different and therefore each has to be addressed and analysed separately using a systematic approach. In this case, a fault-tree analysis has been employed.

The fault-tree was first divided into two branches: (i) failures that result in no treated greywater being produced or supplied, such as the result of a pump breakdown; and (ii) failures that result in the production of partially treated greywater of poor or non-satisfactory quality for reuse (Figure 10.4a; A-1). It should be stressed that in this example the focus is on failures that result in treated-greywater of poor quality, since they are the ones that pose health and environmental risks. The second branch of the fault-tree was subsequently divided into two further compliance failure categories: (i) due to low quality in 'chemical' parameters (COD, BOD, TSS) (Figure 10.4a; B-1); and (ii) due to low quality in microbial indicators (high counts of pathogenic and/or indictor microorganisms) (Figure 10.4b; B-2). In relation to (ii), microbial indicators (B-2), the fault-tree analysis further reveals that five factors can lead to failure in the biological treatment process (C-1), which can ultimately lead to compromised public health and to negative environmental effects. These five factors are: 1) high hydraulic and/or pollutant loads; 2) cross-connection or mixing between raw- and treated-greywater; 3) penetration of toxic or inhibitory substances to the treatment unit *via* the incoming raw greywater; 4) hydraulic problems (short circuiting, clogging); and 5) electrical and mechanical malfunctions (pumps, mixers or blowers, motors).

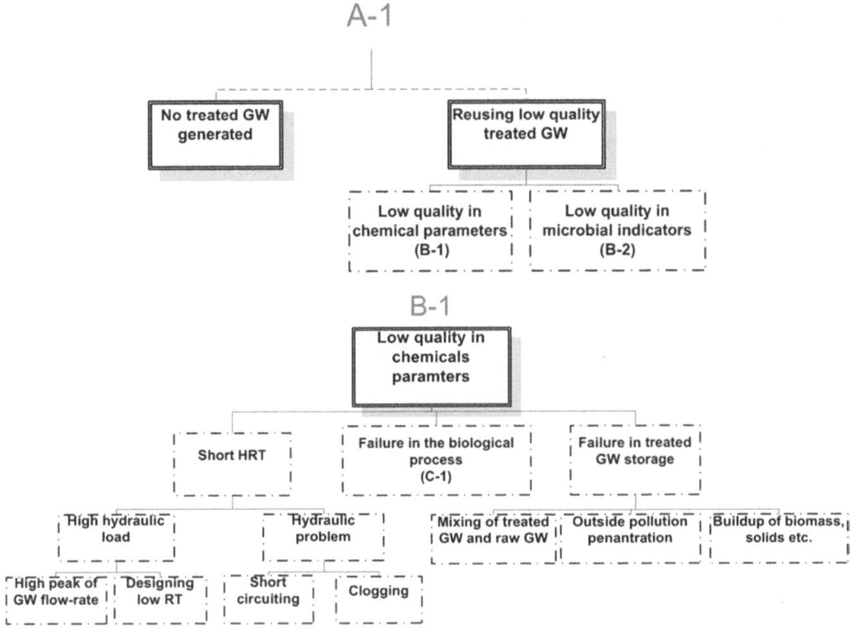

Figure 10.4a General fault-tree for onsite greywater treatment systems (adapted from Alfiya *et al.* 2013).

In relation to 2), the risk of accidental mixing of treated- and raw- greywater can be avoided by installing controls, one-way valves and using different colours for the piping of each stream (e.g., purple colour for pipes conveying treated greywater, which is the international convention for treated effluent pipes). With respect to 1) and 3), raw greywater inherently exhibits high variability of flows, pollutant loads and temperature (Friedler, 2004). By installing an equalisation tank for collecting the raw greywater before the treatment process, these shock-loads can be smoothed considerably and the flow into the biological treatment stage can be kept relatively constant. It should be noted that fixed-film biomass process can cope with this high variability more effectively than suspended biomass, since high hydraulic loads can wash out the suspended biomass from the reactors. Finally, to respond to 1), 4) and 5), a maintenance program should be implemented to eliminate failures caused by hydraulic problems and mechanical and electrical malfunctions.

10.5.2 Using a fault tree analysis to redesign the system

The fault-tree approach outlined in the previous section was adopted in designing recirculating vertical flow constructed wetland (RVFCW) systems and measures were taken in order to reduce potential risks from system failure. Controls were added in order to ensure that raw- and treated- greywater did not mix and that the hydraulic retention time would be sufficient. These adaptations improved the reliability of the units and ensured the production of high quality treated greywater, which is demonstrated in the following case study.

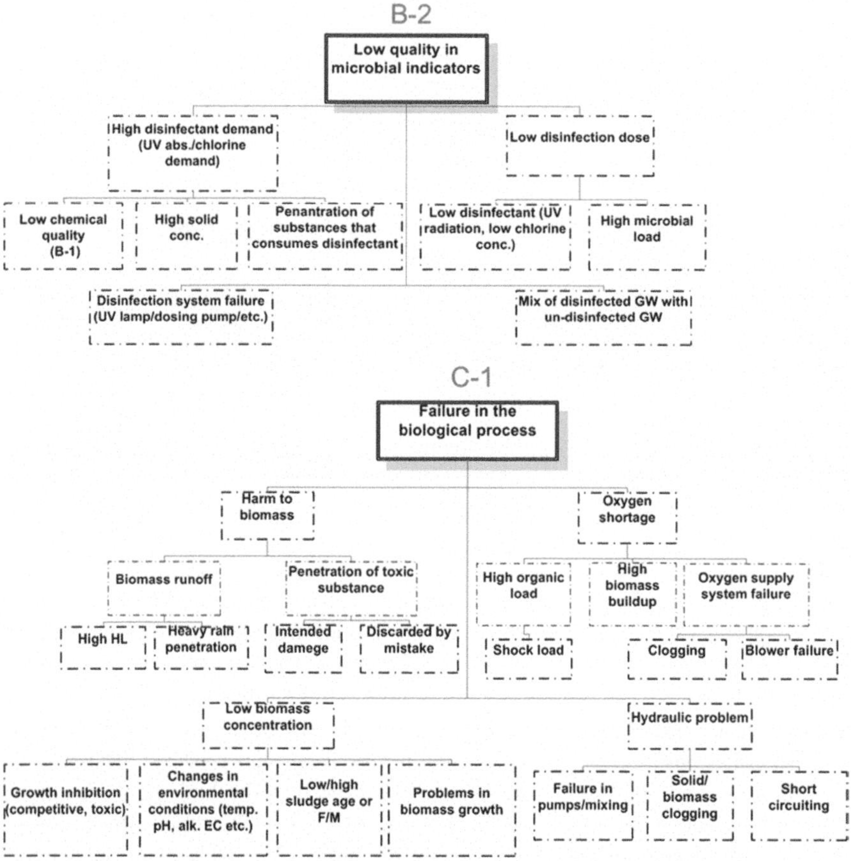

Figure 10.4b General fault-tree for onsite greywater treatment systems.

Each RVFCW consisted of two 500 litre plastic containers (1.0 m × 1.0 m × 0.5 m), one placed on top of the other (Figure 10.5). The upper container holds a planted three-layer bed, while the lower one functions as a reservoir. The bed in the upper container consists of a 5 cm top layer of woodchips, followed by a 35 cm middle layer of tuff gravel and a 10 cm bottom layer of limestone pebbles. Greywater is pumped from a 200 litre collection tank, that also acts as an equalisation/sedimentation tank, and is spread on the top of the bed. From there, greywater trickles through the bed into the reservoir (the lower container) through the perforated bottom of the upper container. From the reservoir, the greywater is recirculated to the top of the upper bed at a rate of about 300 l/h. Further details about the systems can be found in Gross *et al.* (2007).

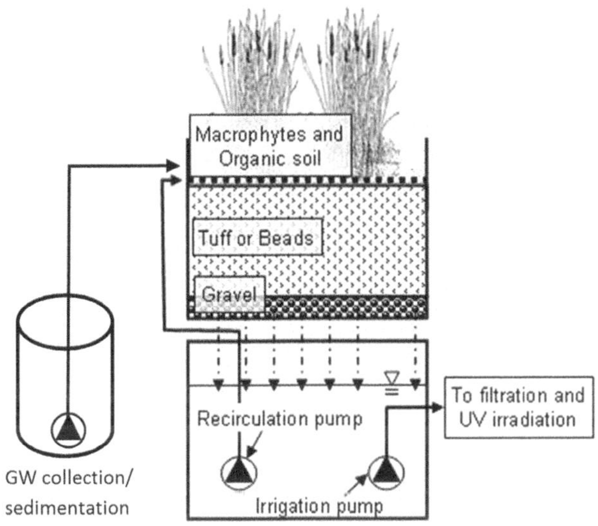

Figure 10.5 Schematic of the onsite recirculating vertical flow constructed wetland (RVFCW) for greywater treatment.

10.5.3 Reliability of a full-scale onsite system – Case study

Twenty of the RVFCW units outlined in the previous section were installed, operated and monitored for several years in three regions of Israel, differing in climatic conditions as follows: nine in the Northern and Central parts of Israel, which are characterised by Mediterranean climate with warm and dry summers, and cool and wet winters (average annual precipitation of 500–600 mm); three in the South Jordan Rift Valley, a semi-arid to arid climate (average annual precipitation of ca. 200 mm); and eight in the central Negev desert, which is an arid region (annual precipitation of less than 80 mm).

The major quality parameters of the raw- and treated-greywater entering and leaving the systems are given in Table 10.5. The quality varied between households and over time, as expressed by the high standard deviations of the raw greywater quality. For example, the systems operating in the North & Central region and the Central Negev region received raw greywater of comparable quality; however, the systems operating in the South Jordan Rift Valley received raw greywater of significantly higher pollutant loads (about 95, 85, 160, 40 and 95% higher, for turbidity, TSS, COD, BOD and MBAS, respectively). However, the large variability in the raw greywater did not have any significant effect on the quality of the treated greywater, which was much more uniform for the duration of the experiment and usually complied with the Israeli effluent quality requirements (Halperin & Aloni, 2003; Inbar, 2007).

Table 10.5 Performance of 20 RVFCW systems installed at sites in different climatic regions of Israel.

	Avg. flow (l/d)	pH (–)	EC (mS/cm)	Turb. (NTU)	TSS (mg/l)	COD (mg/l)	BOD_5 (mg/l)	MBAS (mg/l)
Raw greywater	139(87)*	7.7(0.6)	1.18(0.35)	80(91)	81(98)	299(326)	167(161)	7.7(7.5)
Treated greywater		8.3(0.5)	1.26(0.41)	6.1(6.8)	8.8(7.2)	31(36)	2.7(5.0)	0.38(0.46)
Israeli guidelines**		6.5–8.5	1.4–1.8	5	10	100	10	2

*Values are long term averages; Values in brackets represent standard deviation.
**Based on the Israeli guidelines for unrestricted urban water reuse (Halperin & Aloni, 2003) and regulations for unrestricted effluent reuse in irrigation (Inbar, 2007).

10.5.3.1 Reliability of a greywater biological treatment system

Reliability is a characteristic of an item that is expressed by the probability that the item will perform as specified under given conditions for a stated time interval. Quantitatively speaking, reliability defines the probability that no operational interruptions will occur during a stated time interval (Birolini, 2010). Therefore, the reliability of a greywater biological treatment system should be represented by a probability that the system will produce treated greywater effluent of satisfactory quality during a stated time interval.

A failure is defined as an event where a system stops performing as required (Birolini, 2010), or for the specific focus of this discussion, when the quality of the treated greywater effluent is not satisfactory (e.g., does not meet the required standards/regulations) or when greywater effluent is not produced (no greywater effluent is available for reuse). In the case of a greywater treatment system, most failures are considered repairable and it can be assumed that following the repair of a certain failure, the system is '*as good as new*'. Failure is a random variable and can be described with statistical tools. The relationship between the reliability function $R(t)$, the probability density function (PDF) $f(t)$ and the cumulative distribution function (CDF) $F(t)$, can be formulated as shown in Equation 10.4 (Lazzaroni *et al.* 2011):

$$F(t) = 1 - R(t)$$
$$F(t) = \int_0^t f(t)dt \qquad (10.4)$$

The Mean Time Between (consecutive) Failures (MTBF) can be calculated by integrating the reliability function $R(t)$ (Equation 10.5), and the distribution of failures can be described by models such as normal, exponential, log-normal and Weibull distributions (Lazzaroni *et al.* 2011).

$$\text{MTBF} = \int_0^\infty R(t)dt = \int_0^\infty [1 - F(t)]dt \qquad (10.5)$$

In all of the twenty systems, merely 39 failures occurred during the monitoring period (1.5 years). Only four of the failures (~10%) resulted in irrigation with poor-quality under-treated greywater, which could have led to some transient negative effects on human health and/or the environment. The remaining 90% did not result in any potential negative effects, since they did not affect the quality of the treated greywater nor resulted in halting the irrigation. Two out of the twenty systems encountered seven different failures, each making them responsible for 36% of the overall number of failures. In as many as nine units (45% of the units), no failures at all occurred during the whole period. Figure 10.6 details the failures that occurred, categorised by 14 types. The most frequent causes of failure were clogging or breakdown of the influent pump that conveyed raw greywater from the equalisation tanks to the treatment systems (each one occurred 6 times during the monitored period). Another cause of failure was due to unexplained or un-recognised electrical shutdown ('other electrical failure'), which also occurred 6 times during this period. Interestingly, technical and/or mechanical failures occurred more often than failures of the biological process (treatment), which occurs in media clogging by biomass or sludge accumulation in the lower tank. This suggests that the process itself is much less sensitive than the equipment and that the microbial community in the treatment unit (attached growth biomass) can withstand and overcome short-term failures in the equipment. This observation coincides with a previous study that demonstrated the resilience of the system to withstand disturbances such as high and low pH, high organic load and high doses of cleaning agents, as well as mechanical failures such as pump malfunction (Zapater *et al.* 2011). It should be emphasised that the systems were not serially manufactured in a factory, but custom made in a small workshop, and as such, it can be expected that the number of technical failures would decrease significantly with serial production.

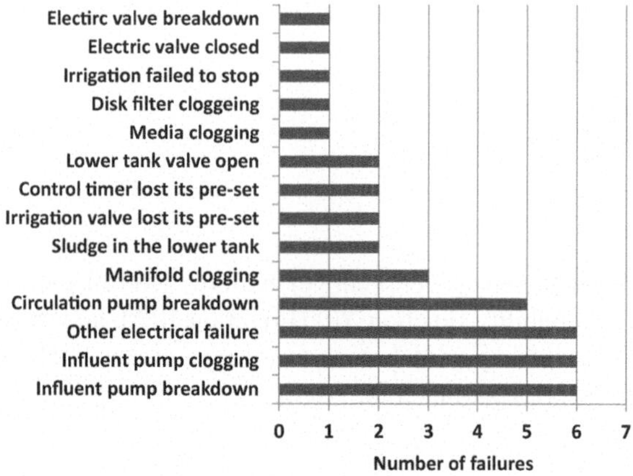

Figure 10.6 Failures recorded in the studied RVFCW system categorised by type of failure (20 units, 542 days of monitoring; based on Alfiya *et al.* 2013).

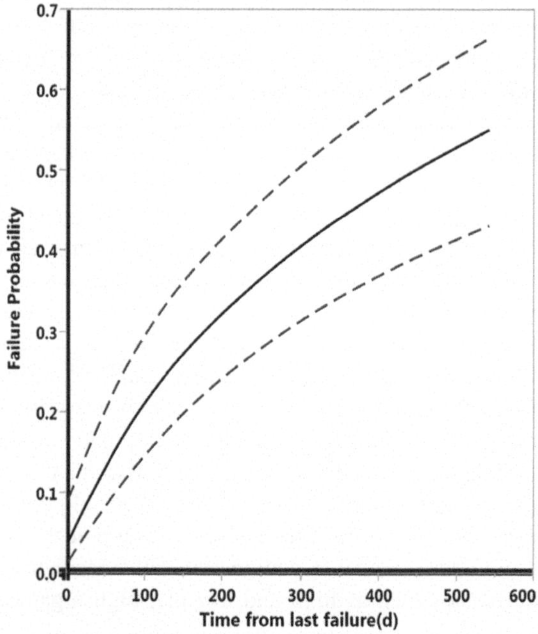

Figure 10.7 CDF plots and fitted Weibull model of time to malfunction event for the 20 RVFCW systems monitored. The solid line represents the Weibull model. The dotted lines represent the model's confidence interval (based on Alfiya *et al.* 2013).

10.5.3.2 Cumulative Distribution Function (CDF) of failures, reliability and Mean Time Between Failures (MTBF)

From the data collected, described above, the cumulative distribution function (CDF) of failures in the systems was plotted using JMP® statistical software (SAS Institute) (Figure 10.7). The Weibull model, which is widely used in reliability engineering, best represents this case study. The MTBF (mean time between failures) was calculated by numerical integration of the CDF plot according to Eq. 10.5 and was found to be 305 days with a relatively low standard deviation. This MTBF is quite long and more than satisfactory when considering systems of this type. As stated above, the MTBF serves as a guideline for deriving a maintenance programme for operating systems. Hence, being so long, it enables the derivation of an extensive (rather than intensive) maintenance schedule (e.g., once in 2–4 months), which is very suitable for single-family units as it should not be very costly.

10.6 SUMMARY AND OUTLOOK

Existing regulations are the basis for creating a more advanced regulatory system that may protect public and environmental health, while encouraging the use of greywater, which is an important yet largely untapped water resource. It is postulated that as long as basic regulatory rules are maintained, greywater reuse poses limited and acceptable risk to public health and the environment, and that its benefits outweigh the associated potential risks. In this chapter, the quantitative microbial risk assessment (QMRA) methodology was demonstrated as an efficient tool for assessing the health risk associated with greywater reuse based on a single pathogen, rotavirus. The QMRA methodology should be extended to other pathogens potentially present in greywater, in order to rationalise the risk and hence derive proper regulations and requirements for treated greywater quality. Since greywater treatment systems are used onsite, their maintenance cannot be very intensive; hence, they should be very reliable. The fault-tree methodology presented in this chapter appeared efficient in identifying and highlighting potential failure causes that can lead to under-performance of the treatment process. Accordingly, measures were incorporated in the design and construction phase of a greywater treatment system, in order to avoid these crucial failures. Additionally, the long-term performance of twenty recirculating vertical flow constructed wetland (RVFCW) greywater treatment systems, that were constructed based on the fault-tree analysis methodology, were evaluated under real-life conditions. The twenty units were proven to be highly reliable with a mean time between failures of about 10 months. This enabled a rather relaxed maintenance schedule to be derived that should be affordable and achievable for those tasked with operating and maintaining small-scale greywater reuse schemes. Risk quantification and management approaches are expected to contribute to safer and more reliable reuse of greywater, which is an important alternative water source. The methods and results presented in this chapter would appear to support the appropriate utilisation of such approaches.

REFERENCES

Adel M., Shmueli L., Bas Y. and Friedler E. (2012). Onsite greywater recycling and its potential future impact on desalination and wastewater reuse in Israel. *4th International Conference on Drylands, Deserts & Desertification*, Ben Gurion Univ., Sede Boqer, Israel, Nov. 2012.

Aizenchtadt E., Ingman D. and Friedler E. (2008). Quality Control of Wastewater Treatment: A New Approach. *European Journal of Operational Research*, **189**, 445–458.

Alfiya Y., Damti O., Stoler-Katz A., Zoubi A., Shaviv A. and Friedler E. (2012). Potential impacts of on-site greywater reuse in landscape irrigation. *Water Science and Technology*, **65**(4), 757–764.

Alfiya Y., Gross A., Sklarz M. and Friedler E. (2013). Reliability of on-site greywater treatment systems in Mediterranean and arid environments – a case study. *Water Science and Technology*, **67**, 1389–1395.

Arizona (USA) Department of Environmental Quality (2001). Title 18, Chapter 9, Article 7. Direct Reuse of Reclaimed Water Phoenix, Arizona. p. 11.

Australian Bureau of Statistics (2007). Environmental Issues: People's Views and Practices (cat no. 4502.0), p. 88.

Birolini A. (2010). Reliability Engineering: Theory and Practice. Springer, Berlin.

Christova-Boal D., Eden R. E. and McFarlane S. (1996). An investigation into greywater reuse for urban residential properties. *Desalination*, **106**, 391–397.

Dekel Oz A. (2011). Study of Bacterial Communities by Molecular Methods as a Tool for Developing an Attached Growth Treatment Process for Greywater. MSc Thesis, Fac. of Civ. and Env. Eng., Technion – Israel Inst. of Technol. (Heb), Haifa, Israel.

Dixon A. M., Butler D. and Fewkes A. (1999). Guidelines for greywater re-use: health issues. *Water and Environment Journal*, **13**(5), 322–326.

Eriksson E., Auffarth K., Henze M. and Ledin A. (2002). Characteristics of grey wastewater. *Urban Water*, **4**(1), 85–104.

Friedler E. (2004). Quality of individual domestic greywater streams and its implication on on-site treatment and reuse possibilities. *Environmental Technology*, **25**(9), 997–1008.

Friedler E., Butler D. and Alfiya Y. (2013). Wastewater composition. Chapter 17, In: Wastewater Treatment: Source Separation and Decentralization, T. Larsen, K. Udert, J. Lienert (eds), IWA Publishing, London, New York, pp. 241–257.

Friedler E. and Hadari M. (2006). Economic feasibility of on-site greywater reuse in multi storey buildings. *Desalination*, **190**, 221–234.

Friedler E., Kovalio R. and Ben-Zvi A. (2006). Comparative study of the microbial quality of greywater treated by Three on-site treatment systems. *Environmental Technology*, **27**(6), 653–663.

Friedler E., Yardeni A., Gilboa Y. and Alfiya Y. (2011). Disinfection of greywater effluent and regrowth potential of selected bacteria. *Water Science and Technology*, **63**(5), 931–940.

Gerba C. P., Rose J. B., Haas C. N. and Crabtree K. D. (1996). Waterborne Rotavirus: a risk assessment. *Water Research*, **30**(12), 2929–2940.

Gross A., Azulai N., Oron G., Ronen Z., Arnold M. and Nejidat A. (2005). Environmental impact and health risks associated with greywater irrigation: a case study. *Water Science and Technology*, **52**(8), 161–169.

Gross A., Shmueli O., Ronen Z. and Raveh E. (2007). Recycled vertical flow constructed wetland (RVFCW) – a novel method of recycling greywater for irrigation in small communities and households. *Chemosphere*, **66**(5), 916–923.

Gross A., Wiel-Shafran A., Bondarenko N. and Ronen Z. (2008). Reliability of small commercial greywater treatment systems and its potential environmental importance. *International Journal of Environmental Studies*, **65**, 41–50.

Gross A., Maimon A., Alfiya Y. and Friedler E. (2012). Safe Greywater Reuse. Report to the Israel Water Authority (Heb).

Haas C. N., Rose J. B. and Gerba C. P. (1999). Quantitative Microbial Risk Assessment, John Wiley & Sons, New York.

Halperin R. and Aloni U. (2003). Guidelines for Urban, Recreational and Industrial Effluent Reuse. IL Minist. of Health (Heb), Jerusalem, p. 15.

Homedes N. (1996). The Disability-Adjusted Life Year (DALY) Definition, Measurement and Potential Use. HCD, World Bank.

Inbar Y. (2007). New standards for treated wastewater reuse in Israel. In: Wastewater Reuse – Risk Assessment, Decision Making and Environmental Security, M. K. Zaidi (ed.), Springer, pp. 291–296.

Lazzaroni M., Cristaldi L., Peretto L., Rinaldi P. and Catelani M. (2011). Reliability Engineering: Basic Concepts and Applications in ICT, Springer, Berlin.

Maimon A., Tal A., Friedler E. and Gross A. (2010). Safe on-site reuse of greywater for irrigation—a critical review of current guidelines. *Environmental Science & Technology*, **44**, 3213–3220.

Mara D. D. (2003). Faecal indicator organisms. In: Handbook of Water and Wastewater Microbiology, D. D. Mara and N. Horan (eds), Academic Press, Elsevier, London, pp. 105–112.

Mara D. D., Sleigh P. A., Blumenthal U. J. and Car R. M. (2007). Health risks in wastewater irrigation: comparing estimates from quantitative microbial risk analyses and epidemiological studies. *Journal of Water and Health*, **5**(1), 39–50.

Morel A. and Diener S. (2006). Greywater Management in Low and Middle-Income Countries, Review of Different Treatment Systems for Households or Neighbourhoods, Swiss Fed. Inst. of Aquatic Sci. & Tech. (Eawag) Dubendorf, Switzerland.

National Water Commission (2008). Requirements for Installation of Rainwater and Greywater Systems in Australia, Waterlines report: Master Plumber and Mechanical Services Association of Australia, Canberra.

Natural Resource Management Ministerial Council (NRMMC), Env. Protection and Heritage Council and Australian Health Ministers' Conf. (2006). National Guidelines for Water Recycling: Managing Health and Environmental Risks (Phase 1).

New South Wales (Au.) Department of Energy, Utilities and Sustainability (2007). NSW Guidelines for Greywater Reuse in Sewered Single Household Residential Premises. Australia, p. 52.

Northern Territory (Au.) Dept. of Health and Community Services (2007). Greywater Reuse in Single Domestic Premises. *Environmental Health*, Information Bulletin No. 2, Darwin, Au., p. 4.

O'Toole J., Sinclair M., Malawaraarachchi M., Hamilton A., Barker S. F. and Leder K. (2012). Microbial quality assessment of household greywater. *Water Research*, **46**, 4302–4313.

Ottoson J. and Stenstrom T. A. (2003). Faecal contamination of greywater and associated microbial risks. *Water Research*, **37**(3), 645–655.

Queensland Government (2003). On-Site Sewerage Facilities: Guidelines for the Use and Disposal of Greywater in Unsewered Areas. Brisbane, Queensland, Australia, p. 26.

Radcliffe J. (2004). Water Recycling in Australia, Australian Academy of Technological Sci. and Eng., Parkville, Vic.

Roesner L., Qian Y., Criswell M., Stromberger M. and Klein S. (2006). Long Term Effects of Landscape Irrigation Using Household Graywater – Literature Review and Synthesis, WERF, SDA.

Shuval H., Lampert Y. and Fattal B. (1997). Development of a risk assessment approach for evaluating wastewater reuse standards for agriculture. *Water Science and Technology*, **35**, 15–20.

South Australia Department of Health (2006). Draft Guidelines for Permanent Onsite Domestic Greywater Systems: Greywater Products and Installation.

State of California (2009). Non potable water reuse systems, emergency graywater regulations. In: *2007 Plumbing Code*, Title 24, Part 5, Chapter 16A, Part 1, p. 10.

Travis M. J., Weisbrod N. and Gross A. (2008). Accumulation of oil and grease in soils irrigated with greywater and their potential role in soil water repellency. *Science of the Total Environment*, **394**(1), 68–74.

Victoria (Au.) Environmental Protection Agency (2006). Reuse Options for Household Wastewater. In EPA, Vic.

Victoria (Au.) Environmental Protection Agency (2008). Code of Practice: Onsite Wastewater Management. In EPA, Vic.

Wiel-Shafran A., Ronen Z., Weisbrod N., Adar E. and Gross A. (2006). Potential changes in soil properties following irrigation with surfactant-rich greywater. *Ecological Engineering*, **26**(4), 348–354.

World Health Organisation (2006). Guidelines for the Safe Use of Wastewater, Excreta and Greywater, Vol. IV, p. 204.

World Health Organisation (2008). Guidelines for Drinking-Water Quality. Vol. 1, 3rd edition incorporating 1st and 2nd addenda.

Zapater M., Gross A. and Soares M. I. M. (2011). Capacity of an on-site recirculating vertical flow constructed wetland to withstand disturbances and highly variable influent quality. *Ecological Engineering*, **37**(10), 1572–1577.

Chapter 11

Greywater recycling: Guidelines for safe adoption

Melissa Toifl, Gaëlle Bulteau, Clare Diaper and Roger O'Halloran

11.1 INTRODUCTION

Concern about the adequacy of potable water supplies has lead to the current international focus on water saving measures, along with more effective management of water supplies and the implementation of policies to reduce wastewater discharges to receiving waters. Together with recycled wastewater, rainwater and stormwater, greywater is often proposed as a potential alternative water source in the domestic setting, both for individual houses and low and high rise multiple occupancy dwellings.

When compared to alternative water sources that rely on rainfall, greywater provides a much more reliable supply of recycled water. Research into the public perception of greywater reuse has also shown a wider acceptance of this water source than recycled wastewater. However, a significant disadvantage of greywater is the large variability in source water quality and flow volume, and consequently it has found limited acceptance as a viable alternative water supply in the wider water industry.

The most basic form of greywater reuse involves diversion of untreated water, whilst more sophisticated greywater systems incorporate treatment using a range of commercially available technologies based on biological, chemical and/or physical processes. The use of untreated greywater has been widely accepted in some developed countries, for example over 40% of households in Melbourne, Australia reported greywater use (Australian Bureau of Statistics, October 2011). However, the use of treated greywater is still not currently widespread, although there is an increasing number of larger scale applications around the world. In these cases, the treated greywater has a variety of end uses, with garden irrigation and toilet flushing being the most common.

A potential barrier to the widespread uptake of greywater treatment systems is the lack of thorough, robust and reproducible testing procedures that can reliably assess the human health and environmental risks. A comprehensive risk assessment of greywater use also needs to be undertaken that incorporates all possible environmental end points, including the potential impacts on plants and soils, groundwater and surface waters.

This chapter presents an overview of the current status of greywater recycling systems around the world, as well as including information about greywater quality and the different types of greywater treatment systems that are commonly used. This is followed by a detailed discussion about current international regulations and guidelines, including the key aspects that need to be considered for the safe and successful implementation of small scale greywater reuse.

11.2 GREYWATER QUALITY

The quantity and quality of greywater has been extensively researched since the late 1970s when work on this water source was first undertaken in the United States (Siegrist, 1978). Recent research has shown that there is a large variability in greywater quality, depending on the source, the household products used and householder behaviour (Eriksson *et al.* 2002). Research has focused on the performance of a variety of treatment systems, where this assessment relies on measuring traditional water quality parameters such as biological oxygen demand (BOD), total suspended solids (TSS), chemical oxygen demand (COD), total Kjeldahl nitrogen (TKN), nitrate, total phosphorus (Tot-P), conductivity, pH and turbidity. Table 11.1 provides a summary of the research efforts into greywater quality.

Other research has focused on the potential human health risks of greywater where microbial water quality has been investigated (Birks & Hills, 2007; Casanova *et al.* 2001; O'Toole *et al.* 2012; Winward *et al.* 2009). More recent research examined the greywater quality in terms of Priority Hazardous Substances identified in the European Union Water Framework Directive (Eriksson *et al.* 2010), xenobiotic substances (those foreign to the natural biological system), organic compounds (Boyjoo *et al.* 2013; Eriksson *et al.* 2002) and other organic micropollutants (Gulyas *et al.* 2011). Xenobiotic substances include surfactants, fragrances, preservatives, UV-filters, and solvents, and so are likely to be present in greywater.

Risk assessment approaches are widely applied to the development of water quality standards (Fewtrell & Bartram, 2001) and have been used to assess the adequacy of the current guidelines for greywater use (Maimon *et al.* 2010). In order to understand the rationale for the different greywater quality measurements and to develop robust greywater guidelines or protocols, an understanding of

Table 11.1 Summary of research examining greywater quality.

Reference	Greywater source	End use	Endpoints	Example parameters
Ottoson and Stenström (2003)	All greywater, multiple houses	Pond (infiltration)	Humans	*E.coli*, somatic coliphages, coprostanol and cholesterol
Rose *et al.* (1991)	Shower, bath and washing machine, multiple houses	Not stated	Humans	Standard plate count, total and faecal coliforms
O'Toole *et al.* (2012)	Washing machine and bathroom	Not stated	Humans	Pathogenic *E coli*, noro-, entero- and rota-virus
Albrechtsen (2002)	Shower and hand wash basin	Toilet flushing	Humans	*E. coli* and Enterococcus
Jamrah *et al.* (2008)	Showers, laundries, kitchens and sinks	Irrigation, car washing, toilet flushing	Non specific	SS, TOC, COD, BOD, Total coliforms and *E.coli*
Christova-Boal *et al.* (1996)	Baths, showers, laundry troughs and washing machines	Toilet flushing and irrigation	Soil and humans	Physical, chemical and microbiological
Mohamed *et al.* (2013)	Showers, laundries, bathtub, and sinks	Irrigation	Soil and plants	Salinity, SAR, organic content, BOD, TSS, pH
Casanova *et al.* (2001)	Various (single houses)	Irrigation	Soil and water	Faecal coliforms and *E.coli*
Gerba *et al.* (1995)	Washing machine and bathroom	Landscape irrigation and toilet flushing	Treatment process	Faecal and total coliforms, nitrate, SS and turbidity
Jefferson *et al.* (2004)	Showers, bathtub and sinks	Not stated	Treatment process	COD/BOD
Birks and Hills (2007)	Baths, showers and washbasins (Halls of residence)	Not stated	Treatment process	Temperature, flow, COD, Salmonella
Friedler (2004)	All greywater	Various	Treatment process	TSS, COD, and BOD, sodium and phosphate
Brandes (1978)	All greywater	None	Treatment process & land	P, TKN, Total and faecal coliforms

risk assessment is required. A risk is the combination of the frequency and the consequence of a particular hazard. The main hazards of greywater use can be identified through a water quality analysis and are often grouped into three categories that need to be considered when assessing the risks. These are:

- Physical – temperature, flow, suspended material, turbidity;
- Chemical – pH, metals, salts, nutrients, organic compounds and xenobiotics;
- Biological – biodegradability, bacteria, viruses and protozoa.

Risks are also related to different end points or end-uses (Table 11.1). As greywater is generally used for irrigation and toilet flushing, these end points are: humans (ingestion, contact, inhalation), plants, soils and groundwater. The robustness or resilience of the greywater treatment system can also be considered as an end point in this risk assessment methodology, so that operational risks are also included. The effect of greywater use upon wastewater flows and concentrations and any subsequent effects on the operation of the sewer network and wastewater treatment plants should also be considered (Marleni *et al.* 2012; Penn *et al.* 2013; Revitt *et al.* 2011).

Incorporating the impact of the different hazards through toxicity and persistence or biodegradability information, allows the primary hazards for different end points to be identified. This approach, referred to as fugacity modelling, is used to predict the likely environmental partitioning and fate of the substances in soil, aquatic (river/lake) and treatment environments. Fugacity modelling was used to predict the behaviour of EU WFD (EU Water Framework Directive) Priority Substances (PS) and Priority Hazardous Substances (PHS) found in greywater, and it was found that the majority will partition into the solid phase (Donner *et al.* 2010).

The understanding and awareness of the different impacts of greywater on human health and the environment is improving. For example, recent work has shown that certain antibacterial greywater components (e.g., triclosan) can impact on soil microbiology (Harrow & Baker, 2010) and others have found that bacteria in different sources of greywater (kitchen, shower and washbasin) survive for different lengths of time in the soil (Abu-Ashour & Jamrah, 2008).

From a microbiological perspective, studies have found increased risks to human health from accidental direct contact with viral, but not bacterial, pathogens when using greywater for sports field irrigation and groundwater recharge (Ottoson & Stenstroem, 2003). Greywater irrigation has also been found to cause a statistically significant increase of faecal coliforms in soil when compared to irrigation with potable water. The presence of children in the collection area has also been found to produce a statistically significant increase in faecal coliform levels in greywater. Others have found that despite high levels of pathogenic indicator organisms, pathogen presence in greywater was undetected (Birks & Hills, 2007) and the presence of *E. coli* was not associated with the presence of human enteric viruses in greywater (O'Toole *et al.* 2012).

Consequently, a wide range of water quality parameters and end points, along with many complex interactions and variability must be considered when developing guidelines, protocols and regulations for greywater use.

11.3 GREYWATER TREATMENT SYSTEMS

There are a significant number of different greywater treatment systems currently available on the market, with new and innovative technologies being developed at a rapid rate. Some of these are aimed at cost reduction so they can provide safe alternative water solutions in low income and water stressed regions (Kariuki *et al.* 2011), while others utilise innovative approaches to reduce chemical usage (Gulyas *et al.* 2009). However, most technologies still include one or a combination of the more common biological, chemical or physical treatment processes discussed below, often followed by disinfection, to treat the greywater to the required standard. Disinfection methods can vary between systems, with the most common types being UV or chemical (typically chlorination or bromination).

The level of treatment is commonly classed into primary, secondary, tertiary and advanced; and will vary depending on the intended end-use of the recycled water. Primary treatments include removal of solids (hair, lint, grit and grease) and suspended solids. Secondary treatments also remove biodegradable organic material. Tertiary treatments further remove nutrients from the recycled water and include disinfection in addition to the treatment process. More advanced treatments may also be applied to further remove material not captured in the initial treatments, however this is not commonly used in small scale greywater treatment technologies. The performance of some selected treatment systems is also discussed in Chapter 10.

11.3.1 Biological systems

Biological processes are commonly used in greywater treatment systems, as they are generally less energy intensive than many of the physical treatment processes available (with the exception of membrane bioreactors (MBR), which use a combination of biological and filtration processes). Biological processes can also have less environmental impact than some of the chemical methods. Biological methods can be aerobic or anaerobic and include trickling filters, MBR (Boyjoo *et al.* 2013; Merz *et al.* 2007), biological aerated filters (BAF) (Ray *et al.* 2012), rotating biological contactors (RBC) (Friedler *et al.* 2005), upflow anaerobic sludge blankets (UASB) (Ellmitwalli & Otterpohl, 2007; 2011), sequencing batch reactors (SBR) (Ghaitidak & Yadav, 2013; Lamine *et al.* 2007) and chambers with suspended or fixed media (Gross *et al.* 2007). Each of these biological treatment processes has the proven ability to adequately treat greywater and reduce nutrients and organic compounds. However, microorganisms are not necessarily removed and so they all require a disinfection step to make the recycled water safe for reuse

(MBR is an exception as the membranes in the filtration stage are able to remove microorganisms). Figure 11.1 shows an example diagram of a biological treatment system. It should be noted that most greywater treatment systems use pumps to move the water between tanks even though these have not been included in the diagram.

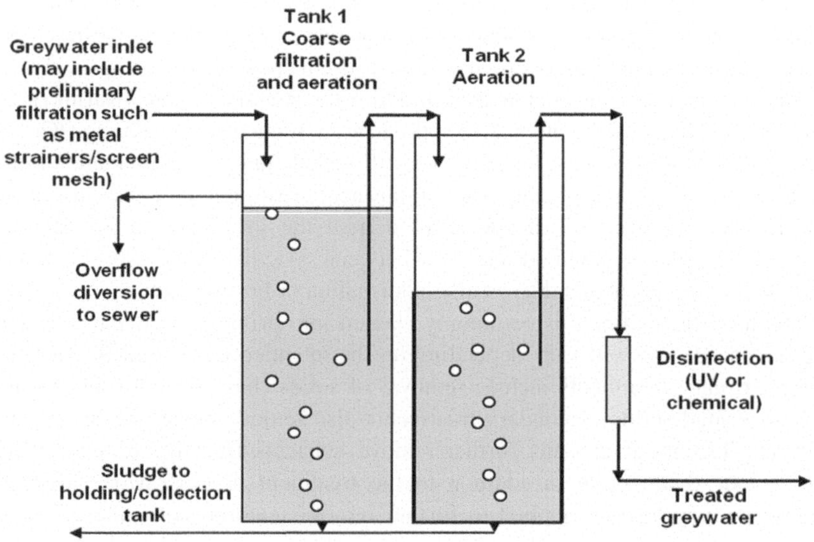

Figure 11.1 An example of a biological greywater treatment technology.

11.3.2 Chemical systems

Chemical treatment processes include activated carbon, coagulation and flocculation, ion exchange resins, and advanced oxidation processes (Sostar-Turk *et al.* 2005; Ciabattia *et al.* 2009). Where chemicals must be added as part of the treatment process, there may be ongoing costs associated with their supply. Additional problems may also arise if these chemicals eventually end up at centralised wastewater treatment plants. Figure 11.2 shows an example of a chemical-based treatment technology without pumps and other accessories such as control panels that are typically fitted.

11.3.3 Physical systems

Physical treatment processes focus on filtration and sedimentation. Filtration is often only a preliminary step and may involve metal strainers, screen meshes and multimedia such as gravel and sand beds. Sand and multimedia filtration methods have also been used as the main treatment for greywater, but these can have problems with clogging (Friedler & Alfiya, 2010). More advanced membrane filtration methods including microfiltration, ultrafiltration and nanofiltration (Ramon *et al.* 2004; Kim *et al.* 2009; Hourlier *et al.* 2010) are known to be very effective for treating greywater and wastewater. However, these methods are also energy intensive and have more maintenance requirements than some of the other treatment processes discussed. Membrane-based filtration processes are discussed in detail in Chapter 12. Filtration also extends to natural systems that are constructed to rely on plants, soil and sand layers to filter and degrade biological material. Known as constructed wetlands (Sundaravadivel & Vigneswaran, 2001; Liehr & Kruzic, 2007; Hsu *et al.* 2011), these systems have traditionally been used in low income countries. Studies have shown that they can treat greywater successfully to a primary or secondary level. However, disinfection is still required if the water is to be used for purposes with high potential for human exposure. The disadvantages of constructed wetlands include the requirement of a large footprint compared to other treatment methods, possible odour and aesthetic issues and their potential as breeding grounds for insects and other pests.

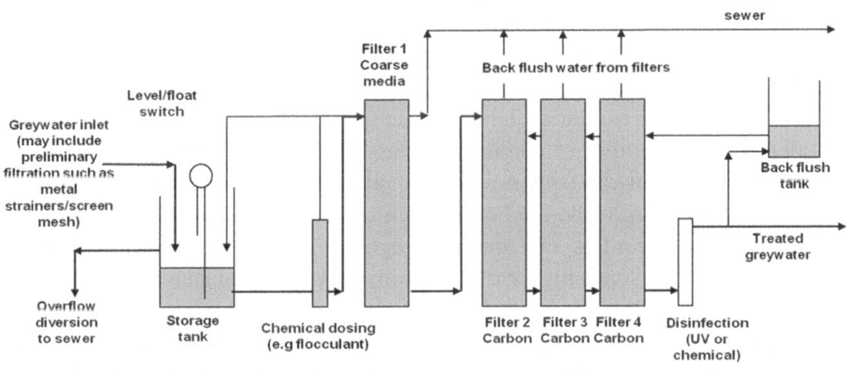

Figure 11.2 An example of a chemical-based greywater treatment technology.

Each of the treatment systems discussed has particular advantages and disadvantages that make them suited for treating particular types of greywater. With more systems constantly being developed and marketed, there has also been an increase in the development of regulations and guidelines available to ensure safety in relation to not only human health but also the environment. Some of the regulations and guidelines currently available are discussed in Section 11.4.

11.4 INTERNATIONAL REGULATIONS AND GUIDELINES

During the last decade, the United States Environmental Protection Agency (USEPA, 2004) and the World Health Organization (WHO, 2006) have formulated guidelines for water reuse, including greywater. An important objective was to guide the process of designing, installing and maintaining greywater systems in a manner that aims to protect human health, plants, soil and the environment. Similarly, Australia, Japan, China, Canada, along with several American states and various European countries, have also developed policies and guidelines designed to regulate greywater recycling within individual households or public premises without compromising public health or environmental quality. Some of these international documents are mandatory regulations, whilst others provide guidelines that can be adapted depending on the specific requirements of the intended water reuse scheme. The documents aim to provide a management framework to guarantee safe water reuse while allowing the use of non-potable water for many purposes that do not require drinking water quality. The existing documents mainly focus on the:

- type of greywater that can be reused;
- permitted uses for reclaimed water;
- treated water quality criteria (parameters and threshold values) depending on domestic end-use.

Some regulatory documents also give details and additional information on the technical requirements and approval processes required to implement a recycling scheme.

Table 11.2 summarises the reclaimed water quality guidelines for domestic end-use adopted by a number of countries. Some of these are specific to greywater recycling and use, whilst others apply to all domestic/municipal wastewater. Greywater coming from the kitchen is often excluded for recycling purposes as this water may contain fats, oils and food particles that are more difficult to treat and consequently present a higher risk. Within an overall management framework, the guideline values are intended to enhance treatment reliability and disinfection effectiveness, thus protecting public health. However, the rationale used to select parameters for water quality monitoring is often not included. Consequently, it is possible that particular local issues/constraints and the available analytical methods could explain the often wide variation of guideline values.

Parameters frequently selected to characterize domestic reclaimed water quality include pH, five-day biochemical oxygen demand (BOD_5), total suspended solids (TSS), turbidity, *E. coli*, thermotolerant coliforms, and chlorine residual. These parameters are monitored and regulated for both health and aesthetic reasons. For example, low BOD_5 levels (<10 mg/l) for toilet flush water helps to ensure that aerobic conditions are maintained in the sewerage system, whilst excessive BOD_5 can also lead to aesthetic and nuisance problems (odour and colour). Organic compounds can be broken down by microorganisms, causing a decrease in the oxygen content of the water, and can also adversely affect disinfection processes (Health Canada, 2010). Because organic pollutants and heavy metals can be adsorbed on particulates, most of the guidelines recommend a low TSS level (<10 mg/l) for indoor end-uses. Turbidity is equally important.

Table 11.2 Comparison of international guidelines and regulations for reclaimed water quality based on the domestic end-uses.

Country/ Organization	Water origin	Parameters and threshold values[a] for water quality criteria depending on the end-use						References
		Toilet flushing	Cold water supply for clothes washing	Car washing	Surface irrigation	Sub-surface irrigation	Garden watering	
WHO (Guideline)	Greywater[b]	$BOD_5 \leq 10$ $TSS \leq 10$ *FC* ≤ *10*		$BOD_5 \leq 10$ $TSS \leq 10$ *FC* ≤ *10*				WHO (2006)
US-EPA (Guideline)	Domestic wastewater	pH: 6–9 $BOD_5 \leq 10$ Turbidity ≤ 2 *FC*: *ND* Res. Cl ≥ 1		pH: 6–9 $BOD_5 \leq 10$ Turbidity ≤ 2 *FC*: *ND* Res. Cl ≥ 1				USEPA (2004)
Australia – ACT (Guideline)	Greywater[b]	$BOD_5 \leq 20$ SS ≤ 30 *FC* ≤ *10*	$BOD_5 \leq 20$ SS ≤ 30 *FC* ≤ *10*	$BOD_5 \leq 20$ SS ≤ 30 *FC* ≤ *10*		$BOD_5 \leq 20$ SS ≤ 30		ACT Heath (2007)
Australia – NSW (Guideline)	Greywater	$BOD_5 \leq 10$[c] SS ≤ 10[c] *FC* ≤ *10*[c] 0.5 < Res. Cl < 2.0	$BOD_5 \leq 10$[c] SS ≤ 10[c] *FC* ≤ *10*[c] 0.5 < Res. Cl < 2.0		$BOD_5 \leq 20$[c] SS ≤ 30[c] *FC* ≤ *30*[c] 0.2 < Res. Cl < 2.0	$BOD_5 \leq 20$[c] SS ≤ 30[c]		NSW Health (2005)
Australia – VIC (Guideline)	Greywater[b]	$BOD_5 \leq 10$ SS ≤ 10 *E. coli* ≤ *10*	$BOD_5 \leq 10$ SS ≤ 10 *E. coli* ≤ *10*	Prohibited	$BOD_5 \leq 20$ SS ≤ 30 *E. coli* ≤ *10*	$BOD_5 \leq 20$ SS ≤ 30 *E. coli* ≤ *10*	$BOD_5 \leq 20$ SS ≤ 30 *E. coli* ≤ *10*	EPA Victoria (2013)

(*Continued*)

Table 11.2 Comparison of international guidelines and regulations for reclaimed water quality based on the domestic end-uses (*Continued*).

		Parameters and threshold values[a] for water quality criteria depending on the end-use						
Country/ Organization	**Water origin**	**Toilet flushing**	**Cold water supply for clothes washing**	**Car washing**	**Surface irrigation**	**Sub-surface irrigation**	**Garden watering**	**References**
China (Guideline)	Municipal wastewater	$BOD_5 < 10$ TDS < 1500 Turbidity < 5 NH_4-N < 10 *FC* < 3 Detergents (anionic) < 1 Res. Cl > 1[d] Res. Cl > 0.2[e]		$BOD_5 < 10$ TDS < 1000 Turbidity < 5 NH_4-N < 10 *FC* < 3 Detergents (anionic) < 0.5 Res. Cl > 1[d] Res. Cl > 0.2[e]			$BOD_5 < 20$ TDS < 1000 Turbidity < 20 NH_4-N < 20 *FC* < 3 Detergents (anionic) < 1 Res. Cl > 1[d] Res. Cl > 0.2[e]	Ernst *et al.* (2007)
Japan (Regulation)	Wastewater	pH: 5.8–8.6 BOD: N/A Odour and appearance not unpleasant Turbidity not unpleasant *TC* ≤ *1000* Res. Cl: trace amount					pH: 5.8–8.6 BOD: N/A Odour and appearance not unpleasant Turbidity not unpleasant *TC* ≤ *ND* Res. Cl: trace amount	Ogoshi *et al.* (2001)

United Kingdom (Guideline)	Greywater[b]	pH: 5–9.5 Turbidity < 10 Res. Cl < 2 Res. Br < 5 *E. coli* < 25 *Int. enterococci* < 10	pH: 5–9.5 Turbidity < 10 Res. Cl < 2 Res. Br < 5 *E. coli. ND* *Int. enterococci: ND*	pH: 5–9.5 Turbidity < 10 Res. Cl < 2 Res. Br: 0.0 *E. coli: ND* *Int. enterococci: ND*	pH: 5–9.5 Turbidity < 10 Res. Cl < 2 Res. Br: 0.0 *E. coli: ND* *Int. enterococci: ND*	pH: 5–9.5 Turbidity: N/A Res. Cl < 0.5 Res. Br: 0.0 *E. coli* < 25 *Int. enterococci* < 10	BSI (2011)
Germany (Guideline)	Greywater[b]	$BOD_7 < 5$ O_2 sat. > 50% *TC* < 10^4 *FC* < 10^3 *P.aeruginosa* < 10^2	$BOD_7 < 5$ O_2 sat. > 50% *TC* < 10^4 *FC* < 10^3 *P.aeruginosa* < 10^2		Class 1 (unrestricted area) *F. streptococci: ND* *E. Coli: ND* *Salmonella: ND/1 L* *Intestinal nematodes Taenia: ND/1 L*	Class 1 (unrestricted area) *F. streptococci: ND* *E. Coli: ND* *Salmonella: ND/1 L* *Intestinal nematodes, Taenia: ND/1 L*	fbr (2005)
Spain (Regulation)	Municipal wastewater	TSS ≤ 10 Turbidity ≤ 2 *E. Coli: 0* *Intestinal nematodes* ≤ *1 egg/10 L*				TSS ≤ 10 Turbidity ≤ 2 *E. Coli: 0* *Intestinal nematodes* ≤ *1 egg/10 L*	Ministerio de la Presidencia (2007)

(*Continued*)

Table 11.2 Comparison of international guidelines and regulations for reclaimed water quality based on the domestic end-uses (*Continued*).

Country/ Organization	Water origin	Parameters and threshold values[a] for water quality criteria depending on the end-use						References
		Toilet flushing	Cold water supply for clothes washing	Car washing	Surface irrigation	Sub-surface irrigation	Garden watering	
							Legionella sp. ≤ *100CFU/L (if spray irrigation)*	
Canada (Guideline)	Domestic wastewater, greywater	$BOD_5 \leq 10$[f] $TSS \leq 10$[f] Turbidity ≤ 2[f] *E. Coli: ND*[f] *FC: ND*[f] 0.5 < Res. Cl < 2.0						Health Canada (2010)

(ND: not detected, FC: faecal or thermotolerant coliforms, *TC*: total coliforms, Res. Cl: residual chlorine, Res. Br: residual bromine, $BOD_{5,7}$: biochemical oxygen demand for 5 or 7 days, (T, D) SS: (total, dissolved) suspended solids, O_2 sat.: O_2 saturation, Int. Enterococci: intestinal enterococci)

[a]Units for physico-chemical parameters are in mg/l (except pH, and turbidity in NTU) and units for *microbiological parameters are in CFU/100 ml (unless otherwise stated)*

[b]Kitchen greywater is excluded

[c]90% of samples

[d]After 30 min

[e]At point of use

[f]Median value

Turbidity is typically limited to between 2 and 10 NTU, as excessive turbidity can interfere with disinfection and decrease its efficiency. Major risks associated with greywater reuse are related to the presence of disease-causing microorganisms. A well-designed and well-operated treatment system should be capable of greatly reducing the levels of pathogens, particularly where a disinfection unit is installed. Most of the international guidelines recommend monitoring of either *E. coli* or faecal coliforms to assess microbiological quality of reclaimed water in order to minimise sanitary risks. As an example, *E. coli* threshold values vary from 'not detected' to 25 CFU/100 ml for toilet flushing end use. It is interesting to note that the guidelines for faecal coliforms range between 3–1000 CFU/100 ml, with the highest value recommended being in Germany. Disinfection is an essential step in greywater treatment, and a chlorine residual of 0.5–2 mg/l is commonly stipulated for domestic recycled greywater to control bacterial regrowth in storage tanks and the recycled water distribution system.

Toilet flushing and garden watering/irrigation are the main permitted uses for which threshold values have been set for greywater recycling. It is important to emphasise that the regulations and guidelines distinguish between high and low exposure end-uses. Toilet flushing, use in washing machines, garden irrigation and car washing are considered to have a high potential for public exposure. This is due to the likelihood of close personal contact and possible inhalation of aerosols and consequently more stringent threshold values are applied. Regulatory processes including certificates of approval, monitoring and auditing, and technology testing protocols are also key management techniques that are used to guarantee safe greywater recycling.

11.5 COMPARISON OF INTERNATIONAL STANDARDS AND TESTING PROTOCOLS

As greywater recycling becomes more popular, codes of practice, standards and testing protocols have been defined to protect the public and to ensure that reliable non-potable water systems are designed, installed and maintained. Some of these are discussed in detail in this section.

11.5.1 British standards BS 8525

Two standards for greywater systems were recently introduced in the United Kingdom by the British Standards Institution (BSI, 2010; BSI, 2011). Part 1 of BS 8525 is a code of practice giving recommendations on the design, installation, alteration, testing and maintenance of greywater systems utilising bathroom greywater to supply non-potable water. It covers systems supplying greywater for domestic uses (in domestic, commercial, industrial or public premises) such as laundry, toilet and urinal flushing and garden watering. In these standards, bathroom greywater is defined as greywater from domestic baths, wash and hand basins, showers and clothes washing machines. The Code of Practice guidelines specify different approaches to the design of greywater systems based on the:

- Determination of demand and yield;
- Water quality guidelines for the intended uses;
- Peak capacity treatment rate.

Technical requirements stipulate that greywater collection pipework should be identified and dedicated to bathroom greywater and should minimise the generation of foam. Concerning storage tanks and cisterns, it is recommended that storage of untreated greywater should be avoided. A back-up water supply should be sized to allow it to meet the full demand requirements.

To prevent the non-potable greywater from entering the drinking water supply, the back-up water supply must be fitted with backflow prevention providing an air gap between the drinking water and reclaimed water. Incorporation of a monitoring unit is strongly recommended to ensure that users are aware of whether the system is operating effectively.

To differentiate the greywater system pipework from the potable water system pipework, a contrasting colour (green or black and green) is recommended and all pipework and fittings should be labelled. The installation and commissioning of greywater systems is an important step to ensure safe water reuse. Accordingly, BS 8525 requires that dye-testing of recycled greywater pipework connections should be carried out before final connections are made to the potable water system. The minimum maintenance requirement recommended is for an annual check of the system components.

Part 2 of BS 8525 specifies requirements and test methods for packaged and/or site-assembled domestic greywater treatment equipment. The test procedures (for a nominal treatment capacity of up to 10 m^3 per day) are carried out on greywater treatment equipment under controlled conditions using public mains water and synthetic greywater. The test methods for hydraulic functions aim to assess the following technical specifications: water tightness and overflow; acceptance flow rate and acceptance volume; controls and failsafe provisions; and treated greywater quality. The protocol to control the discharge of stored water includes tests on the automatic dump facility and on the failsafe conditions in case of interruption to power supply (for electrically powered equipment) and disinfection failure. Their major objective is to ensure that: i) the automatic dump facility discharges the stored treated greywater to drain once the maximum storage period has been exceeded; ii) the water pressure is detected and water is supplied by the back-up water supply detection device; and iii) an alarm(s) gives a visual and/or audible indicator of electrical or disinfection failure. The protocol to control the treated water quality aims to check if it complies with the guidelines based on the end-use. Testing is carried out with synthetic greywater made from de-chlorinated public mains water, shampoo and/or liquid soap, sunflower oil and an inoculant of bacteria from settled treated sewage effluent. All the ingredients are thoroughly mixed and kept at 30°C. Table 11.3 gives the composition of the synthetic greywater recommended for the test.

Table 11.3 Composition of synthetic greywater recommended in the British Standard for greywater.

Parameter	Acceptable range
E. coli (CFU/100 ml)	10^5–10^6
Chemical Oxygen Demand (mg/l)	180 ± 40
Biochemical Oxygen Demand (mg/l)	110 ± 40
NO_3 nitrogen (mg/l)	7.2 ± 0.8
pH	7.0–8.0
Temperature (°C)	30 ± 2.5

Source: BSI (2011).

Three samples are collected during the test. Sample A corresponds to raw greywater and is analysed for microbiological quality (*E. coli*, intestinal enterococci). Samples B and C are collected after treatment (and storage where applicable) from the same sampling point at the same time. Sample B is analysed directly after sampling while sample C is maintained at room temperature during the maximum storage period set for the greywater system before being analysed. Both samples are analysed for microbiological and physico-chemical (turbidity, pH, chlorine, bromine) quality. The results of the tests performed on samples B and C are compared with the results from sample A in order to calculate the difference in *E. coli* levels and to assess the water quality of the treated water. The results from sample A are also used to validate the test cycles and these have to be repeated 10 times to validate the whole test procedure. It is interesting to note that this standard suggests an assessment of only the initial (short-term) performance and that it does not recommend analysis of parameters such as BOD_5 that are good indicators of organic matter content.

11.5.2 New South Wales accreditation guidelines

Accreditation guidelines for domestic greywater treatment systems (DGTS) were introduced by the Department of Health of New South Wales (Australia) in 2005. They set the minimum requirements for accreditation of a manufactured DGTS that may be specifically designed to treat greywater from individual domestic premises for end-uses limited to surface and sub-surface irrigation, toilet flushing and laundry purposes (NSW Health, 2005). The guidelines indicate that an independent agency is to be engaged by the manufacturer to conduct experimental tests on the greywater system and to prepare an evaluation report for submission to the NSW Department of Health to obtain accreditation.

Unlike the British standard, which recommends testing with synthetic greywater, the NSW testing protocol suggests the tests should be performed in premises that are representative of a domestic greywater source, including all greywater source components such as bath, shower, hand basins, laundry and kitchen. Even though no specifications are given for the raw greywater quality, the selection of the test site must comply with several requirements. In particular, the average flows should range from 720 to 900 l/day in order to be representative of an 8 to 10 person rated DGTS (based on a minimum daily flow of 90 l per person per day). Another difference compared to the British standard, is the test period duration that must be 26 weeks from the date of commissioning. This long-term performance assessment allows greater feedback on operational conditions and thus on the process reliability to supply safe reclaimed water over time. Grab samples of influent and effluent should be collected every 12 and 6 days respectively, thus representing 15 and 30 samples during the whole monitoring period. The following prescribed parameters must be analysed during the tests: thermotolerant coliforms (FC), BOD_5, SS and free chlorine. As an example, where the reclaimed water is to be used in toilet flushing and washing machines, the treated water quality should comply with the following criteria:

- $BOD_5 \leq 10$ mg/l for 90% of the samples, with no sample greater than 20 mg/l;
- SS ≤ 10 mg/l for 90% of the samples, with no sample greater than 20 mg/l;
- FC < 10 CFU/100 ml for 90% of the samples, with no sample greater than 30 mg/l;
- Where chlorine is used as a disinfectant, the free residual chlorine concentration shall be ≥ 0.5 mg/l and < 2.0 mg/l in all samples.

The four possible outcomes once the tests are performed are presented in Table 11.4.

Table 11.4 Possible outcomes for DGTS testing protocol in NSW.

Outcome	Action
Pass	DGTS accredited for a five year term (subject to conditions)
Failure due to errors or mishaps in testing procedures or analysis	Extend test period
Failure due to component failure	Retest commencing from initial commissioning
Failure	Rejection – No accreditation

Source: NSW Health (2005).

Warranty and guaranteed service life is another interesting topic addressed in the NSW accreditation guidelines. A service life of at least 15 years is recommended for all metal fittings, fasteners and components of the DGTS (other than pumps and motors), while a minimum service life of 5 years is required for all mechanical and electrical parts. A minimum warranty period of 3 years from the date of delivery is also suggested by the NSW Department of Health. The NSW guidelines aim to guarantee that manufacturers provide long-term reliability of DGTS technologies in order to protect public health.

11.5.3 Commonwealth Scientific and Industrial Research Organisation greywater technology testing protocol

With much of Australia facing water shortages in recent years, there has been an increased focus in both major cities and regional areas on water saving measures. This focus sits alongside changes and improvements in managing water supplies, as well as new policies to reduce the quantity and improve the quality of wastewater discharges to receiving waters. Government agencies and regulatory authorities supported numerous strategies including the use of greywater treatment systems (VGDSE, 2004; Melbourne Water Resources Strategy, 2002). This led to an increase in the types of greywater recycling technologies available on the market, with a number of these systems also trying to minimise environmental impacts by using environmentally friendly agents rather than chemical treatments. However, because responsibility for appropriate regulations rests with the various state governments, a lack of consistency in greywater regulations and testing requirements for greywater treatment technologies made it difficult for manufacturers to get their products to market and impeded the uptake of these systems.

The CSIRO (Commonwealth Scientific and Industrial Research Organisation), Australia's national science organisation, developed a greywater technology testing protocol complete with a synthetic greywater formulation in response to the need for a consistent approach to the Australian situation (Diaper *et al.* 2008; Toifl *et al.* 2008). Although several testing protocols and synthetic greywater formulations have been developed internationally over the past decade for assessing the performance of greywater treatment systems (Brown & Palmer, 2002), greywater composition varies significantly between countries and regions. This is due to variability in household and personal care products, differences in water quality and water usage, and variability in the composition of greywater due to the inclusion or exclusion of various waste streams. For example, in Australia wastewater from the laundry is almost always included in greywater, whereas in Europe the laundry component is generally excluded (Jefferson, 2004). Therefore, whilst providing useful background information, research such as that conducted by Brown and Palmer (2002) and Jefferson (2004) could not simply be applied to Australian conditions.

The CSIRO testing protocol was developed using three small scale greywater treatment systems (Table 11.5) that used combinations of different chemical, physical and biological processes to achieve performance requirements. The treatment technologies were selected on the basis of these unit processes in order to ensure the protocol was appropriate for the different process types.

Table 11.5 Treatment technologies tested during protocol development.

Technology	Process type	Treatment process	Disinfection process
A	Semi batch	Biological with suspended media (SBR)	UV
B	Batch	Chemical flocculant dosing, UV and four stage filtration	UV
C	Semi batch	Settling, biological with fixed media	Chemical(Cl/Br)

The treatment systems were fed with a synthetic greywater developed to mimic an average combined laundry and bathroom greywater from an Australian domestic dwelling. The greywater components included a range of market share household products, some laboratory grade chemicals and secondary sewage effluent sourced from a local wastewater treatment plant (Eastern Treatment Plant, Melbourne, Australia). The parameter ranges for suspended solids, biological oxygen demand (BOD), temperature, pH, turbidity, sodium, zinc, total phosphorous, total Kjeldahl nitrogen (TKN), conductivity, chemical oxygen demand (COD), total organic carbon (TOC), total coliforms and *E. coli* were selected following a review of Australian and international literature and an analysis of data collected from Australian case studies. Whilst calcium and magnesium were analysed in the synthetic greywater, parameter ranges were not specified as these will vary depending on mains water quality. Aluminium was also measured but has no specific parameter range, as this will be highly dependent on the household products used.

The protocol involved 3 stages of testing: i) a tracer study; ii) chemical testing; and iii) microbiological testing. The tracer study provided a profile of the hydraulic flow conditions of the treatment technology and was used to develop flow and dosing regimes for chemical and microbiological testing. The tracer used in the development of the protocol was sodium chloride, as it can be monitored simply using electrical conductivity (EC). The concentrations used did not affect biological treatment, however other suitable chemicals or salts could also be used.

The parameters selected for chemical testing were based on a literature review of greywater components and an investigation of their likely detrimental impacts on soils, plant life and water bodies. Water quality parameters analysed in the feed and product streams were the same as those for the synthetic greywater with the addition of nitrate and *F. Enterococci* and the exception of temperature. Basic microbial analysis was carried out during chemical testing because secondary effluent was added to the synthetic greywater and the performance of the technology could be determined prior to dosing high concentrations of microorganisms.

The purpose of microbiological testing was to prove the log removal of bacterial, protozoan and viral surrogates. The microorganisms selected were in accordance with those suggested in National Water Recycling Guidelines (Environment Protection and Heritage Council *et al.* 2006). The technologies tested during the development of the protocol were challenged with repeated high feed concentrations of the different microorganism surrogates, with the number of repetitions and product sample analysis depending on the technology and the results of the tracer study. Collection of proportional volume feed and product samples, rather than grab samples, was recommended. The three stages of testing in the protocol were designed to provide:

- hydraulic integrity testing of the technology;
- a check of performance in removal of greywater components that are harmful to the environment;
- proof of performance in the removal of a range of surrogate microorganisms;
- some assessment of any operational issues.

As such, the protocol is robust, repeatable and uses standard methods. Therefore, it is suitable for testing treatment systems with high exposure risk end uses, such as domestic dual reticulation, multi-unit dwellings and unrestricted access urban irrigation, as outlined in the National Water Recycling Guidelines (Environment Protection and Heritage Council *et al.* 2006). Since its development the protocol and synthetic greywater formulation have been widely cited and used in the development of several international standards and guidelines for greywater recycling.

11.6 CONCLUSION

Greywater recycling provides an opportunity to reduce the demand on potable water supplies for domestic uses such as toilet flushing, garden irrigation and other non-potable applications. As outlined in this chapter, there are currently many different methods available that can successfully treat varying qualities of greywater. As a consequence, there are a wide variety of commercial greywater treatment systems available for purchase ‘off the shelf’. As the interest in greywater recycling has continued to increase, so has the demand for improved guidelines and regulations for this valuable alternative water source. In recent years, this has led to many countries developing regulations or guidelines to meet their own requirements (Sections 11.4 and 11.5). Ongoing research into greywater characteristics is continually improving the understanding and awareness of the impacts on both human health and the environment. This plays an important role in improving the treatment technologies available, as well as in updating regulations, guidelines and standards to reflect advances in the current state of knowledge.

There are several key aspects that must be considered to allow successful small scale greywater recycling. It is essential to employ treatment systems that are appropriate for each individual situation. In some scenarios a chemical system may be a better option than a biological system, whereas in other situations a low maintenance biological system could be the best choice. Furthermore, a thorough evaluation of the requirements for each installation is necessary, both to ensure the correct size and type of system is selected and to guarantee successful ongoing greywater recycling. This includes an assessment of water usage patterns, greywater volumes generated and the intended end use of the water, as well as an assessment of the ability and commitment of those responsible for the day to day maintenance of the system. In addition, as part of best practice performance monitoring (Toifl *et al.* 2011), a Hazard Analysis Critical Control Point (HACCP) risk management methodology (refer to Chapter 10 for more information) or similar should be completed by the manufacturer to identify hazards in different stages of the treatment system and to develop appropriate risk management strategies. Finally, the importance of any regular maintenance and testing schedule provided by the technology manufacturer should not be underestimated and should always be put into practice.

ACKNOWLEDGEMENT

Permission to reproduce extracts from BS 8525-1:2010 and BS 8525-2:2011 was granted by BSI. British Standards can be obtained in PDF or hard copy formats from the BSI online shop: www.bsigroup.com/Shop or by contacting BSI Customer Services for hardcopies only: Tel: +44 (0)20 8996 9001, Email: cservices@bsigroup.com.

REFERENCES

Abu-Ashour J. and Jamrah A. (2008). Survival of bacteria in soil subsequent to greywater application. *International Journal of Environmental Studies*, **65**(1), 51–56.

ACT Health (2007). Greywater use – Guidelines for Residential Properties in Canberra. Publication No. 07/1380, Second edition, ACT, Australia, http://www.health.act.gov.au/publications/fact-sheets/greywater-use (accessed 9 July 2014).

Albrechtsen H. (2002). Microbiological investigations of rainwater and graywater collected for toilet flushing. *Water Science and Technology*, **46**(6–7), 311–316.

Australian Bureau of Statistics (2011). Household Water and Energy use, Victoria. No. 4602.2, October 2011. ABS, Australia.

Birks R. and Hills S. (2007). Characterisation of indicator organisms and pathogens in domestic greywater for recycling. *Environmental Monitoring and Assessment*, **129**(1–3), 61–69.

Boyjoo Y., Pareek V. K. and Ang M. (2013). A review of greywater characteristics and treatment processes. *Water Science and Technology*, **67**(7), 1403–1424.

Brown R. and Palmer A. (2002). Water Reclamation Standard: Laboratory Testing of Systems Using Greywater. BSRIA, UK.

BSI (2010). Greywater Systems – Part 1: Code of Practice. British Standard Institute Standards Publication BS 8525–1:2010, ISBN 978 0 580 63475 8.

BSI (2011). Greywater Systems – Part 2: Domestic Greywater Treatment Equipment – Requirements and Test Methods. British Standard Institute Standards Publication BS 8525-2:2011, ISBN 978 0 580 63476 5.

Brandes M. (1978). Characteristics of effluents from gray and black water septic tanks. *Journal of the Water Pollution Control Federation*, **50**(11), 2547–2559.

Casanova L. M., Little V., Frye R. J. and Gerba C. P. (2001). A survey of the microbial quality of recycled household greywater. *JAWRA Journal of the American Water Resources Association*, **37**(5), 1313–1319.

Christova-Boal D., Eden R. E. and McFarlane S. (1996). An investigation into greywater reuse for urban residential properties. *Desalination*, **106**(1–3), 391–397.

Ciabattia I., Cesaro F., Faralli L., Fatarella E. and Tognotti F. (2009). Demonstration of a treatment system for purification and reuse of laundry wastewater. *Desalination*, **245**(1–3), 451–459.

Diaper C., Toifl M, and Storey M. (2008). Greywater Technology Testing Protocol. Water for a Healthy Country National Research Flagship, CSIRO.

Donner E., Eriksson E., Revitt D. M., Scholes L., Lützhøft H.-H. and Ledin A. (2010). Presence and fate of priority substances in domestic greywater treatment and reuse systems. *Science of the Total Environment*, **408**(12), 2444–2451.

Elmitwalli T. A. and Otterpohl R. (2007). Anaerobic biodegradability and treatment of grey water in upflow anaerobic sludge blanket (UASB) reactor. *Water Research*, **41**, 1379–1387.

Elmitwalli T. A. and Otterpohl R. (2011). Grey water treatment in upflow anaerobic sludge blanket (UASB) reactor at different temperatures. *Water Science and Technology*, **64**(3), 610–617.

Environment Protection and Heritage Council and Natural Resource Management Ministerial Council. (2006). Australian Guidelines for Water Recycling: Managing Health and Environmental Risks (Phase 1). National Water Quality Management Strategy, Australia.

EPA Victoria (2013). Code of Practice – Onsite Wastewater Management. Publication number 891.3, EPA Victoria, Melbourne, Australia, http://www.epa.vic.gov.au/our-work/publications/publication/2013/february/891-3 (accessed 9 July 2014).

Eriksson E., Auffarth K., Henze M. and Ledin A. (2002). Characteristics of grey wastewater. *Urban Water*, **4**(1), 85–104.

Eriksson E., Donner E. and Ledin A. (2010). Presence of selected priority and personal care substances in an onsite bathroom greywater treatment facility. *Water Science and Technology*, **62**(12), 2889–2898.

Ernst M., Sperlich A., Zheng X., Gan Y., Hu J., Zhao X., Wang J. and Jekel M. (2007). An integrated wastewater treatment and reuse concept for the Olympic Park 2008, Beijing. *Desalination*, **202**, 293–301.

fbr (2005). Greywater Recycling – Planning Fundamentals and Operation Information. *Fachvereinigung Betriebs- und Regenwassernutzung e.V.* (German Assoc for Rainwater Harvesting and Water Utilisation), fbr Info Sheet H 201.

Fewtrell L., and Bartram J. (Eds) (2001). Water Quality: Guidelines, Standards and Health: Assessment of Risk and Risk Management for Water-Related Infectious Disease on Behalf of WHO. IWA Publishing, London.

Friedler E. (2004). Quality of individual domestic greywater streams and its implication for on-site treatment and reuse possibilities. *Environmental Technology*, **25**(9), 997–1008.

Friedler E., Kovalio R., Galil N. I., Matthew K. and Ho G. (2005). On-site greywater treatment and reuse in multi-storey buildings. *Water Science and Technology*, **51**(10), 187–194.

Friedler E. and Alfiya Y. (2010). Physicochemical treatment of office and public buildings greywater. *Water Science and Technology*, **62** (10), 2357–2363.

Gerba C. P., Straub T. M., Rose J. B., Karpiscak M. M., Foster K. E. and Brittain R. G. (1995). Quality study of graywater treatment systems. *JAWRA Journal of the American Water Resources Association*, **31**(1), 109–116.

Ghaitidak D. M. and Yadav K. D. (2013). Characteristics and treatment of greywater – a review. *Environmental Science and Pollution Research*, **20**, 2795–2809.

Gross A., Shmueli O., Ronen Z. and Raveh E. (2007). Recycled vertical flow constructed wetland (RVFCW) – a novel method of recycling greywater for irrigation in small communities and households. *Chemosphere*, **66**, 916–923.

Gulyas H., Choromanski P., Muelling N. and Furmanska M. (2009). Towards chemical-free reclamation of biologically pretreated greywater: solar photocatalytic oxidation with powdered activated carbon. *Journal of Cleaner Production*, **17**, 1223–1227.

Gulyas H., Reich M. and Otterpohl R. (2011). Organic micropollutants in raw and treated greywater: A preliminary investigation. *Urban Water Journal*, **8**(1), 29–39.

Harrow D. I. and Baker K. H. (2010). Impact of triclosan in Greywater on Antibiotic-Resistance in Soil Microbial Communities. Proceedings of the World Environmental and Water Resources Congress 2010: Challenges of Change, Providence, RI, pp. 2939–2949.

Health Canada (2010). Canadian Guidelines for Domestic Reclaimed Water for Use in Toilet and Urinal Flushing. Document H128-1/10-602E, ISBN 978 1 100 15665 1.

Hourlier F., Masse A., Jaouen P., Lakel A., Gerente C., Faur C. and Le Cloirec P. (2010). Membrane process treatment for greywater recycling: investigations on direct tubular nanofiltration. *Water Science and Technology*, **62**(7), 1544–1550.

Hsu C., Hsieh H., Yang L., Wu S., Chang J., Hsiao S., Su H., Yeh C., Ho Y. and Lin H. (2011). Biodiversity of constructed wetlands for wastewater treatment. *Ecological Engineering*, **37**, 1533–1545.

Jamrah A., Al-Futaisi A., Prathapar S. and Harrasi A. A. (2008). Evaluating greywater reuse potential for sustainable water resources management in Oman. *Environmental Monitoring and Assessment*, **137**(1–3), 315–327.

Jefferson B., Palmer A., Jeffrey P., Stuetz R. and Judd S. (2004). Grey water characterisation and its impact on the selection and operation of technologies for urban reuse. *Water Science and Technology*, **50**(2), 157–164.

Kariuki F., Kotut K. and Nganga V. (2011). The potential of a low cost technology for the greywater treatment. *The open Environmental Engineering Journal*, **4**, 32–39.

Kim J., Song I., Oh H., Jong J., Park J. and Choung Y. (2009). A laboratory-scale graywater treatment system based on a membrane filtration and oxidation process – characteristics of graywater from a residential complex. *Desalination*, **238**, 347–357.

Lamine M., Bousselmi L. and Ghrabi A. (2007). Biological treatment of greywater using sequencing batch reactor. *Desalination*, **215**, 127–132.

Liehr S. and Kruzic A. (2007). Natural treatment and onsite processes. *Water Environment Research*, **79**(10), 1451–1473.

Maimon A., Tal A., Friedler E. and Gross A. (2010). Safe on-site reuse of greywater for irrigation - A critical review of current guidelines. *Environmental Science and Technology*, **44**(9), 3213–3220.

Marleni N., Gray S., Sharma A., Burn S. and Muttil N. (2012). Impact of water source management practices in residential areas on sewer networks – A review. *Water Science and Technology*, **65**(4), 624–642.

Melbourne Water Resources Strategy (2002). Final Report: Stage 3 in Developing a Water Resources Strategy for the Greater Melbourne Area. *Water Resources Strategy Committee*. Melbourne Water, Victoria, Australia.

Merz C., Scheumann R., El Hamouri B. and Kraume M. (2007). Membrane bioreactor technology for the treatment of greywater from a sports and leisure club. *Desalination*, **215**, 37–43.

Ministerio de la Presidencia (2007). *Real Decreto 1620/2007, de 7 de diciembre, por el que se establece el régimen jurídico de la reutilización de las aguas depuradas.* (Royal Decree 1620/2007 of 7th December establishing the legal framework for treated wastewater reuse) [in Spanish].

Mohamed R. M. S. R., Kassim A. H. M., Anda M. and Dallas S. (2013). A monitoring of environmental effects from household greywater reuse for garden irrigation. *Environmental Monitoring and Assessment*, **185**(10), 8473–8488.

NSW Health (2005). Domestic Greywater Treatment Systems – Accreditation Guidelines.

Ogoshi M., Suzuki Y. and Asano T. (2001). Water reuse in Japan. *Water Science and Technology*, **43**(10), 17–23.

O'Toole J., Sinclair M., Malawaraarachchi M., Hamilton A., Barker S. F. and Leder K. (2012). Microbial quality assessment of household greywater. *Water Research*, **46**(13), 4301–4313.

Ottoson J. and Stenström T. A. (2003). Faecal contamination of greywater and associated microbial risks. *Water Research*, **37**(3), 645–655.

Victorian Government Department of Sustainability and the Environment (2004). Our Water Our Future: Securing Our Water Future Together. VGDSE, Australia.

Penn R., Schütze M. and Friedler E. (2013). Modelling the effects of on-site greywater reuse and low flush toilets on municipal sewer systems. *Journal of Environmental Management*, **114**, 72–83.

Ramon G., Green M., Semiat R. and Dosoretz C. (2004). Low strength graywater characterization and treatment by direct membrane filtration. *Desalination*, **170**, 241–250.

Ray R., Henshaw P. and Biswas N. (2012). Effects of reduced aeration in a biological aerated filter. *Canadian Journal of Civil Engineering*, **39**, 432–438.

Revitt D. M., Eriksson E. and Donner E. (2011). The implications of household greywater treatment and reuse for municipal wastewater flows and micropollutant loads. *Water Research*, **45**(4), 1549–1560.

Rose J. B., Sun G., Gerba C. P. and Sinclair N. A. (1991). Microbial quality and persistence of enteric pathogens in graywater from various household sources. *Water Research*, **25**(1), 37–42.

Siegrist R. L. (1978). Management of Residential Greywater. University of Wisconsin, Madison. www.soils.wisc.edu/sswmp/pubs/2.23.pdf (accessed 03 June 2013).

Sostar-Turk S., Petrinic I. and Simonic M. (2005). Laundry wastewater treatment using coagulation and membrane filtration. *Resources, Conservations and Recycling*, **44**(2), 185–196.

Sundaravadivel M. and Vigneswaran S. (2001). Constructed wetlands for wastewater treatment. *Critical Reviews in Environmental Science and Technology*, **31**(4), 351.

Toifl M., Diaper C. and O'Halloran R. (2008). Assessing the performance of small scale greywater treatment systems under controlled laboratory conditions. *Water Practice and Technology*, **3**(3), 1–7.

Toifl M., O'Halloran R. and Diaper C. (2011). Best Practice Performance Monitoring for Small Scale Wastewater Treatment Systems. CSIRO: Water for a Healthy Country National Research Flagship, Australia.

US-EPA (2004). Guidelines for Water Reuse. Report EPA/625/R-04/108, US EPA, Washington DC, http://nepis.epa.gov/Adobe/PDF/30006MKD.pdf (accessed 9 July 2014).

WHO (2006). Overview of Greywater Management – Health Considerations. Report WHO-EM/CEH/125/E, Regional Office for the Eastern Mediterranean Centre for Environmental Health Activities, Amman, Jordan, http://applications.emro.who.int/dsaf/dsa1203.pdf (accessed 9 July 2014)

Winward G. P., Avery L. M., Stephenson T., Jeffrey P., Le Corre K. S., Fewtrell L. and Jefferson B. (2009). Pathogens in urban wastewaters suitable for reuse. *Urban Water Journal*, **6**(4), 291–301.

Chapter 12

Membrane processes for greywater recycling

Marc Pidou

12.1 INTRODUCTION

The benefits of water recycling, particularly in regions affected by water scarcity due to an increase in population or simply due to arid conditions, are now widely accepted. As opposed to rainwater harvesting, which is weather dependant, wastewater treatment for reuse offers an alternative source of water constantly available and directly proportional in volume to the population. Of the different wastewater sources available for reuse, greywater has attracted great attention in the past decades. Indeed, total greywater, defined as all domestic wastewaters (bathroom, kitchen and laundry) excluding that used for toilet flushing, has been shown to account for up to 70% of the total domestic wastewater flow with only 30% of the organic load (Kujawa-Roeleveld & Zeeman, 2006) and consequently represents a very attractive source for recycling. By definition, the potential for greywater recycling is focused on residential areas with possible recycling schemes at scales ranging from individual household, to multi-storey buildings or blocks of buildings in urban environments, to isolated communities without centralised wastewater treatment. However, the potential for greywater recycling is also present in commercial settings such as hotels, office buildings, sports facilities and cruise ships. A wide variety of treatment systems including physical (sedimentation, filtration), chemical (disinfection, coagulation, photo-catalysis) and biological (biological aerated filter, rotating biological contactor, sequencing batch reactor, membrane bioreactor) technologies, individually or in combination, have been investigated for greywater recycling (Pidou *et al.* 2007).

This chapter discusses some of the membrane-based technologies for greywater treatment. Membrane processes have the advantage of consistently producing high water quality, as they are a physical barrier to a wide range of pollutants including microorganisms, but require a small footprint for their implementation. For these reasons, membrane processes have significant potential to be used for greywater recycling applications.

12.2 GREYWATER QUALITY AND REUSE STANDARDS

As discussed in the previous chapter, greywater varies greatly in terms of quantity and quality, because its production is as directly affected by householder/user behaviour as it is by their geographical, social and economic situation. However, the shower, bath and wash basin components of greywater flows generally contain less organics, solids and nutrients than kitchen and laundry wastewaters (Table 12.1). To illustrate, average biochemical oxygen demands (BOD) concentrations ranging between 100 and 129 mg/l have been reported in the literature for bathroom greywaters compared to between 286 and 499 mg/l for laundry and kitchen effluents, respectively (Table 12.1). Similarly, higher levels of suspended solids (SS), turbidity and phosphate have been measured in kitchen and laundry wastewaters as compared to bathroom sources (Table 12.1). For practical reasons, it has often been preferred to exclude the more polluted sources from treatment and to treat only the bathroom-sourced greywater, also referred to as 'light' or 'low load' greywater. It should also be emphasised that although toilet wastewater is excluded from greywater, significant levels of microbial indicators have been measured in all types of greywater source. Indeed, total and faecal coliforms counts of 10^2 to 10^8 and 10^1 to 10^7 cfu/100 ml, respectively, have been reported in various greywater components.

The choice of technology for greywater recycling will primarily be driven by the water quality to be achieved for the reuse application. The most common applications reported for recycled greywater are toilet flushing and irrigation (gardens, parks, sports fields), but other applications including clothes or car washing, fire safety, street cleaning or air conditioning have also been considered. Since no international water quality standards for reuse exist, countries have individually set their own guidelines or standards (Chapter 11, Table 11.2). As the primary aim of the standards is to limit health risks to humans, different standards can generally be found depending on the application and the proximity of the users to the reused greywater. However, water reuse standards do not only focus on the microbial contamination of the water, but also on its organic and solids content. A comparison of the typical characteristics of greywater (Table 12.1) and the standards for reuse (Table 11.2) clearly demonstrates the need for treatment before greywater can be reused for any application. A number of treatment technologies can be employed. However, the selection of a particular technology is a function of several factors including influent quality, capital and operational cost, available space and its ability to cope with variations in the influent quantity and quality.

Table 12.1 Greywater characteristics for individual appliances and combined sources.

Parameter	Shower	Bath	Wash basin	Laundry	Kitchen	Combined
BOD_5 (mg/l)	100 (51–212)	129 (59–216)	109 (33–252)	286 (180–472)	499 (453–657)	121 (5–902)
COD (mg/l)	170 (109–501)	210	230 (263–298)	–	–	240 (23–1583)
SS (mg/l)	125 (29–353)	59 (43–304)	167 (36–505)	259 (68–465)	235 (209–245)	54 (11–207)
Turbidity (NTU)	29 (21–375)	60 (33–92)	133 (102–164)	232 (108–144)	–	42 (12–240)
TKN (mg/l)	15.2	8.7 (7–11)	6.8 (4–9.6)	29 (18–40)	–	4.6 (0–27)
NH_4-N (mg/l)	1.2 (0.4–12)	1.3 (1.0–3.6)	–	2.0 (0.7–11)	3.2 (0.3–5.3)	4.0 (0.1–17)
PO_4-P (mg/l)	1.0 (0.3–19)	1.3 (0.4–10)	23 (0.4–49)	61 (13–171)	14 (10–26)	2.3 (0.4–10)
Total coliforms (TC) (cfu/100 ml)	10^2–10^4	10^2–10^6	10^3–10^6	10^2–10^8	–	10^2–10^8
Faecal coliforms (FC) (cfu/100 ml)	10^2–10^6	ND[a]–10^5	10^1–10^2	10^1–10^3	–	10^2–10^7
pH	7.2–7.6	7.5–7.6	7.1–8.1	8.1–10	–	6.5–8.6

[a]ND: not detectable.
Source: Jefferson, (2013).

12.3 TREATMENT PERFORMANCE

Membrane processes applied to greywater recycling can be separated into two categories based on their configuration: direct filtration units and hybrid systems. The following section presents a review of the different configurations and their respective treatment performances for greywater recycling.

12.3.1 Direct filtration

Direct filtration refers to the application of membranes as standalone treatment systems. All membrane types, including low pressure membranes (i.e., microfiltration (MF) and ultrafiltration (UF)) and high pressure membranes (i.e., nanofiltration (NF) and reverse osmosis (RO)), and materials (i.e., organic (polymeric) and inorganic (ceramic, metal)) have been studied and applied to greywater recycling processes (Table 12.2). However, not all membrane systems reported in the literature provided high quality effluent, as may be expected from this technology. Indeed, as shown in a study by Kim *et al.* (2007), only a limited fraction of the organics from a very low strength greywater was removed by metal microfiltration membranes. To illustrate, the metal membranes achieved only 45, 45 and 70% COD removal from an initial 22.9 mg/l feed concentration with pore sizes of 5, 1 and 0.5 μm, respectively. Similarly, the turbidity was only reduced from 12.6 NTU to, respectively, 5.9, 4.8 and 3.2 NTU for the same membranes. Although metal membranes provide an interesting alternative due to their robustness and longevity, the low performance reported hinders their full scale implementation. Overall, Kim *et al.* revealed the limited potential of membranes with larger pore sizes for this application, but also demonstrated improved performance for the tighter membranes with smaller pore sizes. These results are supported by the findings of Ahn *et al.* (1998), Nghiem *et al.* (2006), Li *et al.* (2009) and Bhattacharya *et al.* (2013). All four studies report significant organics and turbidity removal with ultrafiltration membranes (Table 12.2). For example, effluent turbidity values below 1 NTU were consistently measured in these studies. It is worth noting that no differences in performance were observed between the use of ceramic (Ahn *et al.* 1998; Bhattacharya *et al.* 2013) and polymeric (Li *et al.* 2009; Nghiem *et al.* 2006) membranes. The direct impact of the raw greywater composition and in particular of the suspended and soluble fractions on the treatment performance of membrane filtration was further demonstrated by Ramon *et al.* (2004) in their work investigating ultrafiltration and nanofiltration membranes for the treatment of shower water from a sports centre. Indeed, with a particle size distribution mostly between 0.04 and 0.10 μm, only a limited fraction of the organics contained in the shower water was removed. COD removal of 53, 56 and 70% were measured for ultrafiltration membranes with molecular weight cut-offs (MWCO) of 400, 200 and 30 kiloDalton (kDa), respectively. In contrast, a 200 Da nanofiltration membrane achieved a 93% COD removal and complete removal of the suspended

Table 12.2 Direct filtration units for greywater recycling.

Membrane				Operational conditions				Influent source	Treatment performance				Reference
Type	MWCO[a] (kDa)	Material	Surface area (m^2)	CFV[b] (m/s)	Pressure (bar)	Permeability ($l/(m^2 \cdot h \cdot kPa)$)	Test duration (h)		COD In Out (mg/l)	Turbidity In Out (NTU)	SS In Out (mg/l)	FC In Out cfu/100 ml	
MF	5[c]	Stainless steel	0.0098	–	–	219 114 14	0 0.25 0.5	Water from floor cleaning	**22.9** 12.5	**12.6** 5.9	–	–	Kim *et al.* (2007)
MF	1[c]	Stainless steel	0.0098	–	–	195 38 14	0 0.25 0.5	Water from floor cleaning	**22.9** 12.6	**12.6** 4.8	–	–	Kim *et al.* (2007)
MF	0.5[c]	Stainless steel	0.0098	–	–	186 24 14	0 0.25 0.5	Water from floor cleaning	**22.9** 6.8	**12.6** 3.2	–	–	Kim *et al.* (2007)
MF	1[c]	Ceramic γ-alumina with clay-alumina support	0.0026	–	2	324	6.7	Dish-washing water from canteen in office building	**920–2960** 403–525	**89–115** 0.34–0.51	**84–345** 16–29	**2.2×10^5–9.1×10^5** 2.8×10^4–3.5×10^4	Bhatta-charya *et al.* (2013)
UF	0.02[c]	Ceramic γ-alumina with clay-alumina support	0.0026	–	2	202	6.7	Dish-washing water from canteen in office building	**920–2960** 250–291	**89–115** 0.10–0.23	**84–345** ND[d]	**2.2×10^5–9.1×10^5** ND[d]	Bhatta-charya *et al.* (2013)
UF spiral wound	0.0062[c]	–	8.2	–	0.12	–	2160	Combined greywater pre-treated in septic tank	**161**[e] 28.6	**140** 0.5	–	–	Li *et al.* (2009)

(Continued)

Table 12.2 Direct filtration units for greywater recycling (*Continued*).

Membrane				Operational conditions				Influent source	Treatment performance				Reference
Type	MWCO[a] (kDa)	Material	Surface area (m^2)	CFV[b] (m/s)	Pressure (bar)	Permeability ($l/(m^2 \cdot h \cdot kPa)$)	Test duration (h)		COD In Out (mg/l)	Turbidity In Out (NTU)	SS In Out (mg/l)	FC In Out cfu/100 ml	
UF tubular 7 channels 4.5 mm diameter	0.1[c]	Ceramic TiO_2 and ZrO_2 with Al_2O_3-TiO_2 support layer	0.08	4	1.20	–	12	Combined greywater from hotel	**75.4** 8.5	**8.2** 0	–	–	Ahn *et al.* (1998)
	300		0.08	4	1.33	–			**42.2** 5.6	**6.9** 0	–	–	
	15		0.08	1	1.67	–			**48.8** 15.6	**6.2** 0.2	–	–	
				2.5	1.53	–			**83.0** 14.3	**18.1** 0.7	–	–	
				4	1.47	–			**74.0** 7.7	**15.1** 0	–	–	
				4	2.27	–			**41.0** 9.7	**8.9** 0	–	–	
				4	3.66	–			**85.9** 7.2	**11.5** 0	–	–	
UF hollow fibre	0.04[c]	PVDF[f]		–	64[g]	–	6	Synthetic greywater	–	**140** 0.4	–	–	Nghiem *et al.* (2006)

UF flat sheet dead end	400	PAN[h]	0.047	–	1–2	–	–	Shower water from sports centre	**170** 80	**23** 1.4	–	–	Ramon *et al.* (2004)
	200	PAN[h]		–		–	–		**170** 74.3	**23** 1	–	–	
	30	PES[i]		–		–	–		**170** 50.6	**23** 0.8	–	–	
NF tubular	0.2	PA[j]	0.014		6–10	–			**226** 15	**29.5** 0.6	**27.6** 0	–	
NF tubular	0.2	PA[j]	0.033	2.5	20	–	3	Synthetic greywater	**454** 160	**24.1** <1	–	**9.6 × 10³** ND[d]	Hourlier *et al.* (2010)
	0.2				35	–			**454** 261	**24.1** <1	**71.8** 1.3	–	
	0.3				20	–			**454** 138	**24.1** <1	**71.8** ND[d]	**9.6 × 10³** ND[d]	
	0.3				35	–			**454** 151	**24.1** 2.5	**71.8** ND[d]	**9.6 × 10³** ND[d]	
	<0.2				20	–			**454** 170	–	–	**9.6 × 10³** ND[d]	
	<0.2				35	–			**454** 26	**24.1** <1	–	**9.6 × 10³** ND[d]	
	<0.2				35	–		MBGW[k]	**252** <25	**62** <1	**42.2** ND[d]	**1.5 × 10⁵** ND[d]	
NF tubular	0.2	PA[j]	0.033	2.5	35	–	2	Washing machine waste water	**1340** 134	**120** ND[d]	**78** ND[d]	**2.0 × 10³** ND[d]	Guilbaud *et al.* (2010)
	0.3				35	–			**1340** 116	**120** ND[d]	**78** ND[d]	**2.0 × 10³** ND[d]	
	<0.2				35	–			**1340** <25	**120** ND[d]	**78** ND[d]	**2.0 × 10³** ND[d]	

(*Continued*)

Table 12.2 Direct filtration units for greywater recycling (*Continued*).

Membrane				Operational conditions				Influent source	Treatment performance				Reference
Type	**MWCO[a] (kDa)**	**Material**	**Surface area (m^2)**	**CFV[b] (m/s)**	**Pressure (bar)**	**Permeability (l/(m^2·h·kPa))**	**Test duration (h)**		**COD In Out (mg/l)**	**Turbidity In Out (NTU)**	**SS In Out (mg/l)**	**FC In Out cfu/100 ml**	
UF tubular	0.05[c]	Ceramic Al_2O_3, TiO_2, ZrO_2	0.13	–	3	–	3	Laundry waste water	**280** 130	–	**35** 18	–	Sostar-Turk *et al.* (2005)
RO spiral wound	–	PES[i]	1.5	–	20–30	–	2	Laundry waste water	**130** 3	–	**18** 8	–	Sostar-Turk *et al.* (2005)
UF hollow fibre + NF tubular + RO spiral wound	100 + 1 + –	–	15 + 8.9 + 280	–	0.84 + 6.3 + 7.5	–	3 years	–	2.3–5.5	ND[d]	–	ND[d]	Troquet *et al.* (2009)

[a]MWCO: molecular weight cut-off; [b]CFV: cross flow velocity; [c]pore size in μm; [d]ND: not detectable; [e]as TOC; [f]PVDF: polyvynilidene fluoride; [g]flux in L · m^{-2} · h^{-1}; [h]PAN: polyacrylonitrile; [i]PES: polyethersulfone; [j]PA: polyamide; [k]MBGW – Mixed bathrooms greywater.

solids for the same raw water. Similarly, other studies by Hourlier *et al.* (2010) and Guilbaud *et al.* (2010) have confirmed the high performance achieved by high pressure membranes with effluents containing very low turbidity and undetectable levels of suspended solids and faecal coliforms (FC) that meet some of the most stringent standards for greywater reuse (Table 11.2, Chapter 11). Expectedly, such high performances were also reported for reverse osmosis membranes (Table 12.2).

As treatment with direct membrane filtration relies essentially on the physical separation of the pollutants, as shown above, selection of the membrane to be used will depend on the composition of the greywater to be treated and more specifically, the fractions of pollutants in suspended and dissolved forms. Importantly, any solids and colloids present in the greywater will not only have an impact on treatment, but also on operation with a direct influence on membrane fouling. Ultrafiltration membranes have demonstrated good performance and can meet some of the standards for reuse at least for organics and solids. As only limited information is available for their potential to reject the microbial content of greywater and based on experience with these membranes for other applications, some breakthrough of microorganisms in the effluent is possible. Consequently, it should be anticipated that these membranes will have to be combined with a disinfection stage. The dense high pressure membranes (nanofiltration and reverse osmosis) have been shown to produce very high quality effluents. However, as for all water recycling schemes, the effluent quality produced should be considered through a fit for purpose approach. Consequently, considering the water quality produced by these dense membranes, they should at least be considered for unrestricted applications and explored further as a potential option for (indirect or direct) potable applications.

12.3.2 Hybrid membrane systems

Hybrid membrane systems refer to units combining a treatment stage, chemical or biological, and separation by membrane filtration. Two main configurations of hybrid systems have been studied and implemented for greywater recycling to date, those being photo-catalytic membrane reactor (PMR) and membrane bioreactor (MBR), with the latter being the most commonly applied membrane technology for greywater recycling.

Only a few studies have looked at the application of PMRs to greywater recycling (Pidou, 2007; Pidou *et al.* 2009; Rivero *et al.* 2006). PMR systems involve photo-catalytic treatment via titanium dioxide and UV. Powdered titanium dioxide is maintained in suspension in a reactor to ensure contact between the catalyst and the water to be treated. Photo-catalytic treatment is then triggered by the release of photons by the UV lights inserted in the reactor. The catalyst is then separated from the treated effluent by the membrane. The development of the technology for greywater recycling is at an early stage but the treatment performance reported is promising. Indeed, PMRs have achieved effluent concentrations below 10 mg/l for BOD_5, 2 NTU for turbidity, 1 mg/l for suspended solids and non-detectable levels of total and faecal coliforms when used to treat combined and shower only greywaters (Pidou, 2007; Pidou *et al.* 2009; Rivero *et al.* 2006). These systems use micro- or ultra-filtration membranes and when compared to the performance described above for similar membranes operated as direct filtration, the photo-catalytic treatment stage significantly improves the treatment of the greywater producing effluent that meets some of the most stringent standards for reuse (Table 11.2). The presence of the UV lights in the systems is also an advantage, since it enables a complete removal of the micro-organisms and therefore no additional disinfection stage would be required. However, since UV does not offer long lasting germicidal residual effect, microbial regrowth remains a possibility if storage of the treated greywater is required before use.

Alternatively, MBRs combine a biological treatment stage including activated sludge and separation of the biomass from the treated effluent with membranes. Again, for this application, both MF and UF membranes are being used. Although MBRs are implemented at full scale for many greywater recycling applications, most of the literature available is for small scale studies (Table 12.3). Interestingly, only one type of MBR configuration with the membrane immersed in the biological tank, in comparison to side-stream systems, has been studied in any detail. In all cases, either hollow fibre (HF) or flat sheet (FS) membranes were used. Immersed systems are more commonly implemented for municipal wastewater treatment applications particularly because they are less energy intensive than their side-stream counterparts. A review of the available literature also revealed that the biological reactors were operated mostly in two ways, continuously and as sequencing batch reactors (SBR). Overall, the evaluation of the treatment performance from all the studies reported in the literature clearly showed that MBRs achieved high treatment performance irrespective of the greywater source or the system's configuration and operation. To illustrate, effluent concentrations of at most 13 mg/l and generally below 6 mg/l for BOD_5, turbidity mostly below 1 NTU, suspended solids (SS) below 4 mg/l and total coliforms (TC) levels below 100 cfu/100 ml were measured (Table 12.3). These results confirm the high performance reported for the technology for other applications and demonstrate its ability to treat greywater to meet the strictest criteria for reuse (Table 11.2). Furthermore, a wide range of hydraulic retention times (HRT) were investigated with values between 2 and 63 hours (Table 12.3). The data revealed that excellent treatment performance was obtained even for systems operated with short retention times. For example, Young and Xu (2008) reported BOD and SS effluent concentrations of below 6.1 and 4.0 mg/l, respectively, for an MBR operated at HRTs of 2.5–5.5 hours. In the specific case of the MBR, it is well documented that the biofilm formed on the surface of the membrane, as a result of the contact with the biomass, increases the selectivity of the membrane. This leads to improved rejection of solutes smaller than the pore size

Table 12.3 Membrane bioreactors for greywater recycling.

System					Membrane							Source	Treatment performance				Reference
Type	Volume (m^3)	SRT[a] (d)	HRT[b] (h)	MLSS[c] (g/l)	Type	Surface area (m^2)	Material	Pore size (μm)	Pressure (bar)	Flux (l/($m^2 \cdot h$))	SADm[d] (m^3/($m^2 \cdot h$))		BOD_5 In Out (mg/l)	Turbidity In Out (NTU)	SS In Out (mg/l)	TC In Out (cfu/100ml)	
iMBR[e]	0.6	–	18	–	FS[f]	5	PE[g]	0.4	–	6.6	–	Combined source from apartment building	**250–270**[h] 14	**68–70** 0.5	**50–60** 2.5	–	Baban *et al.* (2011)
iMBR[e]	0.7	–	21–37	5–7	FS[f]	4	–	–	–	4–8	–	Office building (hand wash, dish washer, kitchen)	**257** <2.4	–	**11** ND[i]	**3.3 × 10^7** <100	Paris (2009)
iMBR[e]	2.8	–	15–22	3–4.5	FS[f]	21	–	–	–	6–9	–	University dormitory (kitchen, showers, hand wash)	**151** <4.2	<1	**63** ND[i]	**4.7 × 10^7** <100	Paris (2009)
iMBR[e]	1.5	–	28–63	2.1	FS[f]	21	–	–	–	1–3	–	Hotel (hand wash, showers)	<1.7	<0.7	–	ND[i]	Paris (2009)
iMBR[e]	0.0036	∞	2.8–3.1	4.5–6.4	HF[j]	0.047	PVDF[k]	0.04	0.071	25–28	5.2	Combined source from office building	**339**[h] 50	96.6 5.0	**78** ND[i]	**4.2 × 10^5** 39	Smith and Bani-Melhem (2012)

(*Continued*)

Table 12.3 Membrane bioreactors for greywater recycling (*Continued*).

System					Membrane							Source	Treatment performance				Reference
Type	Volume (m^3)	SRT[a] (d)	HRT[b] (h)	MLSS[c] (g/l)	Type	Surface area (m^2)	Material	Pore size (μm)	Pressure (bar)	Flux (l/($m^2 \cdot h$))	SADm[d] (m^3/($m^2 \cdot h$))		BOD_5 In Out (mg/l)	Turbidity In Out (NTU)	SS In Out (mg/l)	TC In Out (cfu/100ml)	
iMBR[e]	0.003	–	9–18	0.4–1.9	HF[j]	0.04	–	0.1	0.07–0.4	7–11	8.0	Sport centre (showers)	**59** 4	**29** 0.5		**1.4×10^5**[l] 68	Merz *et al.* (2007)
iMBR[e]	1.2	–	19.5	–	FS[f]	7	PES[m]	0.05	–	19.2	–	Office building (showers and hand wash)	**138** 1	**68** 0.2	**57.5** <1	**3.3×10^4**[n] <5	Santas-masas *et al.* (2013)
iMBR[e]	0.0654	48–65	2.5–5.5	5.9–7.8	HF[j]	2	PVDF[k]	0.2	–	14.9–18.3	0.3	Synthetic greywater	**65.6** 3.0–6.1	–	**75** 3.4–4.0	–	Young and Xu (2008)
–	0.035	4–20	2	2–10	–	–	–	–	–	–	–	Bathrooms and kitchen	**493**[h] 24	–	**90** <1	–	Lesjean and Gnirss (2006)
Gravity iMBR[e]	0.066	–	31.5	–	FS[f]	0.48	PS[q]	0.4	0.06	–	2.5	Synthetic greywater	**120**[h] 9.6	<1	**52.2** 3.9	ND[i]	Jefferson *et al.* (2001)
Gravity iMBR[e]	0.010	–	13.6	10–25	FS[f]	0.1	PO[r]	0.4	0.015–0.030	9	6.0	Synthetic kitchen sink and washing machine	**675**[h] 26.3	–	–	–	Huelgas and Funa-mizu (2010)
Gravity iMBR[e]	0.0018–0.0025	15–∞	4.5–24	11–13	HF[j]	–	PAN[s]	100 kDa	0.05	–	–	University restaurant (dish washer)	**770–2050**[h] 8–72	–	–	–	Huelgas *et al.* (2009a)

Gravity iMBR[e]	–	–	4–12	9–11	HF[j]	–	PAN[s]	100 kDa	0.05	–	–	Kitchen sink and washing machine	**540–1200**[h] 28–124	–	–	–	Huelgas *et al.* (2009b)
	–	–	4.5–12	11–13	HF[j]	–	PAN[s]	100 kDa	0.05	–	–	Kitchen sink	**770–2050**[h] 11–73	–	–	–	
Gravity iM–SBR[o]	0.017	–	13	4	FS[f]	0.15	PE[g]	0.4	0 0046	7	8.0	–	**97.3** 12.3	–	**33** ND[i]	**1.1 × 10^5** ND[i]	Lamine *et al.* (2012)
iM-SBR[o]	0.5	>360	24–33	2.5–4.5	FS[f]	7.8	PP[p]	0.4	0.1–0.4	5–8	–	Synthetic greywater	**230**[h] 18.9	–	–	–	Scheu-mann and Kraume (2009)
	0.5	>360	12	4.5–7.8	FS[f]	15.6	PVDF[k] PES	0.4	0.1–0.25	9–12	–			–	–	–	
iM-SBR[o]	0.5	>360	8–33	–	FS[f]	7.8	–	–	–	10–35	–	Synthetic greywater	17.5[h]	–	–	–	Kraume *et al.* (2010)
	0.6	>50	18	–	–	5	–	–	–	8–38	–	Combined source from house holds	13.7[h]	–	–	–	
	0.003	>50	18	–	–	–	–	–	–	–	–	Sport centre (showers)	21.0[h]	–	–	–	
iM-SBR[o]	1	–	–	–	HF[j]	–	PP[p]	0.2	–	–	–	Combined source (bathroom, kitchen, laundry)	**79**[h] 20	–	**185** <2	–	Shin *et al.* (1998)

[a]SRT: sludge retention time, [b]HRT: hydraulic retention time, [c]MLSS: mixed liquor suspended solids, [d]SADm: specific aeration demand, [e]iMBR: immersed membrane, [f]FS: flat sheet, [g]PE: polyethylene, [h]as COD, [i]ND: not detectable, [j]HF: hollow fibre, [k]PVDF: polyvinylidene fluoride, [l]as faecal coliforms, [m]PES: polyethersulfone, [n]as *E. Coli*, [o]sM-SBR: submerged membrane sequencing batch reactor, [p]PP: polypropylene, [q]PS: polysulfone, [r]PO: polyolefin, [s]PAN: polyacrylonitrile.

of the membrane and can explain the better performance of MBRs compared with that achievable by direct filtration. In addition, as highlighted previously (Pidou *et al.* 2007), greywater and in particular sources from the bathroom and kitchen, may be deficient in the nutrients necessary for optimum biological treatment (Table 12.1). However, this apparently does not hinder MBRs and other aerobic biological systems from treating greywater to very high standards. The versatility of MBRs for this application is further demonstrated by the excellent treatment performance achieved when treating high organic strength sources such as kitchen and washing machine greywaters. Huelgas and co-workers (2009a, b; 2010) reported removal of organics of over 90% for MBRs treating wastewaters from kitchen sinks and dish-washers with an initial COD content as high as 2050 mg/l (Table 12.3). This is also supported by similar results obtained for MBRs applied to the treatment of effluents from industrial laundries (Andersen *et al.* 2002; Hoinkis & Panten, 2008). Andersen *et al.* (2002) reported average effluent BOD_5 and COD concentrations of 2 and 50 mg/l, respectively, for an MBR treating the wastewater from an industrial laundry in Denmark with influent concentrations of 680 and 1700 mg/l for BOD_5 and COD, respectively.

With reports of total coliforms removal of up to 5 log (Table 12.3), MBRs are also very efficient at altering the microbial content of greywater. However, Merz *et al.* (2007) and Jefferson *et al.* (2001) highlighted the potential for micro-organism regrowth in effluent pipes. Indeed, they observed that when the effluent pipe was disinfected with chlorine, faecal coliforms would not be detected in the effluents for more than a month but eventually numbers would start increasing after that. Although these observations may be due to the specific way these small scale research units were installed and operated, this suggests that a disinfection stage might still be required to meet the standards for reuse, especially those requiring non detectable levels. Finally, it should be noted that no clear differences could be observed between the performance of continuous and sequencing batch reactor operations in terms of removal of the organics, solids and micro-organisms. However, sequencing batch reactors (SBR), having the possibility to alternative anoxic and aerobic conditions, benefited from the removal of nutrients. For example, Kraume *et al.* (2010) reported total nitrogen removal of 72–88% and ammonium removal of 84–96% by three membrane-SBRs treating a synthetic greywater, shower water from a sports centre and a combined source from households, respectively.

12.4 OPERATION, MAINTENANCE AND COSTS

As stated in the introduction, the implementation of a treatment technology for greywater recycling will not only rely on good treatment performance, but also on other crucial aspects such as footprint, operation and maintenance requirements and, of course, associated costs. In the next section, some of the key advantages and disadvantages of membrane technologies are discussed.

12.4.1 Operation and maintenance

12.4.1.1 Fouling control measures

The main operational challenge with membrane technologies that has a direct impact on performance and maintenance is fouling. Membrane fouling is defined as the deposition and accumulation of materials of organic, inorganic and/or biological nature on the surface and in the pores of the membrane. Membrane fouling may impact both the hydraulic and treatment performance of the system and creates additional operation and maintenance costs to maintain and/or recover acceptable performance levels. Strategies for fouling control include physical methods such as relaxation (briefly stopping filtration), cross flow shear with liquid and/or air and backwash. It should be noted that not all methods can be applied to all types and configurations of membranes. For dense membranes, nanofiltration and reverse osmosis, which are most commonly used as spiral wound modules, fouling formation can only be controlled by adjusting the liquid cross flow velocity. However, with these membranes, fouling control can also be achieved chemically with the addition of antiscalants and biocides in the feed water to limit the formation of inorganic and bio- fouling respectively. It should also be noted that in most greywater reuse applications dense membranes will require a pre-treatment stage, often in the form of low pressure membranes. This removes solids from the feed water, as they may have a major impact on the operation of the system by blocking the feed channels in the membrane modules. The addition of another technology in the treatment train will then add to the complexity and costs of the system. The physical methods to be applied for fouling control in micro- and ultra- filtration membranes will depend on the membrane module configuration, whereas relaxation and liquid or air cross flow can be applied to all configurations (tubular, flat sheet and hollow fibre). Backwashing, defined as reversing the flow of filtration through the membrane to remove the accumulated foulants from its surface, is almost exclusively used with hollow fibre membranes. These fouling control methods can however be used individually or in combination. For example, in direct filtration applications, the backwash of hollow fibre membranes is often coupled with air scouring to optimise the removal of the fouling layer.

As discussed above, these methods will help limit and remove reversible fouling and extend the duration of performance of the membrane units. However, it is likely that over time irreversible fouling (i.e., fouling not removed by these physical methods) will accumulate and affect the performance of the membranes. When the performance drops to unacceptable levels, it can be recovered by chemical cleaning of the membranes. Typical chemical cleaning procedures will often be based on a sequence of cleanings with a caustic solution or sodium hypochlorite for the removal of organics and biofouling, and an acid (hydrochloric acid, citric acid, nitric acid) to remove inorganics. Other cleaning agents such as surfactants and chelating agents are also used (Judd, 2011). It should also be noted that strong oxidants such as sodium hypochlorite have a damaging effect on the structure of dense membranes with a polyamide active layer. In order to avoid damage and premature ageing of the membranes, the use of weaker oxidants such as chloramines is preferred (Donose *et al.* 2013).

For dense membranes, the cleaning chemicals are pumped into the feed channels of the membrane modules, left to soak and then flushed. Low pressure membranes can be similarly cleaned by soaking in chemicals in situ (cleaning in place – CIP) or ex situ. However, this is not always practical especially for applications such as membrane bioreactors and therefore a chemically enhanced backwash (CEB) method is often preferred. In this case, the cleaning chemical is pumped into the membrane modules through the permeate line (Judd, 2011). The volume of chemical required is consequently smaller and the membrane tank does not have to be emptied or the membrane modules taken out. Generally, CEBs are performed regularly as a preventive action and maintenance soak cleanings are used occasionally to restore performance. During cleaning of the membranes, operation of the system has to be stopped. In order to avoid disruption, it is very important that this maintenance step is as short as possible and its frequency is reduced to a minimum.

12.4.1.2 Direct filtration

Most studies focused on direct filtration of greywater are based on short term trials and only limited information is available on fouling over long periods of operation. Moreover, for this particular application, fouling will be directly linked to the feed water quality (and the membrane characteristics and operation) and will so be site specific. Li *et al.* (2009) operated UF membranes for the treatment of combined greywater in batches. The membrane was operated in cycles of 10 minutes of filtration followed by a 30-second backwash over a two-week period. In that period, at a constant filtration pressure of 0.12 bar, the flux generally decreased by about 40% from an initial value of 10 $l/(m^2 \cdot h)$. The membranes were then chemically cleaned after each batch on a fortnightly basis and the results demonstrated an excellent recovery of the performance suggesting only very limited irrecoverable fouling. Nghiem *et al.* (2006) reported similar results as they observed full recovery of the performance of UF membranes treating synthetic greywater after cleaning with typical household bleach. Additionally, Ahn *et al.* (1998) evaluated UF membranes with different pore sizes and results further supported these findings and demonstrated the influence of the raw feed water composition on fouling. Indeed, because the greywater investigated in this study contained large particles, all tested membranes displayed similar hydraulic performance characterised by a rapid decrease of the flux in the first hour of operation followed by stable conditions for the remainder of the test (up to 12 days). In these conditions, the fouling observed is essentially cake layer formation with the accumulation of the large particles on the surface of the membrane depending on the hydraulic conditions in the system (cross flow velocity, relaxation and/or backwash sequence). However, raw waters containing small particles and colloids may have a greater impact on the operation of the membranes as they will penetrate the structure of the membrane and may contribute to pore blocking. This type of fouling is often more difficult to control and remove. The studies reported in the literature that investigated the application of dense membranes for greywater recycling essentially assessed their performance for treatment and very limited information is available on fouling. Guilbaud *et al.* (2012) reported the use of NF membranes for the treatment of greywater aboard ships with batch operation at high fluxes (50–60 l/($m^2 \cdot h$)). Such intense operation of the membranes necessitated frequent (daily) cleaning. However, this was found to be a suitable way of operating this specific system because only very limited space was available and continuous operation was not required. Operation with conservative conditions (low flux or pressure) may limit fouling and subsequent requirements for control and cleaning. However, if a low flux is used additional membranes will be needed in the system to meet the desired treatment flow. Consequently, a trade-off between operational and maintenance costs and capital cost will have to be found. Ultimately, the knowledge and experience gained from the operation and fouling control of the many direct filtration systems already implemented for municipal and industrial wastewater treatment applications can be transferred directly to greywater recycling.

12.4.1.3 Hybrid systems

For hybrid systems and in particular MBRs, fouling formation will not only depend on the feed water quality, but also on the operational conditions in the biological system, as the membrane is directly exposed to the biomass. Interestingly, the systems reported in the literature for the treatment of greywater have been studied for a wide range of conditions (Table 12.3). For example, fluxes between 1 and 38 l/($m^2 \cdot h$) are reported (Table 12.3) in comparison to 10–30 l/($m^2 \cdot h$) for typical fluxes for iMBRs in municipal wastewater treatment application (Judd, 2011). Kraume *et al.* (2010) operated an SBR with immersed membranes at fluxes of 10–35 l/($m^2 \cdot h$) with filtration cycles of nine minutes followed by one minute relaxation. They observed that the membranes required cleaning after 3–4 months. In the same study, another SBR with the membrane operated in continuous filtration with more conservative fluxes of 5–12 l/($m^2 \cdot h$) could be operated for six months between chemical cleans. These studies demonstrate that different strategies can be applied to limit fouling and extend the performance of the membranes. The studies by Huelgas *et al.* (2009b) and Huelgas and Funamizu (2010) also demonstrated the impact a chosen filtration sequence may have on the performance of a system. Significant fouling, characterised by a rapid decrease of the flux at constant pressure over forty days, was observed in an MBR operated with a three minute backwash every five days for the treatment of a high strength kitchen and washing machine greywater (Huelgas *et al.* 2009b). In comparison, another MBR also treating high strength greywater maintained good filtration performance for

90 days with operation at 9 l/($m^2 \cdot h$) alternating ten minutes of filtration and two minutes of relaxation. Other work reported by Young and Xu (2008) and Smith and Bani-Melhem (2012) showed that by operating MBRs for the treatment of greywater continuously at high fluxes of 18 and 25–28 l/($m^2 \cdot h$), respectively, weekly chemical cleaning was needed. In contrast, Merz *et al.* (2007) and Lamine *et al.* (2012) demonstrated that the frequency of chemical cleaning could be reduced by operating the system with conservative fluxes and by implementing relaxation. For example, Merz *et al.* (2007) obtained stable operation of an MBR treating shower waters from a sport centre for 140 days with a flux of 8 l/($m^2 \cdot h$) and intermittent operation of 45-minute filtration and 15-minute relaxation.

From the review of the above literature, it can be summarised that the careful selection of filtration sequence enables extended operation (several months) with limited maintenance requirements, but this may still have an influence on cost. As previously discussed, if more conservative fluxes are chosen, more membranes will be needed to treat a given flow and consequently the capital cost of the system will be increased. It should also be noted that for all immersed membranes used in MBR systems, in addition to the possible relaxation and backwash to control fouling, air sparging on the membrane can be used for fouling control. In addition to the air provided to the biomass in the biological tank for its development, air is also injected at the base of the membrane modules. The coarse air bubbles scour the surface of the membranes (or provide vibration of the fibres in hollow fibre membranes) to limit the formation of a fouling layer. It is essential to have a good understanding and control of air sparging, as it is known to have a beneficial effect up to a certain flow level after which no improvement on filtration performance is observed and consequently energy is wasted. With a specific aeration demand (SADm) mostly between 2.5 and 8 $m^3/(m^2 \cdot h)$, the membranes investigated in the studies reported here were found to be over-aerated, as is often the case in small scale research units. Indeed, typical SADm for large scale MBRs in municipal wastewater treatment are typically between 0.2–1.5 $m^3/(m^2 \cdot h)$ (Judd, 2011). This highlights that improvement in the operation and performance of the systems can be achieved when applied at a larger scale.

Another key aspect of MBRs is sludge management. In most cases, the sludge retention time is extended to longer periods to limit the need for sludge disposal. However, this has to be done carefully because at long sludge retention times (SRT) the biomass is concentrated in the system and this may have an impact on the operation of the membrane. This was shown by Young and Xu (2008) who experienced significant fouling in an MBR operated with a long sludge age to limit sludge formation and disposal. Interestingly, Lesjean and Gnirss (2006) opted for a different approach and tested a pilot MBR with short SRT and HRT. Their objective was to produce high sludge quantities to improve carbon and nitrogen recovery within an overall ecosan approach with onsite sludge treatment. However, in most cases, to limit the costs associated with sludge handling and treatment, operational conditions will be set so that biomass production is reduced. This is confirmed by the generally high SRT used with most values reported over 48 days (Table 12.3). In urban environments, sludge wastage can generally be controlled by discharge to sewers with a waste pump operated on a timer, in which case requirement for onsite maintenance will be limited (Sneller, 2009).

Finally, when evaluating membrane systems for greywater recycling especially for application in accommodation, it is important to understand the capability of the system to cope with variations in quantity and quality. The systems may be exposed to rapid changes in the composition of the greywater and also to potentially extended periods of inactivity (e.g., during holiday periods). Membranes used for direct filtration are not expected to be affected by such variations as they are simple physical treatment units and can be turned on and off as required. However, the biological treatment component of an MBR may be affected by such variation and consequently treatment and operation may be impacted. Interestingly, the study by Pidou *et al.* (2004) clearly demonstrated the robustness of MBRs against such variation. An MBR treating greywater was exposed to a stoppage of aeration or feed for 8 hours with no effect to its performance as effluent quality remained stable. The robustness and suitability of MBRs for this application was further demonstrated when the performance of the system was not affected by a 25-day interruption in operation, simulating an extended holiday period.

Membranes are pressure driven processes and consequently can be easily automated by monitoring and controlling pressures and flows in the system. As these systems are operated at constant flow, when fouling builds up on the membrane, the resistance to flow will increase and consequently the pressure required across the membrane to maintain this flow will increase. These changes can easily be monitored by pressure gauges. Fouling control measures can then be triggered by time, in preventive mode and/or by pressure, in corrective mode. Due to the complexity of these technologies, trained personnel may be required for operation and maintenance. However, with automation, the control of the system can be effected remotely enabling reduced maintenance costs. This means that operators are only required to be onsite for limited durations, mainly to refill the chemical tanks for cleaning.

12.4.2 Energy and costs

Membrane systems are known to be energy intensive, which can be a barrier to their implementation. The energy requirements for the operation of NF membranes for greywater treatment have been reported to be between 4.2–14 kWh/m^3 (Humeau *et al.* 2011; Guilbaud *et al.* 2010, 2012). In comparison, energy requirements for MBRs have been reported between 1.7 and 7 kWh/m^3 (Paris *et al.* 2007; Sellner, 2009; Baban *et al.* 2011; Humeau *et al.* 2011; Lamine *et al.* 2012; Santasmasas *et al.* 2013). This is in the range reported for typical small scale MBRs of 3–12 kWh/m^3 and between 0.6–2 kWh/m^3 for large systems (Verrecht *et al.* 2010). Such energy consumption figures are of course higher than those of other technologies potentially used for greywater recycling. In their study,

Baban *et al.* (2011) compared the economics of the implementation of an MBR and a constructed wetland (CW) for greywater recycling. The energy requirement of 1.7 kWh/m^3 reported for the MBR was indeed significantly higher than that for the CW with only 0.02 kWh/m^3, but in a similar range to other biological systems such as an SBR (3.8 kWh/m^3) or an RBC (1.2 kWh/m^3). In contrast, the capital cost of the MBR was lower with £531 (€645) per population equivalent against £988 (€1200) per population equivalent for the CW. In addition, the footprint required for both systems was also found to be a key implementation consideration. The space requirement for the MBR was only 1 m^2 per m^3 of water treated daily against 63 m^2/m^3 for the CW. The small footprint of membrane systems is a significant advantage over other treatment technologies especially when considering that most applications for greywater recycling will be in dense urban areas where space is both limited and often expensive. Additional data show space requirements for MBRs treating flows between 0.8 and 8.5 m^3/d to vary between 1.5 and 7.5 m^2 corresponding to 0.9–1.9 m^2/m^3 (Sellner, 2009). A large MBR installed in a hotel with a treatment capacity of 14.6 m^3/d only required a footprint of 16 m^2 (1.1 m^2/m^3) and was in fact installed in the basement (Paris *et al.* 2007). In this specific case, the investment cost for the unit was £55,599 (€67,553) with an additional £1894 (€2301) per annum for maintenance and £2107 (€2560) per annum for energy. Based on these costs, the payback period was estimated to be 4.5 years (Paris *et al.* 2007). Fletcher *et al.* (2007) reported single house package plant MBRs to have a capital cost in the range of £2535–4609 (€3080–5600) with £889 (€1080) per annum for operation and maintenance. However, as a general rule, more expensive units as an investment are usually cheaper to operate because more control is incorporated in the system. Finally, with overall costs of the recycled water produced by NF membranes reported to be between 3.5 and 6.4 £/m^3 (4.2 and 7.8 €/m^3) (Humeau *et al.* 2011; Guilbaud *et al.* 2010, 2012), significantly higher than drinking water, these systems are expensive especially for small scale applications and would be viable only in specific circumstances. In recent years, significant research and development work has been focused on reducing membrane manufacturing costs and the energy consumption of membrane based systems, which will in turn help decrease the overall costs for greywater reuse applications and improve the long term viability of the technologies.

12.5 CONCLUSION

This chapter has provided an overview of literature relating to different membrane technologies for application in the field of greywater reuse. It has highlighted that the advantages exhibited by membrane technologies provide a strong case for their implementation for greywater recycling. Indeed, direct filtration with dense membranes and hybrid systems, such as membrane bioreactors, have been shown to meet the most stringent standards for reuse regardless of the feed water quality. However, it is important to bear in mind that one technology cannot suit all water recycling applications and every case should be dealt with according to a fit for purpose approach. For example, direct filtration with low pressure membranes is proven to be efficient at removing solids and micro-organisms. However, direct filtration has limited effectiveness in the removal of dissolved organics and nutrients, consequently providing effluent quality suitable for reuse applications such as irrigation. Overall, membrane systems have high energy requirements in comparison to other technologies, but they offer certain advantages including lower capital cost, small footprint, automation and robustness, which make them a very strong contender for greywater recycling applications.

REFERENCES

Ahn K.-H., Song J.-H. and Cha H.-Y. (1998). Application of tubular ceramic membranes for reuse of wastewater from buildings. *Water Science and Technology*, **38**(4–5), 373–382.

Andersen M., Kristensen G. H., Brynjolf M. and Grüttner H. (2002). Pilot-scale testing membrane bioreactor for wastewater reclamation in industrial laundry. *Water Science and Technology*, **46**(4–5), 67–76.

Baban A., Hocaoglu S. M., Atasoy E. and Regelsberger M. (2011). Appraisal of technical options for grey water treatment processes. In *Proceedings of the 3rd International Congress Smallwat11 on "Wastewater in small communities, Towards the Water Framework Directive (WFD) and the Millenium Development Goals (MDG)"*, 25–28 April 2011, Seville, Spain.

Bhattacharya P., Sarkar S., Ghosh S., Majumbar S., Mukhopadhyay A. and Bandyopadhyay (2013). Potential of ceramic microfiltration and ultrafiltration membranes for the treatment of gray water for an effective reuse. *Desalination and Water Treatment*, **51**, 4323–4332.

Donose D. C., Subash S., Pidou M., Poussade Y., Keller J. and Gernjak W. (2013). Effect of pH on the ageing of reverse osmosis membranes upon exposure to hypochlorite. *Desalination*, **309**, 97–105.

Fletcher H., Mackley T. and Judd S. (2007). The cost of a package plant membrane bioreactor. *Water Research*, **41**(12), 2627–2635.

Guilbaud J., Massé A., Andrès Y., Combe F. and Jaouen P. (2012). Influence of operating conditions on direct nanofiltration of greywaters: Application to laundry water recycling aboard ships. *Resources Conservation and Recycling*, **62**, 64–70.

Guilbaud J., Massé A., Andrs Y., Combe F. and Jaouen P. (2010). Laundry water recycling in ship by direct nanofiltration with tubular membranes. *Resources Conservation and Recycling*, **55**(2), 148–154.

Hoinkis J. and Panten V. (2008). Wastewater recycling in laundries-From pilot to large-scale plant. *Chemical Engineering and Processing: Process Intensification*, **47**(7), 1159–1164.

Hourlier F., Massé A., Jaouen P., Lakel A., Gérente C., Faur C. and Le Cloirec P. (2010). Membrane process treatment for greywater recycling: Investigations on direct tubular nanofiltration. *Water Science and Technology*, **62**(7), 1544–1550.

Huelgas A. and Funamizu N. (2010). Flat-plate submerged membrane bioreactor for the treatment of higher-load graywater. *Desalination*, **250**(1), 162–166.

Huelgas A., Nagata H. and Funamizu N. (2009a). Effect of organic loading rate for on-site treatment of wastewater using SubMBR. *Environmental Engineering Science*, **26**(1), 15–23.

Huelgas A., Nakajima M., Nagata H. and Funamizu N. (2009b). Comparison between treatment of kitchen-sink wastewater and a mixture of kitchen-sink and washing-machine wastewaters. *Environmental Technology*, **30**(1), 111–117.

Humeau P., Hourlier F., Bulteau G., Massé A., Jaouen P., Gérente C., Faur C. and Le Cloirec P. (2011). Estimated costs of implementation of membrane processes for on-site greywater recycling. *Water Science and Technology*, **63**(12), 2949–2956.

Jefferson B. (2013). Water Recycling Lecture Notes, Cranfield University, UK.

Jefferson B., Laine A. L., Stephenson T. and Judd S. J. (2001). Advanced biological unit processes for domestic water recycling. *Water Science and Technology*, **43**(10), 211–218.

Judd S. J. (2011). The MBR Book: Principles and Applications of Membrane Bioreactors for Water and Wastewater Treatment. Elsevier, Oxford, UK.

Kim R.-H., Lee S., Jeong J., Lee J.-H. and Kim Y.-K. (2007). Reuse of greywater and rainwater using fiber filter media and metal membrane. *Desalination*, **202**(1–3), 326–332.

Kraume M., Scheumann R., Baban A. and El Hamouri B. (2010). Performance of a compact submerged membrane sequencing batch reactor (SM-SBR) for greywater treatment. *Desalination*, **250**(3), 1011–1013.

Kujawa-Roeleveld K. and Zeeman G. (2006). Anaerobic treatment in decentralised and source-separation-based sanitation concepts. *Reviews in Environmental Science and Bio-Technology*, **5**(2006) 115–139.

Lamine M., Samaali D. and Ghrabi A. (2012). Greywater treatment in a submerged membrane bioreactor with gravitational filtration. *Desalination and Water Treatment*, **46**(1–3), 182–187.

Lesjean B., Gnirss R. (2006). Grey water treatment with a membrane bioreactor operated at low SRT and low HRT. *Desalination*, **199**(1–3), 432–434.

Li F., Gulyas H., Wichmann K. and Otterpohl R. (2009). Treatment of household grey water with a UF membrane filtration system. *Desalination and Water Treatment*, **5**(1–3), 275–282.

Merz C., Scheumann R., El Hamouri B. and Kraume M. (2007). Membrane bioreactor technology for the treatment of greywater from a sports and leisure club. *Desalination*, **215**(1–3), 37–43.

Nghiem L. D., Oschmann N. and Schäfer A. I. (2006). Fouling in greywater recycling by direct ultrafiltration. *Desalination*, **187**(1–3), 283–290.

Paris S. (2009). The importance of membrane technology for greywater recycling. *In Proceedings of the Fbr conference on "Greywater Recycling and Reuse"*. 5 February 2009, Berlin, Germany.

Paris S., Schlapp C. and Netter T. (2007). A contribution to sustainable growth by research and development. *In Proceedings of the IWA Symposium – "Water Supply and sanitation for all"*, 27–28 September 2007, Berching, Germany.

Pidou M. (2007). Hybrid Membrane Processes for Water Reuse. PhD thesis, Cranfield University, UK.

Pidou M., Memon F. A., Stephenson T., Jeffrey P. and Jefferson B. (2007). Greywater recycling: treatment options and applications. *P I Civil Eng: Engineering Sustainability*, **160**, 3859–3867.

Pidou M., Palmer A., Avery A., Jeffery P., Stephenson T., Judd S. and Jefferson B. (2004). Robustness of technology solutions for grey and black water recycling in urban environments. *In Proceedings of the Annual Symposium of the Watereuse Association: "Water Reuse: Bringing Life to the Desert"*, 19–22 September 2004, Phoenix, USA.

Pidou M., Parsons S. A., Raymond G., Jeffrey P., Stephenson T. and Jefferson B. (2009). Fouling control of a membrane coupled photo catalytic process treating greywater. *Water Research*, **43**, 3932–3939.

Ramon G., Green M., Semiat R. and Dosoretz C. (2004). Low strength graywater characterization and treatment by direct membrane filtration. *Desalination*, **170**(3), 241–250.

Rivero M. J., Parsons S. A., Jeffrey P., Pidou M. and Jefferson B. (2006). Membrane chemical reactor (MCR) combining photocatalysis and microfiltration for grey water treatment. *Water Science and Technology*, **53**(3), 173–180.

Santasmasas C., Rovira M., Clarens F. and Valderrama C. (2013). Grey water reclamation by decentralized MBR prototype. *Resources Conservation and Recycling*, **72**, 102–107.

Scheumann R. and Kraume M. (2009). Influence of hydraulic retention time on the operation of a submerged membrane sequencing batch reactor (SM-SBR). for the treatment of greywater. *Desalination*, **246**(1–3), 444–451.

Shin H.-S., Lee S.-M., Seo I.-S., Kim G.-O., Lim K.-H. and Song J.-S. (1998). Pilot-scale SBR and MF operation for the removal of organic and nitrogen compounds from greywater. *Water Science and Technology*, **38**(6), 79–88.

Smith E. and Bani-Melhem (2012). Grey water characterization and treatment for reuse in an arid environment. *Water Science and Technology*, **66**(1), 72–78.

Sneller M. (2009). Greywater treatment in apartment building-Austria. Combined greywater treatment using a membrane bioreactor. *Sust. Sanitation Practice*, **1**, 30–33.

Šostar-Turk S., Petrinić, I. and Simonič, M. (2005). Laundry wastewater treatment using coagulation and membrane filtration. *Resources Conservation and Recycling*, **44**(2), 185–196.

Troquet J., Lasserre J. C., Le Calvez C. and Lamaze B. (2009). Grey Waters Recycling at Concordia Antarctic Station: Feedback from Three-Year Operation. In *Proc. of the European Advanced Life Support Workshop*, 2–4 June 2009, Barcelona, Spain.

USEPA (2012). Guidelines for Water Reuse. Report EPA/600/R-12/618. Washington DC, USA.

Verrecht B., Maere T., Benedetti L., Nopens I. and Judd S. (2010). Model-based energy optimisation of a small-scale decentralised membrane bioreactor for urban reuse. *Water Research*, **44**(14), 4047–4056.

Young S. and Xu A. (2008). Development and testing of a low sludge discharge membrane bioreactor for greywater reclamation. *Journal of Environmental Engineering and Science*, **7**(5), 423–431.

Chapter 13

Energy and carbon implications of water saving micro-components and greywater reuse systems

Fayyaz Ali Memon, Abdi Mohamud Fidar, Sarah Ward, David Butler and Kamal Alsharif

13.1 INTRODUCTION

This chapter briefly introduces the drivers for water efficiency and, for the UK context, discusses domestic water consumption patterns, resulting energy and carbon implications and available policy options, such as the *Code for Sustainable Homes* (CSH) to reduce consumption. The CSH requires reduction in per capita water consumption in households and accordingly provides a rating scheme. Higher water efficiency (e.g., levels 5 and 6) is possible if water efficient micro-components (water using fixtures and appliances) are installed in conjunction with rainwater harvesting (RWH) or greywater reuse (GWR) systems. Greywater is a relatively less polluted stream consisting of wastewater mainly from hand washbasins, showers and baths and after sufficient treatment can be used for non-potable applications (e.g., toilet flushing). A number of studies (including those reported in Chapter 10, 11 and 12) suggest that a well maintained greywater recycling system on its own can reduce mains supply by 25–30%. However, the energy consumption and the carbon loads associated with the onsite GWR require further research.

This chapter describes a methodology and the architecture of an assessment tool to assess the potential energy and carbon implications of meeting Level 5 and 6 of the CSH. The tool application resulted in the generation of numerous composite strategies (i.e., combinations of water saving fixtures and commercially available GWR systems) and their associated energy and carbon footprint. The analysis of the strategies indicated that although greywater at domestic level can offer considerable per capita water demand reduction potential, its effectiveness can decrease if applied in conjunction with water efficient toilets. Furthermore, it can increase the energy consumption and resulting carbon load associated with in-house water use.

13.2 DRIVERS FOR WATER EFFICIENCY

World population is increasing by 6 million people per month. 1.2 billion people live in areas affected by physical water scarcity and 1.6 billion live in areas affected by economic water scarcity. Currently, 1.4 billion do not have sufficient electricity. It is estimated that in 2030, 1.2 billion people will still lack access to electricity. An extra billion tonnes of cereals will be needed annually by 2030. Water is needed to produce both electricity and food. Simply, available freshwater resources are not sufficient enough to meet the present and future demand. According to the World Economic Forum, water supply security is among the top five risks the world is facing currently (WEF, 2012).

Efficient use of water (i.e., doing more with less without compromising the quality of service) is a way forward. However, the uptake of water efficiency measures in many parts of the world, including the UK, is significantly lower than some developed countries (e.g., Australia and Japan), which have higher penetration of water saving technologies and established mandatory policies. Because of fragmented policies and increased frequency of flooding, promoting water efficiency remains an uphill task in the UK. Hosepipe bans imposed by 7 UK water companies in the recent past are nothing but a grim reminder of the fact that the country should enhance its resilience to withstand pressures imposed by the uncertainty in climate change. This resilience building, to swiftly manage the consequences of frequent extreme events (droughts and floods), requires a multi-track approach of influencing water user behaviour, developing 'fit for purpose' supply sources, minimising leakage, reducing demand through water saving appliances and fixtures and implementing appropriate flood mitigation and adaptation measures.

13.3 DOMESTIC WATER CONSUMPTION AND ASSOCIATED ENERGY FOOTPRINT

In UK households, typical per capita water consumption is about 150 litres per day (Memon & Butler, 2006). Of these 150 litres of rigorously treated high quality drinking water, about one third is used for toilet flushing (Figure 13.1). This clearly is an unsustainable and unwise use of a precious resource. In several developing countries, the provision of the World Health Organisation set water requirement of 50 litres, to meet daily basic human needs, still remains a challenge. A typical UK family consumes approximately 500 litres of water daily and almost 95% of this becomes wastewater, requiring extensive treatment before its safe disposal. The provision of safe drinking water and subsequent transport of resulting wastewater and its treatment requires considerable energy input. Using water in our homes, contributes around 35 million tonnes of greenhouse gases a year (EA, 2014). The water industry is the fourth largest energy user in the UK and used approximately 7700 gigawatt hours (GWh) of energy in 2006, which is 1% of the average daily electricity consumption in England and Wales (Caldwell, 2009). At a household level, the second largest use of energy is for water heating and is significantly higher than the actual amount of energy required externally to produce one m^3 of potable water and treat wastewater.

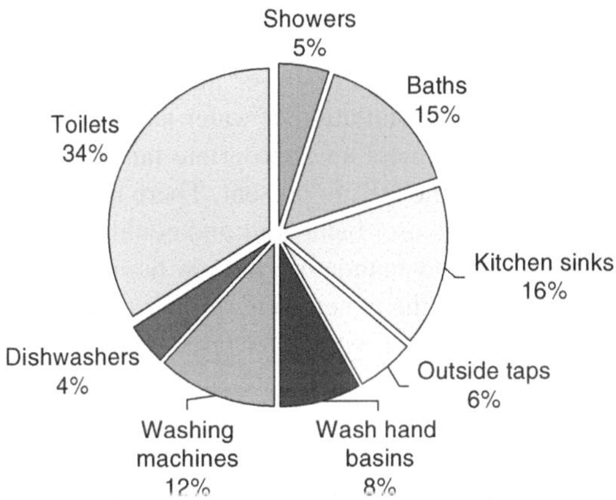

Figure 13.1 Typical share of micro-components in domestic per capita consumption of water (Memon & Butler, 2006).

13.4 WATER EFFICIENCY POLICY AND ENABLING TECHNOLOGIES

Future Water, the Government's water strategy for England (Defra, 2008a), aspires to reduce per capita consumption to 130 litres a day. The Building Regulations in England and Wales (Part G) require that in all new homes per capita consumption must not exceed 130 litres per day. This reduction of 20 litres in the typical consumption of 150 litres can be realised easily through the installation of water saving appliances/fixtures, often referred as micro-components of demand (e.g., water saving taps, low cistern volume toilets, low flow shower heads and water efficient white goods (washing machines and dishwashers)). Considerable research and innovation has resulted in the development and availability of a wide range of products and fixtures, claiming widely varying water saving potentials. Appliance performance is now also reported for water in addition to their energy consumption and fixtures come with a widely varying price tag. Past studies have suggested that there is no direct link between the price and associated water savings from a particular appliance, implying that achieving water efficiency at a lower cost is also possible (Grant, 2006). Water companies, as part of water efficiency promotion campaigns, distribute products including tap magic inserts, low flow shower heads and hippos (a basic water displacement object to reduce WC flush volume). The long term effectiveness of such campaigns requires independent scrutiny. Appropriately chosen appliances can potentially reduce not only the overall water demand, but also associated energy consumption and therefore contribute towards the Government's target of reducing greenhouse gas emission by 80% by 2050. For new households, the uptake of certain micro-components (e.g., low flush toilets) shows an upward trend. However, for the existing housing stock, the pace of retrofitting is much slower and therefore does not translate into substantial water savings. Installation of water meters to record actual water consumption and charge consumers an appropriate tariff is a common practice in many countries other than the UK, at present. There is a broad consensus that metering potentially influences user behaviour and could reduce consumption by 10%. In the UK, metering is now mandatory for new homes. The existing housing stock is mostly unmetered and the penetration rate for meters varies from region to region and is strongly correlated with water price. Owing to cost implications, the UK is unlikely to achieve universal metering in the near future. However, water users can request a meter and water companies are obliged to install one with no charge to customer.

Although the above mentioned water demand management measures, if implemented universally, could reduce consumption significantly, this may alter the hydraulic regime and therefore performance of water supply networks and wastewater drainage systems. Recent investigations suggest that reduced water demand could increase water age, a water quality indicator, in potable water distribution systems (Atkinson *et al.* 2014). On the other hand, reduced wastewater flows to sewers have the potential to reduce self-cleansing velocity and accelerate solids build up within sewers (Blanksby, 2006). These undesirable implications, however, can be overcome through improved infrastructure operation and management strategies. Water efficiency at a universal scale will certainly help to eliminate or delay the construction of additional new infrastructure to meet future additional demand. According to the Environment Agency, by 2020 the demand for water could increase by 800 million extra litres a day (EA, 2014).

The Code for Sustainable Homes (DCLG, 2008), the national standard for the sustainable design and construction of new homes, aims at improving the sustainability of buildings by efficient use of resources such as water and energy. It requires, among other things, reduction in the domestic consumption of potable water by installing water efficient micro-components, rainwater harvesting units and GWR systems. With regard to water consumption, the CSH employs a rating system (levels 1 to 6), based on per capita water consumption and sets mandatory minimum standards for each level (Table 13.1). The CSH offers a 6-star (Level 5 and 6) rating to low/zero carbon homes achieving per capita consumption not exceeding 80 litres per day.

Table 13.1 Water consumption levels of the Code for Sustainable Homes.

Water consumption (litres/cap. day)	Credits	Levels
≤120	1	Levels 1 and 2
≤110	2	
≤105	3	Levels 3 and 4
≤90	4	
≤80	5	Levels 5 and 6

Source: DCLG (2008).

Reducing consumption to achieve CSH Level 5 or 6 through water saving micro-components alone is not viable and therefore some form of recycling of low grade (fit for purpose) water for non-potable applications can help. About 80% of domestic water use is for non-potable applications. RWH and GWR are two established options which can be considered as an alternative non-potable supply source. The British Standards for RWH (BS 8515:2009) and GWR (BS 8525-2:2011) provide guidance for these water reuse systems' design, installation, operation and maintenance. Although rainwater, in comparison to greywater, requires minimal treatment, its uninterrupted supply is uncertain. On the contrary, greywater supply is reliable and fairly consistent all year round.

13.5 GREYWATER TREATMENT AND REUSE SYSTEMS

Depending on the scale of use, a range of technologies (e.g., membrane bioreactors, constructed wetlands, simple coarse filtration before storage followed by some disinfection) can be applied to treat greywater and reuse it for toilet flushing. GWR is not viable for households with a single occupant. Studies conducted in Israel and elsewhere indicate that for a properly designed and maintained system, the risk to human health, relative to the risks associated with other day to day activities, is fairly limited (Chapter 10). Safeguards against any probability of minor risks can be put in place through following the British Standard on GWR systems (BS 8525-2:2011). These are not fit and forget systems and do require a degree of maintenance and depending on the required final quality of the treated effluent and treatment technology employed, there can be considerable cost and energy implications.

A number of conventional and innovative approaches have been investigated extensively in the recent past to establish their operational envelop, robustness and efficiency to treat greywater (Friedler *et al.* 2005; Li *et al.* 2009; Melin *et al.* 2005; Shin & Johnson, 2007; Pidou, 2007; Winward *et al.* 2008; Toifl *et al.* 2008; Wu *et al.* 2009). However, limited attention has been given to their energy consumption and the environmental impacts, including carbon emissions.

In the study presented in this chapter, four commercially available packaged GWR systems have been considered. Since the objective of the study was not to compare the performance of the commercially available systems, details cannot be given of the manufacturers of the considered systems. The systems have been named as A, B, C and D. A schematic of each system is given in Figure 13.2 and their key components are described as below.

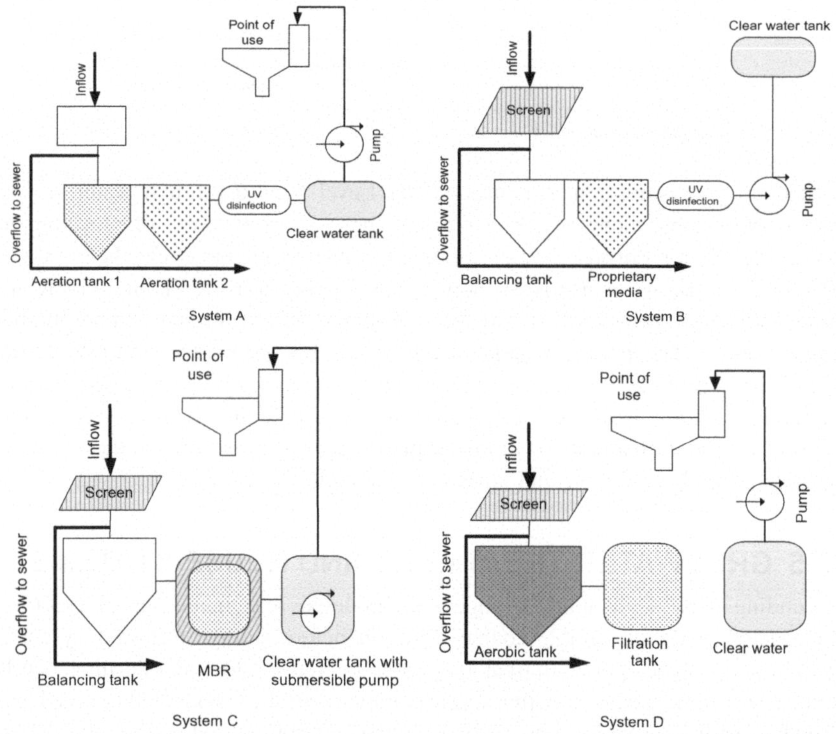

Figure 13.2 Schematic of treatment processes employed in the considered GWR systems.

With System A, the wastewater from the greywater generating micro-components passes through a filter that removes larger particles such as hair and lint. The pre-treated greywater undergoes a sequence of biological treatment processes in which organic pollutants are decomposed. On the way to the storage tank, the effluent flows through a UV-light lamp to disinfect it. When the treated effluent is required (e.g., when a WC is flushed), a booster pump automatically pushes the effluent to the point of use. According to the manufacturer's claim, the system uses 1.2 kWh of energy per m^3 of greywater treated. In addition to this, lifting the effluent to the point of use increases the total energy consumption associated with the system.

System B combines biological and physical treatment processes. The system comprises a balancing/equalisation tank, treatment column and treated water storage tank. Screens remove lint and coarse material before it flows into the balancing tank. The water is pumped to a treatment column where it flows down through a vertical treatment column with a bed of proprietary media. The treated greywater is finally disinfected with UV light. The system uses 1.5 kWh of energy per m^3 of greywater treated. An additional energy is required to pump the effluent from the treatment process to the clear water tank.

System C employs a submerged membrane bioreactor (MBR). The MBR combines an activated sludge process with a membrane separation process. The treatment process starts with passing the greywater through a filter to remove large particles. The removed solids are automatically flushed into the sewer system. The partially treated greywater flows into an aeration tank, where microbes (with the help of externally supplied oxygen) degrade the organic matter in the greywater. After a short interval of sedimentation and flotation the biological treatment continues in the bioreactor. The effluent then passes through a submerged ultrafiltration (UF) to obtain clear water for further applications. The treated effluent is stored in a tank until it is required. When a WC is flushed, a booster pump delivers the effluent to the point of use. According to the system manufacturer, the energy consumption of the system (including the booster pump) is about 3.5 kWh per m^3 of treated greywater.

System D combines biological and physical treatment processes. Greywater is collected in an aeration tank, which encourages natural stabilisation/oxidation of bio-degradable particles. The biologically treated greywater flows into a tank with submerged ultra-filtration membrane designed to filter out the remaining particles. The filtered effluent is transferred and kept into a clean water tank until water is required for non-potable application. A pump begins supplying water to the consumers on demand. According to the manufacturer, the system uses about 2.0 kWh of energy per m^3 of treated greywater. Additional energy is required to lift the effluent from the clean (treated) water tank to where it is being used.

These systems were included in the assessment (Section 13.6) as an illustration to quantify the overall energy and carbon implication of the composite strategies capable of meeting the CSH highest water efficiency target (i.e., Level 5 and 6). The assessment methodology requires the implementation of an assessment tool developed by Fidar (2010). The tool has a flexible architecture to accommodate new GWR systems in addition to the four systems mentioned above.

13.6 ASSESSMENT METHODOLOGY

The methodology includes a multi-step approach to:

- quantify water consumption and resulting greywater generation;
- calculate the energy consumption and carbon load associated with the water use in a household, the operation of the considered GWR systems (described in Section 13.5) and the delivery of the treated water to the point of use; and
- apply a multi-objective based assessment tool to generate optimal composite strategies to reduce consumption to meet CSH water efficiency thresholds.

The water using micro-components considered in this study include WCs, showers, basin taps, kitchen taps, baths, dishwashers and washing machines. Installation of a GWR system is considered to meet the water demand for WC flushing.

13.6.1 Quantification of water volumes

The volume of water consumed per person and resulting greywater generation were calculated using water user behaviour characteristics (i.e., the frequency and duration of use for each water-using device/micro-component) and the extent of water efficiency offered by a range of micro-components commercially available in the market. The water consumption is composed of various end uses. The water consumption characteristics of the micro-components used in the assessment are shown in Table 13.2. These are typical characteristics of the micro-components installed in the majority of households in the UK. The volume of greywater generated is equal to the volume of wastewater produced from three end uses: baths, showers and basin taps.

Table 13.2 Typical water use characteristics of micro-components in the UK.

Micro-component	Use frequency (uses/capita/day)	Usage unit	Event duration (min/use)	Source
Basin tap	7.2	litres/min	0.67	DCLG (2008)
Bath	0.4	litres/use	N/A*	DCLG (2008)
Dishwasher	0.28	litres/use	N/A	MTP (2008b)
Kitchen tap	7.2	litres/min	0.67	DCLG (2008)
Shower	0.6	litres/min	5.0	DCLG (2008)
Washing machine	0.31	litres/use	N/A	MTP (2008a)
WC	4.8	litres/use	N/A	DCLG (2008)

*Not applicable.

13.6.2 Estimation of energy and carbon load

The energy consumption and the carbon emissions associated with any given water consumption level is calculated from the volume of water used, the temperature at which the water is delivered, the percentage split of hot and cold water, the energy use of the GWR system and the standard energy use of white goods (washing machines and dishwashers).

White goods and electric showers considered in this research use electricity to heat up water internally (i.e., they do not use water from household's hot water system). The energy consumption of the white goods used in the assessment is based on the manufacturers' energy efficiency labelling and has been calculated to reflect the energy requirement for different wash programmes (MTP, 2008a; MTP, 2008b). Further details are available in Fidar (2010).

The energy consumption (in kWh) of the hot water using micro-components, such as showers, internal taps and baths is calculated using the following equation (Gettys *et al.* 1989):

$$E = \frac{mc\Delta T}{3.6 \times 10^6 \eta} \tag{13.1}$$

where,

E = energy requirement (kWh)
m = mass of the water used (kg)
c = specific heat capacity of water (4190 J/kg/°C)
ΔT = change in water temperature (°C)
η = efficiency of the heating system

The constant is the conversion factor from joules to kWh. It was assumed that only 50% of water consumed from internal taps was heated. The in-built flexibility in the assessment methodology allows assessing the impact of any user defined hot-cold water split.

Added to the overall energy consumption of the end uses is 1.315 kWh per m^3, which represents the energy required for the treatment and delivery of potable water and resulting wastewater collection, treatment and disposal (Water UK, 2008).

The energy use of the GWR systems described above is summarised in Table 13.3. Calculating the energy use associated with lifting the effluent to the point of use requires determining the hours the booster pump is operated annually and the power rating of the specified pump. The flow rate of the specified pump and the mass (volume) of the water to be lifted determine the hours. Note that the WC flush volume represents the volume of water to be lifted. With System A, the manufacturer specified a 750 Watt booster pump. For Systems B and D, the energy use for lifting the water is calculated with the assumption that the power rate and flow rate of the booster pump are 500 Watt and 6 litres/minute (or the nominal cistern volume of a water saving WC), respectively. In the assessment, the performance efficiency of the pump was assumed to be 80%.

Table 13.3 Energy use of the considered GWR systems.

System	Energy use (kWh/m³)	
	Treatment processes	Booster pump
A	1.2	2.6
B	1.35	1.7
C*	3.5	
D	2.0	1.7

*The booster pump is integrated within the treatment processes of system C.

The energy consumption associated with residential water using end uses is converted to greenhouse gas (GHGs) emissions, based on Defra (2008b) guidelines to GHG conversion factors, taking the source of the energy (in this study, gas and electricity) into consideration. An additional 0.97 kg CO_2 equivalent per m^3 resulting from water supply and wastewater treatment operations is added (Water UK, 2008).

The energy consumption and the carbon emissions were calculated per household per year, based on a household occupancy of 2.4, which is in agreement with Waterwise (2011). In addition to this, a gas boiler with a performance efficiency of 70% was assumed for the water heating system, as this is typical in UK homes (MTP, 2008d).

Since a wide variety of water saving micro-components offering varying claims of water efficiency, associated energy consumption and costs is commercially available, it is possible to develop numerous composite strategies (i.e., combinations of water using appliances/fixtures/micro-components) to deliver a desired water efficiency (CSH level). In order to automatically select the most efficient (in terms of cost and energy and water consumption) optimal composite strategies, a multi-objective based simulation and assessment tool was developed to identify appropriate composite strategies meeting Level 5 and 6 of the CSH.

13.6.3 Application of a multi-objective optimisation based assessment tool

A multi-objective based optimisation simulation tool was developed and applied to facilitate the generation of micro-component based composite strategies, integrate the generated composite strategies with GWR systems and analyse their energy and carbon implications.

The assessment tool consists of several interconnected components including an input module, a database on water efficient fixtures, a composite strategies generator, a filter, an optimisation engine and an analyser. These components are briefly described here.

- *Input module*: This enables the tool user to set the number of composite strategies to be generated, define constraints on water and/or energy consumption, specify hot and cold water split and input micro-component use characteristics. This was programed using Visual Basic for Applications (VBA) code.
- *Water efficient fixtures database*: This is at the centre of the tool and stores the details of the various household water using micro-components. Seven micro-components, as shown in Table 13.2, have been included in the database. For each micro-component, different commercially available types have been considered. In total the database has over 300 different types of micro-components. The database also contains information on flow rates for the different types of showers and taps, flush volume of the WCs, overflow capacity of baths and water and energy use associated with washing machines and dishwashers. The database has been compiled from the brochures provided by the micro-components' manufacturers/suppliers including BMA (2014).
- *Composite strategies generator*: This is used to automatically generate composite strategies for desired per capita water consumption and calculate the associated energy consumption and carbon emissions, based on the approach discussed earlier in Section 13.6.2. An extensive number of permutations can be generated to develop composite strategies delivering the desired levels of water consumption. The calculation of water and energy consumption and carbon loads for such an extensive number of composite strategies requires considerable computational effort. To facilitate smooth data processing, the generator was compiled using VBA.
- *Filter*: The composite strategies produced by the generator are further processed through a filter. The strategies that meet the constraints defined in the input module are passed to the analyser for further analysis.
- *Optimisation engine*: A multi-objective optimisation engine, that being GANetXL (Bicik *et al.* 2008) has been incorporated to process the composite strategies (produced by the generator) to identify the solutions with the minimal resource consumption and environmental impact. This optimisation functionality allows users to achieve a better trade off for a given set of constraints (e.g., savings in water consumption vs. energy consumption vs. cost etc.)
- *Analyser*: This component facilitates investigation of the role of each micro-component in the overall water and energy consumption in a given composite strategy.

The interaction and data flow between these components is explained in Fidar (2010).

13.7 RESULTS AND DISCUSSION

To quantify the implications of on-site GWR systems, numerous composite strategies were generated that can deliver the higher water efficiency levels set out in the CSH (i.e., Levels 3 & 4 and Level 5 & 6, as shown Table 13.1). Of the generated composite strategies, 10 were randomly chosen (five each for CSH Levels 3 & 4 and Level 5 & 6) to analyse in detail the potential water saving that could result from reusing greywater. Strategies A1 – A5 deliver Levels 3 and 4, whereas B1 – B5 can deliver Levels 5 and 6. The details of the selected composite strategies are presented in Table 13.4. The table shows that with on-site greywater treatment and reuse, it is possible to achieve the higher water efficiency levels of the CSH without considerably reducing the *actual* total water consumption.

In the UK, the volume of water used to flush the WC in a typical household is slightly smaller than the volume of water available from showers, baths and basin taps (EA, 2008). It therefore appears that the water demand for WC flushing can readily be met by reusing the greywater. Theoretically, this would provide significant savings since WC flushing represents a relatively large proportion of household water demand (Figure 13.1). However, it has been reported that the proportion of water used for WC flushing is decreasing, while the fraction used for personal washing is increasing (Clarke *et al.* 2009). The implication of this is that the potential water savings that could be achieved through reusing greywater becomes smaller as the water demand for WC flushing reduces, unless additional non-potable applications are considered. In addition to this, smaller water savings can in turn lead to longer payback period, making the GWR options less attractive to consumers. The water savings associated with GWR can be increased if the effluent is used for other water using activities such as washing machines and car washing. However, this might require higher quality effluent.

Figures 13.3 and 13.4 show the influence of the tool generated composite strategies on per capita mains potable water reduction and the associated energy and carbon loads, respectively. In the figures, each point refers to a simulated composite strategy. The figures also compare the performance of the generated composite strategies with and without GWR systems. With regard to energy use and environmental implications, this study confirmed that: while GWR options have the potential to save water, they are more likely to increase the domestic water-related energy consumption and the associated carbon loads. The increase in energy consumption and the associated carbon loads resulting from GWR ranged from 30 to 98 kWh/household/year and between 12 and 34 kg CO_2 eq/household/year, respectively. It is however, important to mention that GWR systems can reduce the potable water consumption by about 11 to 37%, depending on the WC flush volume.

Table 13.4 Selected composite strategies and water-using characteristics of micro-components.

CSH level	Composite strategies	Volume of water consumed by each micro-component*							Total water use by each composite strategy (litres/ capita. day)	
		WC (litres/ use)	Shower (litres/ min)	Basin tap (litres/ min)	Kitchen tap (litres/ min)	Dishwasher (litres/use)	Washing machine (litres/use)	Bath (litres/ use)	Without GW reuse	With GW reuse
Level 3 & 4	A1	6.0	7.0	3.5	5.0	15.0	45.0	155.0	133.9	105.0
	A2	5.0	8.0	3.0	5.0	14.0	49.0	145.0	128.8	105.0
	A3	6.0	11.0	2.5	3.5	15.0	53.0	140.0	133.9	105.0
	A4	4.5	6.0	4.0	5.8	15.0	52.0	120.0	126.5	105.0
	A5	5.0	6.7	3.6	5.8	11.0	49.0	132.0	129.0	105.0
Level 5 & 6	B1	4.0	6.7	2	2.5	14.0	48.0	120.0	99.0	80.0
	B2	6.0	6.0	2.0	3.5	12.0	46.0	110.0	108.7	80.0
	B3	3.75	6.7	2.0	2.5	13.0	50.0	118.0	98.0	80.0
	B4	5.0	8.0	1.8	2.5	12.0	42.0	116.0	103.8	80.0
	B5	6.0	7.0	1.8	2.5	15.0	45.0	125.0	108.8	80.0

*Based on the information provided by the manufacturers of the selected micro-components, GW – Greywater.

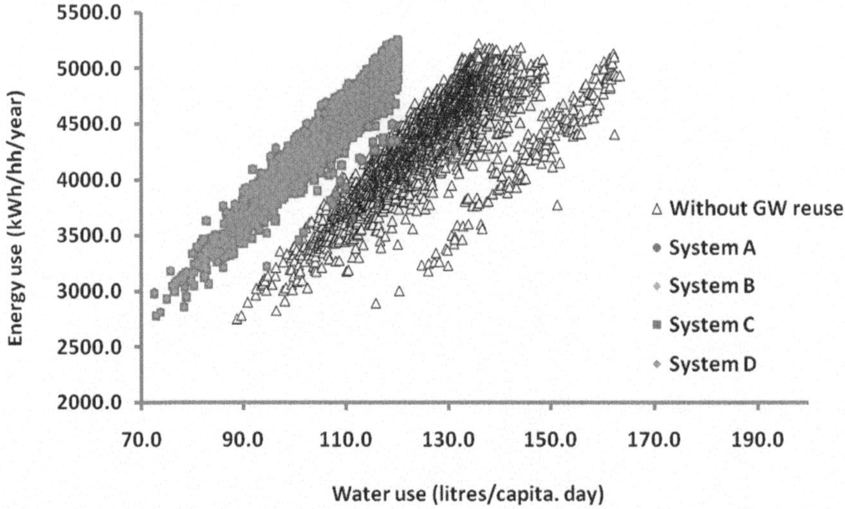

Figure 13.3 Influence of the various GWR systems on the energy use of the simulated water-saving strategies.

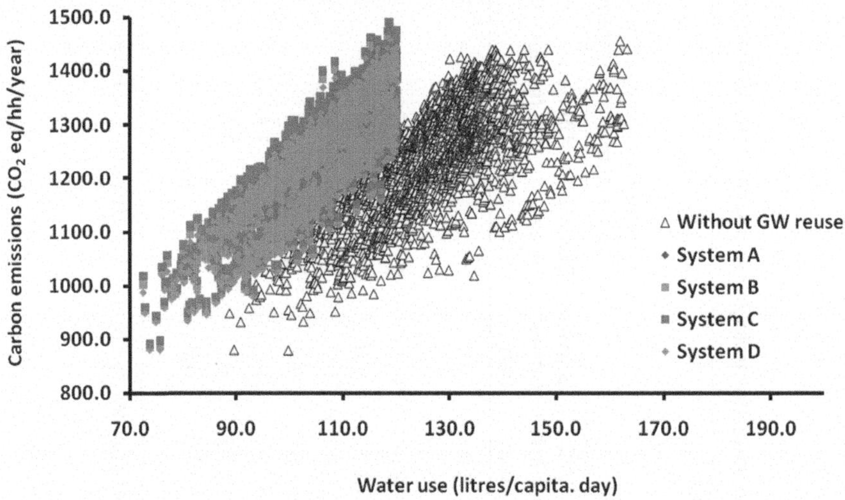

Figure 13.4 Influence of the various GWR systems on the carbon emissions of the simulated water-saving strategies.

Figures 13.5 and 13.6 present the annual household energy consumption and carbon emissions of the selected 10 composite strategies (Table 13.4), respectively. The figures show comparison of the obtained results with the performance of the respective composite strategies, excluding the GWR systems. As discussed above, in all strategies, GWR systems have increased the residential water-related energy consumption. It is important to note that the GWR systems have increased the proportions of the energy use and carbon loads associated with the in-house water use activities. It should also be noted that often a significantly greater fraction of energy is associated with water consumption in households than the combined water supply and wastewater treatment and disposal operations (Hackett & Gray, 2009).

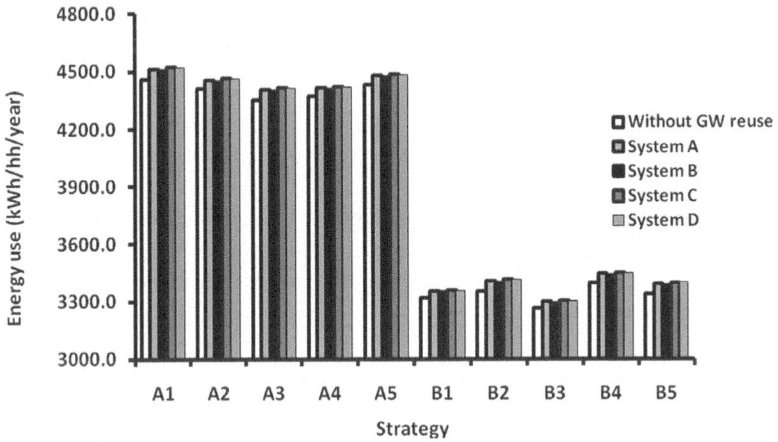

Figure 13.5 The Influence of GWR systems on energy consumption of the selected composite water-saving strategies given in Table 13.4.

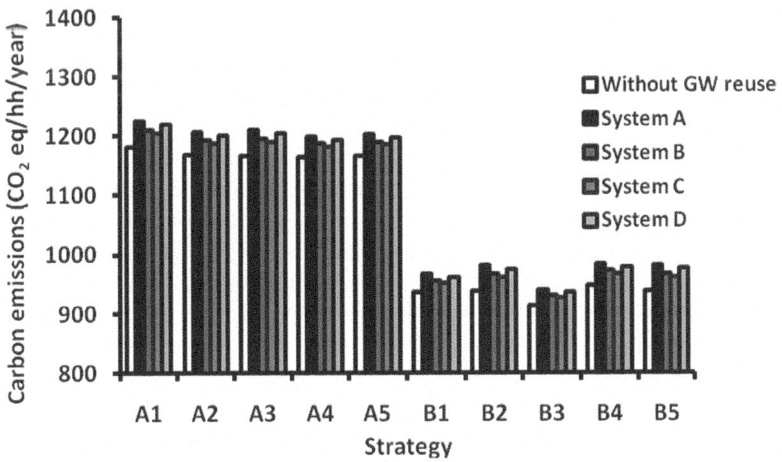

Figure 13.6 the Influence of GWR systems on carbon emissions for the selected composite water-saving strategies given in Table 13.4.

The increase in energy use and carbon emissions resulting from on-site greywater treatment and reuse is determined mainly by the treatment processes employed (by the systems), the volume of effluent treated and the total head that the booster pump has to overcome.

Systems with simple treatment processes tended to have relatively lower energy consumption and carbon emissions. However, according to Friedler (2008), reusing greywater without proper treatment may carry some health risk and have negative aesthetic and environmental effects. It is therefore important to carry out appropriate treatment in order to minimise concerns regarding public health implications. Although, in the UK, greywater used for WC flushing is required to be disinfected (Wheatley & Surendran, 2008), there are commercially available systems that do not disinfect the effluent. Similarly, in the USA, the reclaimed water used for WC flushing is required to undergo eventual filtration and disinfection (Al-jayyousi, 2003).

As stated previously, the variables that determine the actual energy consumption associated with GWR systems include the system flow rate (the volume of greywater processed per unit time). Many studies linked the energy consumption of the GWR systems to the volume of effluent treated (Zhang *et al.* 2010; Ryan, 2007; Cornel & Krause, 2004; Zhang *et al.* 2003). Friedler (2008) and Friedler and Hadari (2006) showed that the energy use of the GWR systems with rotating biological contractor (RBC) increased exponentially with the system flow rate. According to Fletcher *et al.* (2007), the energy consumption of a submerged MBR GWR system arises mainly from a combination of aeration and liquid pumping together with a small fraction required for the control equipment. Aeration is an essential process in the majority of wastewater treatment processes and constitutes the largest fraction of plant energy consumption (Chamber *et al.* 1998; Lekov *et al.* 2009; Rosso & Stenstrom, 2006; Doan & Lohi, 2009). Note that the oxygen requirement of biological treatment activities and converting nitrite to nitrate is determined, among other factors, by the system's flow rate (Ryan, 2007; Fletcher *et al.* 2007).

It is clear from the above results and analysis that achieving water efficiency targets through reusing greywater may be at the cost of energy use and the associated carbon loads. This means that energy consumption and the carbon emissions are likely to rise with the increasing water savings through GWR systems. For example, the implications of GWR systems have been evaluated by comparing the performance of two similar (hypothetical) households, one with a 6-litre flush WC and the other with a 4.5-litre WC. With System A (Figure 13.2), the household with the 6-litre WC would see an annual household water saving of more than 25 m^3, while its annual energy consumption and the resulting carbon emissions would increase by approximately 83 kWh and 20 kg CO_2eq, respectively. In comparison, the household with the 4.5-litre WC, would experience an annual household water saving of about 19 m^3 and an increase in energy consumption and carbon loads of around 62 kWh and 15 kg CO_2eq, respectively. Similarly, as discussed earlier, using the treated greywater for other end uses such as washing clothes can reduce domestic water consumption, but would further increase water-related energy use and the associated carbon emissions. Therefore, an objective-oriented (i.e., whether to reduce water consumption or energy use and carbon emissions) trade off is required.

As previously discussed, pumping the effluent from the treatment unit to the point of use is an important component in the energy consumption associated with the GWR systems. Based on the results obtained from the GWR systems considered in the study presented, the pumping activities consume a significant fraction of the total energy use of the system (Table 13.3). Factors that determine the actual energy consumption of this component include the mass (volume) of water to be lifted, the elevation difference, the head loss in the pipes and the pump efficiency. The mass of water to be lifted is determined by the water demand of the application for which the effluent is used such the WC flush volume. The elevation difference varies with buildings. Friedler (2008) and Friedler and Hadari (2006) considered each storey to be about three metres. Since the GWR system treatment units are mostly situated in the basement of the building, they included an extra three metres in their calculation. For example, for a two-story building, the elevation difference would be 9 metres. The head loss associated in the pipe can be calculated with the Darcy-Weisbach equations (Equation 13.2) and was found to be negligible (less than 2 cm).

$$h_f = \frac{flv^2}{2gD} \tag{13.2}$$

where

h_f = the head loss (m)
f = coefficient of friction
l = length of the pipe (m)
v = velocity of the water through the pipe (m/second)
g = acceleration due to gravity (m/second^2)
D = diameter of the pipe (m).

The value of f can be assumed as 0.005 (Michael, 2003).
The power (in kW) required by the pump can be calculated using Equation 13.3.

$$P = \frac{Q\rho gZ}{3.6 \times 10^6 \eta} \tag{13.3}$$

where

Q = the flow (m^3/h)
ρ = water density (kg/m^3) and
η = the pump overall efficiency (assumed to be 80%)
Z = total head (m) = elevation head + head loss

The equation shows that the power requirement increases with the mass of the liquid and the total head. As a result, recycling more water will result in higher energy consumption and the associated carbon emissions. For example, (based on Equation 13.3), the power required to elevate six litres of water (the maximum allowable WC flush volume in the UK) to twelve metres would be about 15 Watts, whereas lifting 4.5 litres of water would require about 11 Watts. However, as discussed above, recycling less water means less water savings resulting from installing the GWR system, and that makes the option less attractive. It is also important to highlight that most of the commercially available pumps used for lifting the greywater to the point of use appear to be overdesigned. Consequently, innovation in or the utilisation of more energy efficient pump types for GWR, could facilitate a reduction in the energy and carbon implications associated with GWR systems.

Finally, it should be noted that in a typical household, the WC is the only water using micro-component that does not require energy input. However, if a greywaster reuse system is installed to meet the water demand for WC flushing, this end-use (WC flushing) will also then represent an energy input.

13.8 CONCLUSIONS

Globally, particularly in water stressed regions, the wider uptake of water efficiency measures has become possible due to mandatory policies and wide-ranging public acceptability. However, the situation in the UK, although improving, remains very much focused on the new housing stock with little effective impact on the majority of the existing dwellings. To reduce fresh potable water mains supply, water saving micro-components are preferred over rainwater harvesting or reuse of greywater. The available reuse schemes, described elsewhere in this book, can be regarded as demonstration projects providing a fertile platform for further research, identification of issues and strategies to overcome them.

The results of the study presented in this chapter have confirmed that water efficiency measures with a GWR option have the potential to reduce potable water consumption. However, they are more likely to increase the domestic water-related energy consumption and the associated carbon loads. The extent of the increase in energy use and carbon emissions resulting from on-site greywater treatment and reuse is determined mainly by the volume of effluent treated and the total head that the booster pump has to overcome. Therefore, an objective-oriented (i.e., whether to reduce water consumption or energy use and carbon emission) trade-off is required.

The wider uptake of GWR systems utilising conventional energy intensive treatment processes appears unlikely, in the current climate, due to carbon implications, associated costs and social perceptions. However, energy related implications are likely to reduce with further technological innovations and wider uptake of renewable energy sources.

REFERENCES

Al-jayyousi O. (2003). Greywater reuse: towards sustainable water management. *Desalination*, **156**(1–3), 181–192.

Atkinson S., Farmani R., Memon F. A. and Butler D. (2014). Reliability indicators for water distribution system design: a comparison. *ASCE – Journal of Water Resources Planning and Management*, **140**(2), 160–168.

BMA (2014). Bathroom Manufacturers Association (http://www.bathroom-association.org/) (accessed 2 June 2014).

Bicik J., Morley M. S. and Savić D. A. (2008). A Rapid Optimization Prototyping Tool for Spreadsheet-Based Models. *Proceedings of the 10th Annual Water Distribution Systems Analysis Conference WDSA2008*, Kruger National Park, South Africa, pp. 472–482.

Blanksby J. (2006). Water conservation and sewerage systems. In: Water Demand Management, D. Butler and F. A. Memon (eds), ISBN 9781843390787, IWA Publishing Chapter 5, 107–129.

Caldwell P. (2009). Energy Efficient Sewage Treatment: Can Energy Positive Sewage Treatment Works become the Standard Design? *3rd European Water and Wastewater Management Conference*, 22–23 September 2009, http://www.hydrok.co.uk/downloads/CaldwellPaper.pdf (accessed 22 March 2014).

Clarke A., Grant N. and Thornton J. (2009). Quantifying the Energy and Carbon Effects of Water Saving. Environment Agency and Energy Saving Trust. http://www.energysavingtrust.org.uk/Publications2/Housing-professionals/Heating-systems/Quantifying-the-energy-and-carbon-effects-of-water-saving-summary-report (accessed 31 March 2014).

Cornel P. and Krause S. (2004). State of the art MBR in Europe. *Technical Note of National Institute for Land and Infrastructure Management*, **186**, 151–162.

DCLG (2008). Code for Sustainable Homes: Technical Guide. Department for Communities and Local Governments. https://www.planningportal.gov.uk/uploads/code_for_sust_homes.pdf (accessed 31 March 2014).

Defra (2008a). Future Water – the Government's Water Strategy for England. Department for Environment, Food and Rural Affairs, Cm 7319, http://archive.defra.gov.uk/environment/quality/water/strategy/pdf/future-water.pdf (accessed 31 March 2014).

Defra (2008b). Guidelines to Defra's GHG Conversion Factors. Department for Environment, Food and Rural Affairs. http://archive.defra.gov.uk/environment/business/reporting/pdf/passenger-transport.pdf (accessed 31March 2014).

Doan H. and Lohi A. (2009). Intermittent aeration in biological treatment of wastewater. *American Journal of Engineering and Applied Science,* **2**, 260–267.

EA (2014). Save water. Environment Agency, http://www.environment-agency.gov.uk/homeandleisure/beinggreen/117266.aspx (accessed 22 March 2014).

EA (2008). Greywater: an Information Guide. EHO0408BNWQ-E-E. Environment Agency, http://www.highland.gov.uk/NR/rdonlyres/4F44CF29-5BC5-43FD-A68F-EF390F972B45/0/GreywaterRecyclingInfoGuide.pdf (accessed 31 March 2014).

Fidar A. (2010). Environmental and Economic Implications of Water Efficiency Measures in Buildings, PhD Thesis, University of Exeter.

Fletcher H., Mackley T. and Judd S. (2007). The cost of a package plant membrane bioreactor. *Water Research*, **41**(12), 2627–2635.

Friedler E. (2008). The water saving potential and the socio-economic feasibility of GWR within the urban sector – Israel as a case study. *International Journal of Environmental Studies*, **65**, 57–69.

Friedler E. and Hadari M. (2006). Economic feasibility of on-site GWR in multi-storey buildings. *Desalination*, **190**, 221–234.

Friedler E., Kovalio R. and Galil N. (2005). On-site greywater treatment and reuse in multi-storey buildings. *Water Science and Technology*, **51**(10), 187–94.

Gettys W., Keller F. and Skove M. (1989). *Physics: Classical and Modern*. McGrow-Hill Books Company, 380.

Grant N. (2006). Water conservation products. In: *Water Demand Management*, D. Butler and F. A. Memon (eds), ISBN 9781843390787, IWA Publishing Chapter 4, pp. 82–106.

Hackett M. and Gray N. (2009). Carbon dioxide emission savings potential of household water use reduction in the UK. *Journal of Sustainable Development*, **2**(1), 36–43.

Henkel J., Lemac M., Wagner M. and Cornel P. (2009). Oxygen transfer in membrane bioreactors treating synthetic greywater. *Water Research*, **43**(6), 1711–1719.

Lekov A., Thompson A., McKane A., Song K. and Piette M. (2009). Opportunities for Energy Efficiency and Open Automated Demand Response in Wastewater Treatment Facilities in California – Phase I Report. Ernest Orlando Lawrence Berkley National Laboratory. http://drrc.lbl.gov/sites/all/files/lbnl-2572e.pdf (accessed 8 August 2014).

Li S., Wichmann K. and Otterpohl R. (2009). Review of the technological approaches for greywater treatment and reuses. *Science of the Total Environment*, **407**(11), 3439–49.

Melin T., Jefferson B., Bixio D., Thoeye C., De Wilde W., De Koning J., van der Graaf J. and Wintgens T. (2005). Membrane bioreactor technology for wastewater treatment and reuse. *Integrated Concepts in Water Recycling*. Khan, Schäfer and Muston (eds) – ISBN 1741280826.

Michael A. (2003). *Irrigation: Theory and Practice*. Vikas Publishing Ltd, New Delhi.

MTP (2008a). BNW05: Assumptions Underlying the Energy Projections for Domestic Washing Machines. Version 3.1. Market Transformation Programme.

MTP (2008b). BNW07: Assumptions Underlying the Energy Projections for Domestic Dishwashers. Version 3.1. Market Transformation Programme.

MTP (2008c). BNW16: A Comparison of Manual Washing up with a Domestic Dishwasher. Version 2.5. Market Transformation Programme.

Pidou M., Memon F. A., Stephenson T., Jefferson B. and Jeffrey B. (2007). Greywater Recycling: Treatment Options and Applications. *Proceedings of the Institution of Civil Engineers Engineering Sustainability*, **160**, pp. 119–132.

Rosso D. and Sternstrom M. (2006). Aplpha Factors in Full-scale Wastewater Aeration Systems, *Proceedings of the Water Environment Federation*, DOI:10.2175/19386470 6783762940. http://www.environmental-expert.com/Files%5C5306%5Carticles%5C1 2651%5C386.pdf (accessed 31 March 2014).

Ryan O. (2007). Energy Efficiency Solutions for a Full-Scale MBR Plant. MSc Thesis. Cranfield University.

Shin C. and Johnson R. (2007). Permeate characteristics and operation conditions of a membrane bioreactor with dispersing modified hollow fibres. *Journal of Industrial and Engineering Chemistry*, **13**(1), 40–46.

Toifl M., Diaper C. and O'Halloran R. (2008). Assessing the performance of small scale greywater treatment systems under controlled laboratory conditions. *Water*

Practice and Technology, **3**, doi:10.2166/wpt.2008.077. http://www.iwaponline.com/wpt/003/0077/0030077.pdf (accessed 31 March 2014).

WEF (2012). World Economic Forum Global Risk Report. 7th edn, http://www.weforum.org/reports/global-risks-2012-seventh-edition (accessed 22 March 2014).

Wagner M., Cornel P. and Krause S. (2002). Efficiency of Different Aeration Systems in full Scale Membrane Bioreactors. *Proceedings of the Water Environment Federation*, http://www.ingentaconnect.com/content/wef/wefproc/2002/00002002/00000010/art00035 (accessed 31 March 2014).

Water UK (2008). Sustainability Indicators 2007/08. Water UK. http://www.water.org.uk/home/news/press-releases/sustainable-water/sustainability2008.pdf (accessed 31 March 2014).

Waterwise (2011). Evidence Base for Large-scale Water Efficiency: Phase II Final Report, Waterwise, http://www.waterwise.org.uk/resources.php/12/evidence-base-for-large-scale-water-efficiency-phase-ii-final-report#sthash.kYRBDnrY.dpuf (accessed 31 March 2014).

Wheatley A. and Surendran S. (2008). Greywater treatment and reuse results from a 50-person trial. *Urban Water Journal*, **5**(3), 187–194.

Winward G., Avery L., Stephenson T. and Jefferson B. (2008). Chlorine disinfection of grey water for reuse: Effect of organics and particles. *Water Research*, **42**(1–2), 483–491.

Wu Z., Zhou Z., Wang Z., Tian L., Pan Y. and Wang X. (2009). The application of membrane bioreactor technology to the treatment of wastewater from a multifunctional supermarket. *Environmental Progress & Sustainable Energy*, **29**(1), 52–59.

Zhang Q., Wang C., Xiong Q., Chen R. and Cao B. (2010). Application of life cycle assessment for an evaluation of wastewater treatment and reuse project – case study of Xi'an, China. *Bioresource Technology*, **101**(5), 1421–1425.

Zhang S., Houten R., Eikelboom Doddema H., Jiang Z., Fan Y. and Wang J. (2003). Sewage treatment by a low energy membrane bioreactor. *Bioresource Technology*, **90**(2), 185–192.

Section III

Wastewater Reuse Systems

Chapter 14

Introduction to sewer mining: Technology and health risks

Amit Chanan, Saravanamuth Vigneswaran, Jaya Kandasamy and Stuart Khan

14.1 INTRODUCTION

Sewer mining describes the process of extracting valuable water from the sewerage network, by treating raw sewage to very high standards. The Australian Capital Territory Electricity and Water (ACTEW), an Australian utility company, is credited for coining an interchangeable term 'Water mining'. ACTEW used it to describe the process of extracting raw sewage from a sewer and treating it to suitable standards for use as irrigation water for public open space (Butler & McCormick, 1996).

Chanan and Kandasamy (2009) and Chanan (2012) preferred using the term water mining over sewer mining, to highlight that the substance of value being mined in this process is water and not sewage. It can be argued that the term sewer mining is rather ambiguous and perhaps carries a negative connotation.

According to Chanan and Kandasamy (2009), sewer mining operates independently from the conventional centralised sewage treatment facility. A small-scale treatment plant simply taps into a sewer main and extracts the effluent, processing it to a suitable standard. Sludge or any other process residues such as filter backwash water and plant wash-down water are returned to the sewer and treated in the usual manner at the central sewage treatment plant (Phillips, 2004). Figure 14.1 illustrates the typical set-up of a sewer mining scheme.

According to Asano (2007), the sewer mining concept is nearly fifty years old, dating back to the implementation of Whittier Narrows Water Reclamation Plant in Los Angeles County in the 1960s. In the United States, sewer mining schemes are commonly described as Satellite Treatment Systems, referring to their outpost location away from the central sewage treatment plant. The location of sewer mining facilities closer to the point of water use has also been highlighted in some literature, giving sewer mining schemes the title of *Point-of-sale Reuse.*

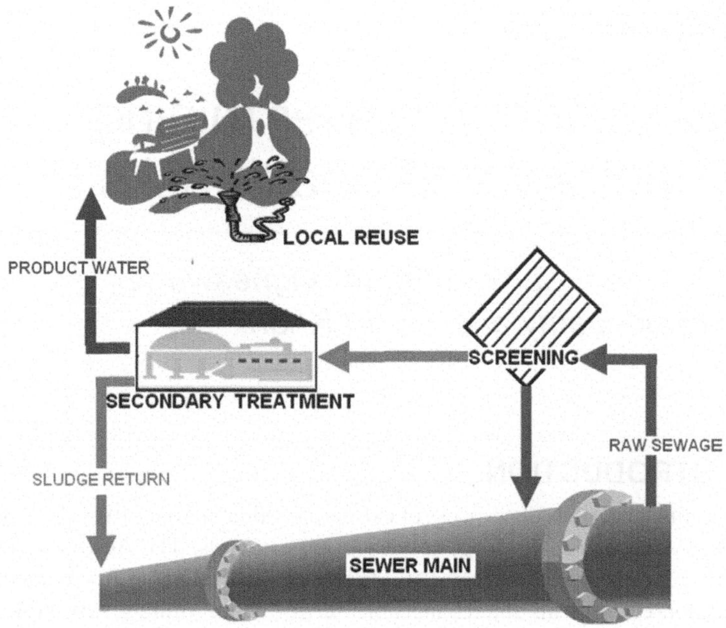

Figure 14.1 Schematic illustration of a water mining facility (adapted from Gikas & Tchobanoglous, 2007).

14.2 ADVANTAGES OF SEWER MINING

There are several advantages of moving away from conventional centralised sewer management systems and adopting innovative concepts such as sewer mining. Some of these crucial advantages have been discussed by Chanan and Kandasamy (2009) and are outlined below.

14.2.1 Reduced transportation costs

Sewer systems are typically designed to convey projected peak wastewater flows; consequently a direct result of the increasing urbanisation process has been a progressive increase in the size of interceptor sewers and tunnels (Chanan & Kandasamy, 2009). Larger sewer conveyance systems simply enable transporting of larger volumes of wastewater over significant distances to the centralised sewage treatment facilities. Constructing this large collection infrastructure network comes with obvious major cost implications.

Andoh (2004) highlighted that municipal wastewater is typically composed of more than 99% water and only less than 1% solids. Increasing the dry solids content of wastewater from 1% to 2% therefore effectively reduces the quantity of water that has to be conveyed by half, providing scope for major costs savings. As highlighted in Figure 14.2, sewer mining schemes, by virtue of extracting valuable water out of the sewage stream, assist in reducing the volume of water needed to be transported to centralised facilities.

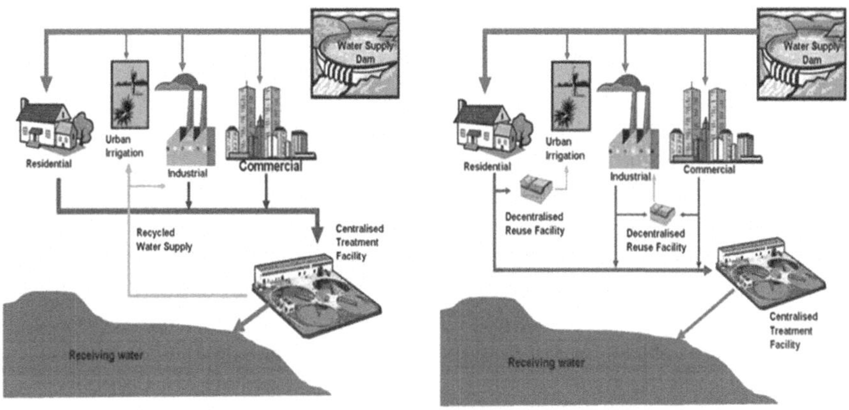

Figure 14.2 Centralised reuse vs sewer mining (Chanan Kandasamy, 2009).

In a comparative study of the three utility sectors, Marsden (2005; cited in Chanan & Kandasamy, 2009), concluded that gas and electricity are expensive to produce, but relatively inexpensive to transport. Bulk water and wastewater services on the other hand, are relatively less expensive to produce than to transport. Table 14.1 shows that the transportation costs in the water and wastewater sector comprise 21% of total cost compared with only 8% and 14% for electricity and gas, respectively.

Table 14.1 Comparison of transportation & production costs in the utility sector.

	Transmission infrastructure, as % of total asset value	Transportation costs, as % of total business costs	Production costs, as % of total business costs
Water	70	21	31
Electricity	50	8	50
Gas	60	14	40

Source: ACIL Tasman (1997).

Transportation costs are also an issue when considering the supply network for treated water. One of the major difficulties in introducing water recycling in urban areas with centralised wastewater systems is the issue of long distance transport of treated water. Sewer mining allows the flexibility of extracting water directly from a sewer practically anywhere and treating it closer to the site of its intended use (Khan, 2007). It therefore eliminates the need to transport water over long distances from centralised treatment facilities at the end of the sewer network to where it is needed.

14.2.2 Improved treatment of organic solids

Treatment of urban wastewater involves a number of unit processes. A conventional wastewater treatment plant involves preliminary/primary treatment processes for large debris, gross solids, floatables and readily settle able solids. Primary treatment processes are typically followed by biological treatment processes for secondary treatment to remove finer solids, dissolved pollutants and nutrients (Andoh, 2004). When water reuse is an objective, additional final polishing commonly includes filtration, either by granular media or synthetic membranes. Figure 14.3 highlights that the size of the particles being treated determines the appropriate treatment process. Looking at the figure, one can make a general observation that more complex unit processes and treatment stages are required with reducing particle sizes.

In most centralised sewerage networks, large organic solids are typically discharged in water closets (WCs; toilets), at the upper reaches of the system. These solids get degraded into smaller sized particles with age and transport through the sewerage network. This process is further assisted by ancillary components such as pumping stations that create high turbulence and shear. Consequently, wastewater at the end of an extensive sewerage network has a higher proportion of smaller sized solids, when compared with wastewater at the top end of the system (Chanan & Kandasamy, 2009).

As outlined in Table 14.2, larger organic solids found at the top end of the collection system settle rapidly and can therefore be easily removed using sedimentation processes. The earlier this separation is implemented in the sewerage system, the easier it is to achieve water quality benefits. Sewer mining facilities located in the upper reaches of a sewer network provide an ideal opportunity to gain from this treatment advantage (Chanan & Kandasamy, 2009).

14.2.3 Enhanced resilience and disaster recovery

Given the heightened terrorism awareness in today's world, it is now well recognised that centralised water infrastructure, including sewerage systems, can become potential targets for terrorist activities. According to Gikas and Tchobanogloukas (2009), damage to centralised sewage treatment facilities whether caused by terrorism or by natural disasters, such as earthquakes or floods, has the potential to cause severe public health and environment impacts.

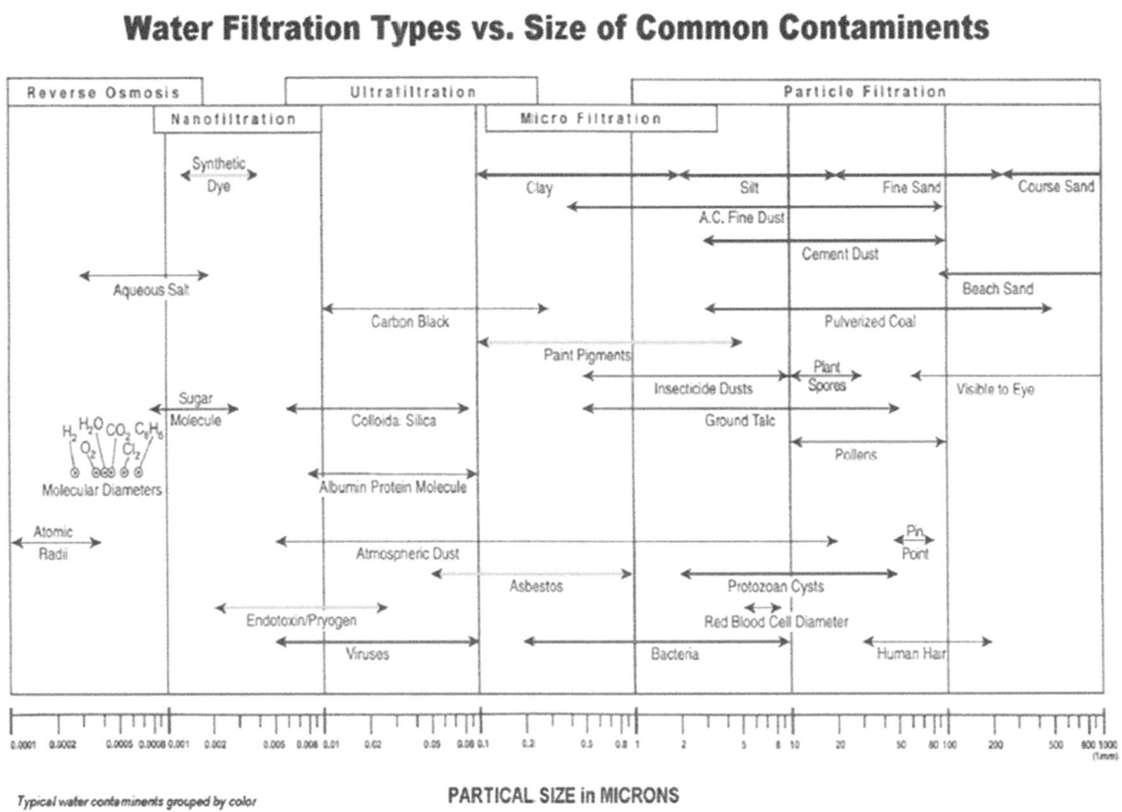

Figure 14.3 Particle sizes and appropriate treatment processes (Advanced Water Filters, 2011).

Table 14.2 Settling times for various solids.

Particle size (microns)	Order of size	Time required to settle (inert sediments, specific gravity = 2.65)	Time required to settle (organic solids, specific gravity = 1.2)
10000	Gravel	0.4 seconds	1.2 seconds
1000	Coarse Sand	3 seconds	9 seconds
100	Fine Sand	34 seconds	5 minutes
10	Silt	56 minutes	8 hours
1	Bacteria	4 days	32 days
0.1	Colloidal	1 year	9 years
0.01	Colloidal	>50 years	>50 years
0.001	Colloidal	>50 years	>50 years

Source: Adapted from Andoh (2004).

The centralised piped network is particularly vulnerable to natural disasters because of the large spatial spread of pipes crossing a variety of environments (Howard & Bartram, 2010). These include, for instance, low-lying areas, where there is increased vulnerability to floods and other risk events. The vulnerability of piped systems also arises from the large numbers of pipe joints, which are often the points of greatest weakness both for breaks and for ingress of contaminated water.

Within the water sector, sewerage systems have particularly limited resilience. Flood events can cause physical damage to sewerage infrastructure, resulting in leakage of sewage into the environment causing environmental contamination and public health risks (CSIRO, 2007). Differential ground settlement that can occur after floods or after prolonged periods of drought can also damage the sewer network (Fehnel *et al.* 2005; cited in Howard & Bartram, 2010).

Security and disaster recovery of a centralised sewerage system can be optimised by incorporating a number of water mining facilities throughout the network, thereby reducing the impact associated with such unforeseen events. A catastrophic failure of a local water mining facility would not cause the whole system to shut down. Similarly, failure in the central treatment facility could fall back on the cumulative capacity of the water mining facilities operating within the network.

14.2.4 Volume stripping and deferred capital investment

According to Chanan and Kandasamy (2009), water mining facilities treat sewage to extract water for local reuse and in doing so these facilities also take some of the load off the centralised treatment plant located at the end of the sewer network.

The phenomenon can be described as Volume Stripping, when considering the volume of wastewater reaching the central facility.

Depending on the number and size of the sewer mining facilities, volume stripping can delay the need for expansion of the central plant. Figure 14.4 illustrates how a well-planned and implemented sewer mining strategy is capable of reducing a water utility's costs through avoiding or deferring the need for a major capital upgrade of the central treatment plant and associated network.

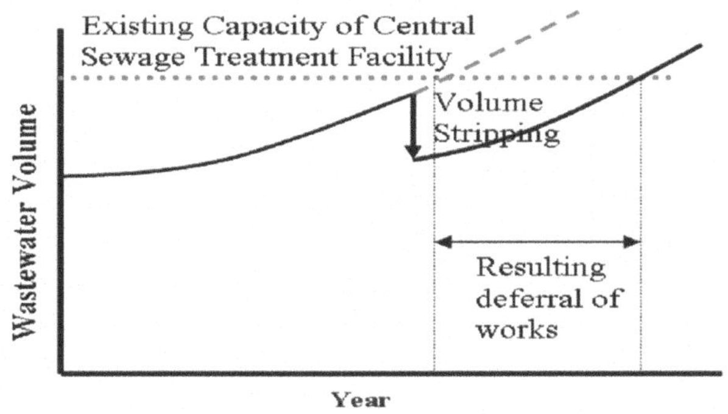

Figure 14.4 Volume stripping benefit of water mining plants (adapted from White, 1998).

The benefits of avoiding or deferring the expansion of existing systems through volume stripping go beyond financial gains. For instance, the expansion of sewer networks invariably involves disruptions in the flow of traffic and other public activities, which is not viewed favourably by most municipal governments and communities (Gikas & Tchobanoglou kas, 2007). Volume stripping helps to avoid these disruptions.

14.2.5 Fit for purpose treatment

Based on the type and level of treatment, sewer mining can be adapted to a wide range of community and industrial wastewater applications ranging from residential development projects to sports facilities and parks, for a range of volumes. As summarised in Table 14.3, the US EPA (2004) provides suggested water quality objectives for sewer mining facilities to safely produce treated water for various non-potable purposes including:

- Irrigation of public parks and recreation centres, including school yards and playing fields, and landscaped public areas;
- Irrigation of landscaped areas surrounding medium and high-density residential developments, general wash down, and other maintenance activities;

Table 14.3 Suggested water quality objectives for recycled water.

Proposed reuse purpose	Suggested treatment	Reclaimed water quality	Comments
Urban Reuse All types of landscape irrigation (e.g., golf courses, parks, cemeteries), vehicle washing, public & commercial toilet flushing, commercial air-conditioners, and other uses with similar exposure	Secondary + Filtration + Disinfection	pH = 6–9 BOD < 10 mg/l NTU < 2 F. Coli/100 ml = Not detectable Residual Cl_2 – 1 mg/l (minimum)	Chemical (coagulant and/or polymer) addition prior to filtration may be necessary to meet water quality requirements. The reclaimed water should not contain measurable levels of viable pathogens. Reclaimed water should be clear and odourless. A higher chlorine residual and/or a longer contact time may be necessary to assure that viruses and parasites are inactive or destroyed. A chlorine residual of 0.5 mg/l or greater in the distribution system is recommended to reduce odours, slime and bacterial growth.
Recreational Impoundments Incidental contact (e.g., fishing and boating) and full-body contact with reclaimed water allowed	Secondary + Filtration + Disinfection	pH = 6–9 BOD < 10 mg/l NTU < 2 F. Coli/100 ml = Not detectable Residual Cl_2 – 1 mg/l (minimum)	Dechlorination may be required to protect aquatic flora and fauna. Reclaimed water should be non-irritating to skin and eyes. Reclaimed water should be clear and odourless. Nutrient removal may be necessary to avoid algal growth in the impoundments. Chemical (coagulant and/or polymer) addition prior to filtration may be necessary to meet water quality requirements. The reclaimed water should not contain measurable levels of viable pathogens. A higher chlorine residual and/or a longer contact time may be necessary to assure that viruses and parasites are inactivated or destroyed. Fish caught in impoundments can be consumed.
Construction Use Soil compaction, dust control, washing aggregate, making concrete	Secondary + Disinfection	BOD < 30 mg/l TSS < 30 mg/l F. Coli/100 ml < 200 Residual Cl_2 – 1 mg/l (minimum)	Worker contact with reclaimed water should be minimised. A higher level of disinfection, for example to achieve <14 faecal coliform/100 ml should be provided when frequent work contact with reclaimed water is likely.

Source: Adapted from US EPA (2004)

- Irrigation of landscaped areas surrounding commercial, office, and industrial developments;
- Irrigation of golf courses;
- Commercial uses such as vehicle washing facilities, laundry facilities, window washing, and mixing water for pesticides, herbicides, and liquid fertilizers;
- Ornamental landscape uses and decorative water features, such as public fountains, reflecting pools and waterfalls;
- Dust control and concrete production for construction projects;
- Toilet and urinal flushing in commercial and industrial buildings.

14.2.6 Right to reclaimed water

There is still significant debate on the ownership of the sewage, or more importantly the valuable water contained within. However, the literature is already discussing the pros and cons of public gain or private profit in what happens to municipal sewage. Yule (2008) questioned the equity in '*our sewage becoming a private property for sale, once it leaves our property*'.

In the United States, this battle has already reached the court of law, in the name of the community's 'Right to Reclaimed Water' (Chanan & Kandasamy, 2009). Allocating all or most of the available reclaimed water from a city to a single user, as is commonly considered in centralised sewer schemes, may not be most equitable. The City of Phoenix and the owners of the Palo Verde Nuclear Power Plant were recently sued by a land developer over the 'rights to the reclaimed water'. The court case resulted in the power plant implementing a demand management program and making excess water (about half of the original demand) available to other non-potable users within the city (Okun, 2000).

Sewer mining plants, unlike large scale centralised recycling facilities, are small in size and capture and reuse water within a local sewer catchment as opposed to the whole city. It can therefore be argued that sewer mining facilitates equity, when it comes to the right to reclaimed water (Chanan & Kandasamy, 2009).

14.3 TREATMENT OPTIONS FOR SEWER MINING

Water treatment processes typically involve biological, chemical and physical removal mechanisms (Chanan *et al.* 2010). These include:

- *Physical Removal* – These treatment technologies rely on physical separation processes such as filtration, sedimentation and flotation to remove pollutants;
- *Chemical Removal* – Chemicals, typically coagulants and flocculants, are used in these treatment techniques to increase the removal rate of pollutants;
- *Biological Removal* – These treatment technologies use biological processes to transform pollutants to more manageable forms for separation.

A range of commercially viable treatment technologies are now available to treat sewage to specified water quality targets. Given the emphasis on decentralised systems, most of these technologies have minimal plant footprint requirements, making them suitable for sewer mining operations. Commercially available sewer mining technologies are categorised into three major types (Holt & James, 2006):

- *Biological Processes*:
  - Sequencing Batch Reactors (SBR);
  - Natural systems such as subsurface wetlands;
  - Rotating Biological Contactors;
- *Physical Processes*:
  - Sand and media filtration;
  - Membrane Filtration (micro-, ultra-, nano-filtration and reverse osmosis);
- *Hybrid Processes (these combine physical, biological and/or chemical processes to achieve optimum results).*
  - Fine Solids Separator (FSS) and Biological process;
  - Membrane bioreactors (MBR).

The SBR technology is based on an activated sludge process, involving four sequential steps of filling, reacting (aeration), settling (sedimentation/clarification) and decanting (removal of clarified water) as shown in Figure 14.5. To reduce the plant footprint, the treatment steps are carried out in the same tank (Asano, 2007).

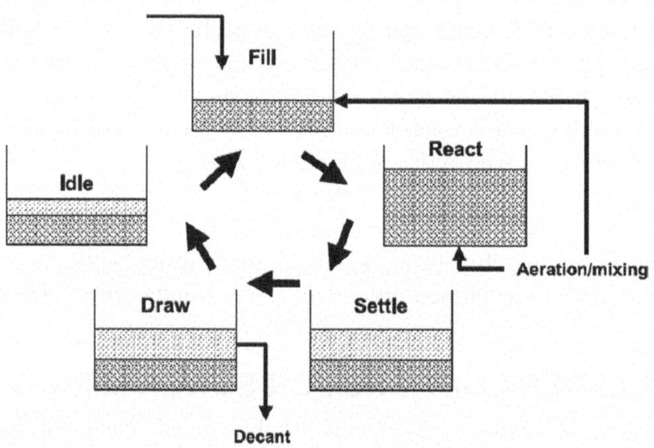

Figure 14.5 Schematic of sequential batch reactor process (adapted from US EPA, 2002).

By combining biological processes and membrane filtration, MBRs are capable of enhanced organics and suspended solids removal. Membrane filtration also allows MBR-type sewer mining facilities to be fairly compact in size, simple to operate, reduce sludge production and produce high quality reclaimed water

(Asano, 2007). The process involved in the MBR sewer mining plant is illustrated in Figure 14.6.

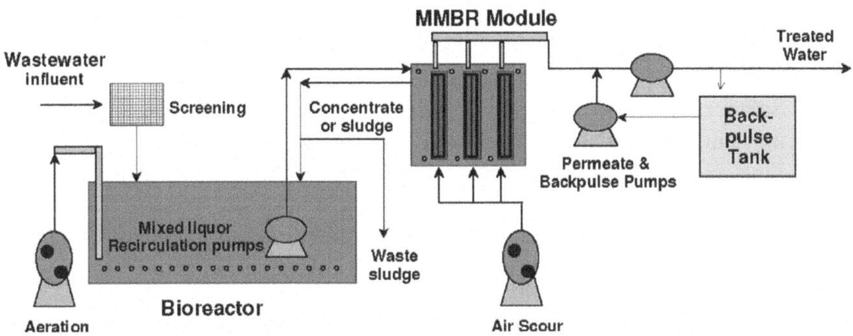

Figure 14.6 Schematic of MBR processes, popularly used in sewer mining schemes (adapted from Holt & James, 2006).

Another hybrid technology that utilises biological processes in conjunction with chemical and physical removal processes of coagulation and filtration, is known as the ReAqua Chemical Assisted Separation (CAS) process. This technology was applied to deliver Sydney, Australia's first sewer mining project at Beverley Park Golf Club. As outlined in Figure 14.7, the ReAqua CAS process involves coagulation to remove fine solids. Coagulation is followed by a submerged aerated filter for biological treatment. Effluent from the biological process is further polished through a sand filter and is disinfected by UV and/or chlorination (Chanan & Kandasamy, 2009).

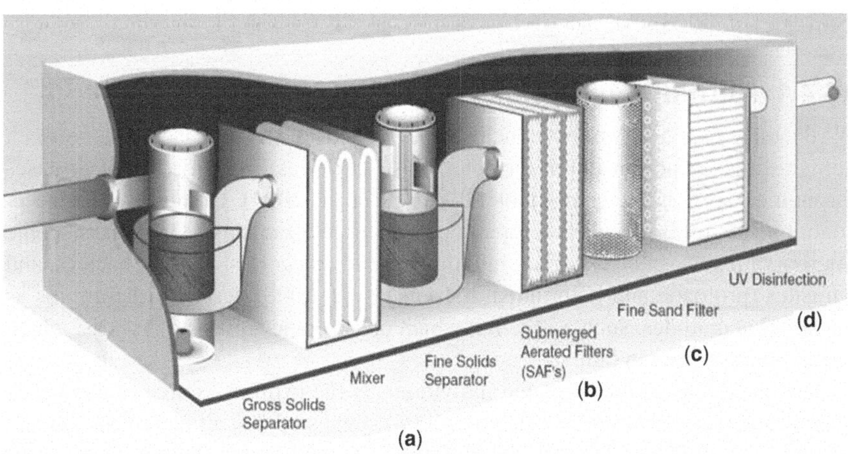

Figure 14.7 Schematic of the ReAqua CAS technology based sewer mining plant (adapted from CDS Technologies, undated).

As outlined in Figure 14.7, this hybrid System combines chemical, physical and biological treatment mechanisms. The system involves the following four key treatment steps:

(a) Chemical assisted fine solid separation (Chemical/Physical) – uses coagulation to remove fine solids;
(b) Submerged aerated filter (Biological) – enables biodegradable organic and nutrient removal;
(c) Fine sand or multimedia filter (Physical) – additional suspended solids removal;
(d) UV disinfection – the last step in the treatment process, commonly coupled with chlorination to meet the residual chlorine requirements.

For domestic sewage, the majority of pollutants are associated with fine particles (<50 μm) and colloidal solids (Levine *et al.* 1985). A compact and cost-effective primary treatment process that can rapidly remove high levels of these fine and colloidal solids has obvious application in wastewater treatment and reuse. To be suitable for sewer mining facilities, such processes should be rapid, compact and low in capital cost.

14.4 SEWER MINING RISKS

Water recycling schemes pose potential public and environmental health risks. A risk management approach is the best way to ensure protection of the public and the environment. A risk assessment needs to be undertaken for recycling water schemes and two key factors that need to be considered in this risk assessment are the standard to which the recycled water has been treated and its intended use (Steven *et al.* 2008).

14.4.1 Human health risks

While there is only limited conclusive evidence to show association between human disease and current wastewater reuse practices, transmission of diseases through reuse is however plausible (Westrell *et al.* 2004). Domestic wastewater contains hundreds of types of pathogenic microorganisms. These pathogenic microorganisms can be classified into three broad categories: viruses, bacteria and parasites (protozoa and helminths). Stevens *et al.* (2008) described helminths as intestinal nematodes, such as Taenia, which causes tapeworm and Ascaris, which causes roundworms in humans.

The faecal material of infected individuals is the primary source of pathogens in domestic wastewater. Westrell *et al.* (2004) suggested that all pathogens that are excreted in faeces could potentially be found in wastewater including:

(a) pathogens that mainly cause gastroenteritis (*Salmonella, Giardia, Cryptosporidium, rotavirus*);

(b) pathogens that cause milder respiratory infections (*adenovirus*);
(c) pathogens that can cause more severe disease, such as haemolytic uremic syndrome (*enterohaemorrhagic E. coli*).

Information on the removal of enteric viruses and protozoa (*Cryptosporidium* and *Giardia*) is considered to be crucial in the context of water reuse because of their low-dose infectivity, their long-term survival in the environment and the difficulties in monitoring these pathogens (Dewettinck *et al.* 2001).

A quantitative microbial risk assessment (QMRA) incorporated into a risk management framework can function as a valuable tool to identify potential human health threats (as detailed in Chapters 10 and 18). Such an approach can help in controlling possible health risks and at the same time encourage the public's level of confidence in different recycling alternatives (Westrell *et al.* 2004).

14.4.2 Environmental risks

Some of the common environmental risks associated with recycled water use include (Stevens *et al.* 2008):

- Salinity, which in high concentrations can degrade soils and impact on freshwater plants and invertebrates in the natural ecosystems;
- Nitrogen and Phosphorous, which can cause eutrophication in land and aquatic ecosystems;
- Organic loads and turbidity, which can have severe impact on the health of aquatic ecosystems.

In addition to the above common environmental risks, there are a number of other risks associated with recycled water use that may impact on the health of receiving flora. For example, sodicity (sodium content) can cause soil dispersion and swelling, and accumulation of Boron in soil can cause plant toxicity in some sensitive plant species (Stevens *et al.* 2008).

14.5 HAZARD ANALYSIS AND CRITICAL CONTROL POINTS (HACCP)

HACCP has been described as a systematic approach, leading to the detection, description and finally the control of hazards. Originally developed by the Pillsbury Company in 1960s to deliver safe food products for NASA's space program, HACCP is now widely used as the quality assurance strategy of choice by the food industry (Dewettinck *et al.* 2001). In 1993 the United Nations Food and Agricultural Organisation and World Health Organisation categorised HACCP as the standard means to assure food and beverage safety (Deere & Davison, 1999).

HACCP is a preventive system that helps to assure that products reaching the consumer are safe for consumption. If a hazard appears out of control, a quick and appropriate intervention can take place thereby minimising the risk on the health of the consumers (Dewettinck *et al.* 2001). Figure 14.8 provides a three step basic overview of the HACCP process as defined by Davison *et al.* (2001).

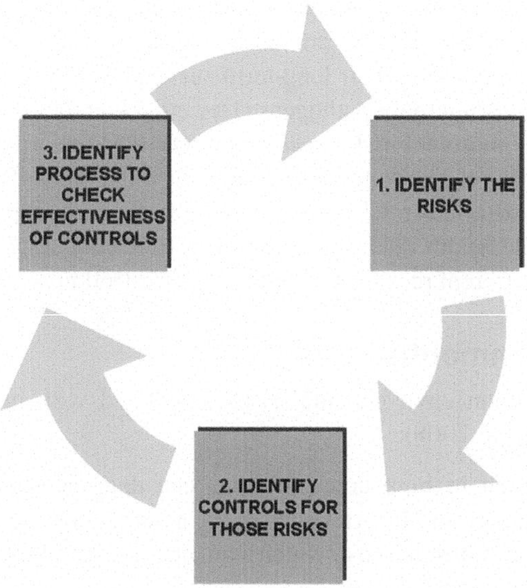

Figure 14.8 A schematic representation of the Hazard Assessment and Critical Control Points process (adapted from Davison *et al.* 2001).

HACCP involves the identification of Critical Control Points (CCPs) within production and distribution systems to control hazards and to maintain best management practices throughout processes. Criteria are established for monitoring at each CCP and corrective actions are established that are to be carried out when critical limits are not met (Westrell *et al.* 2004).

Yokoi *et al.* (2006) described the standard seven-step procedure for introducing the HACCP. Step 1 of hazard analysis involves identifying the health hazards caused by biological, chemical and/or physical means, and researching the likelihood of their occurrence, including the severity of the health hazards. By referring to the outcomes of hazard analysis, the most important hazard prevention measures are determined at the CCP. At each CCP, the monitoring systems and the acceptable range, called the critical limit (CL), are determined and the hazard prevention operations are checked for accuracy. The frequency for monitoring systems at the CCP must be enough to ensure product safety (Yokoi *et al.* 2006). Table 14.4 shows the standard procedures for HACCP introduction.

By identifying the CCPs in the production process, it is possible to control the production environment to ensure that safe products are being delivered. The proper identification of CCPs is an important issue in HACCP because the major efforts in process control and monitoring will be directed towards them. Incorrectly identified CCPs may lead to monitoring effort being wasted on a non-critical part of the production process, meanwhile hazards could be left unchecked at another location within the process.

Table 14.4 Standard procedures for HACCP introduction.

Step		Description
Step 1	Hazard Analysis	Identification of hazards, their respective severity and control measures
Step 2	Determine CCP	Important steps in the production process where hazard can be eliminated
Step 3	Establish CL for each CCP	Acceptable levels for monitoring at each CCP
Step 4	Establish monitoring system for each CCP	Monitoring system with appropriate frequency that satisfies the requirements for likely hazards at each CCP
Step 5	Establish corrective actions	Measures to be carried out at each CCP, in response to monitoring results going above the CL
Step 6	Establish verification procedures	Periodical verification procedures of the HACCP system
Step 7	Documentation and record keeping	Important data such as monitoring results and corrective actions to be recorded

Source: Adapted from Yokoi *et al.* (2006).

14.5.1 HACCP in the water industry

Following the systematization of HACCP by the Codex Alimentarius Commission of the World Health Organisation (WHO), its applicability to water supplies was first confirmed by Havelaar (1994), who considered the major microbiological hazards in the water supply system. Barry *et al.* (1998) applied HACCP for management of protozoan parasites in a compromised catchment. Deere and Davison (1998) and Davison and Deere (1999) discussed the relevance of HACCP to water management in Australia.

The WHO published its 3rd edition of the Guidelines for Drinking Water Quality in 2003. The guidelines include a requirement for the establishment of water safety plans (WSPs). WSPs are an improved risk management tool, which formally introduce risk assessment and risk management approaches in water supply systems. The WSPs primarily refer to the HACCP concept, which as previously discussed, is the quality management method of choice for food and medical industries. HACCP has been formally incorporated as part of the WSPs in the WHO Guidelines (WHO, 2003).

A major difference in producing safe wastewater compared to safe food is that the raw product, being the wastewater, already contains all the hazards, whereas in case of safe food production, the goal is to prevent contamination from entering food. The focus in sewer mining and water reuse must therefore be on controlling the exposure to raw wastewater and on eliminating or reducing the hazards through effective treatment.

As the starting point of any wastewater reuse scheme already contains the hazards, using compliance driven end-point testing is not enough because it would invariably be late in giving feedback. Such monitoring leaves the consumer open to risk and consequently the water provider open to litigation and loss of consumer confidence (Davison *et al.* 1999). The duty of care obligations on the water providers (including recycled water) requires them to demonstrate to any person using the water that the supplied water is fit for purpose (agreed intended use). According to Kogarah Municipal Council (2007), recycled water not meeting the agreed water quality standards could also be captured under the defective goods provisions of a consumer protection legislation.

The benchmark for the above-discussed 'duty of care' should be taken to be the highest applicable standard from the following (Kogarah Municipal Council, 2007):

- Prevailing revisions of the HACCP system for the sewer mining facility;
- Relevant National or Regional Guidelines for the Management of Recycled Water Quality (e.g., Australian Guidelines for Water Recycling: Managing Health and Environmental Risks (Phase 1); NSW Government 2007, NSW Guidelines for Management of Private Recycled Water Schemes);
- Appropriate industry technical guidance and standards.

It is therefore critically important that managers of any water recycling scheme fully understand their legislative obligations. Lack of understanding of one's legislative responsibilities is not an admissible defence in the court of law. On the other hand, demonstrable 'due diligence' by showing an up-to-date HACCP system and compliance with relevant guidelines, can be used as defence in the event of a legal challenge.

14.6 CONCLUSION

This chapter has introduced 'sewer mining' as the process of extracting valuable water from a sewerage network by treating raw sewage to high standards. Sewer mining has the potential to play a significant role in addressing water scarcity in major urban centres throughout the world. A range of sewer mining treatment technologies is now commercially available to treat sewage to specified water quality targets. Most of these technologies have minimal plant footprint requirements, making them suitable for decentralised operations, even in heavily urbanised areas with land availability constraints.

It is imperative to note that water is used for various purposes in our cities, with varying quality and quantity requirements. After use, the wastewater generated also varies in quality and quantity. Sewer mining now allows the water industry to take a major leap forward, by providing the opportunity to treat sewage only to a standard that fits the purpose of its intended use. Sewer mining facilities can be tailored to the local users' requirements thereby moving forward from the current 'one size fits all' philosophy, where all of a city's wastewater is treated to the same high level before wastefully discharging it into the nearest water-body.

Sewer mining forms part of a 'soft path' for water management, which emphasises the optimisation of end-use efficiency, small-scaled management systems, incorporates fit-for-purpose water use, and recommends the use of diverse, locally appropriate and commonly decentralised infrastructures.

REFERENCES

ACIL Tasman (1997). Third Party Access in the Water Industry: An Assessment of the Extent to which Services Provided by Water Facilities Meet the Criteria for Declaration of Access. A final report prepared for the National Competition Council, September 1997.

Advanced Water Filters (2011). Water Filtration Types vs Size of Common Contaminants. https://www.advancedwaterfilters.com/faq-particle-size-chart/ (accessed 29 April 2014).

Andoh R. Y. (2004). Why satellite treatment within collection systems makes sense. *Proceeding of Collection Systems 2004: Innovative Approaches to Collection Systems Management*, August 8–11 2004, Milwaukee, USA.

Asano T. (2007). Water Reuse: Issues, Technologies and Applications. Metcalf and Eddy Inc. McGraw-Hill, p. 1570.

Barry S. J., Atwill E. R., Tate K. W., Koopman T. S., Cullor J. and Huff T. (1998). Developing and implementing a HACCP-based program to control *cryptosporidium* and other waterborne pathogens in the alameda creek watershed: case study. *In Proceeding of American Water Works Association Annual Conference*, 21–25 June 1998, Dallas, Texas. Water Resources Vol. B, pp. 57–69.

Butler R. and MacCormick T. (1996). Opportunities for decentralized treatment, sewer mining and effluent re-use. *Desalination Journal*, **106**, 273–283.

CDS Technologies (undated), ReAqua: Water Mining Process/Reuse Water. Product Brochure.

Chanan A. and Kandasamy J. (2009). Water mining: planning and implementation issues for a successful project. In: Waste Water Treatment Technologies, Encyclopedia of Life Support Systems, V. Vigneswaran (ed.), UNESCO Publishing, Paris.

Chanan A., Saravanamuth V., Kandasamy J. and Shon H. K. (2010). Chemical-assisted physico-biological water mining system. *Proceedings of the ICE – Water Management*, **163**(9), 469–474.

Chanan A. P. (2012). A Case Study Research into Urban Water Reuse. Unpublished PhD Thesis, University of Technology Sydney, August 2012.

CSIRO (2007). Infrastructure and Climate Change Risk Assessment For Victoria. Report to the Government of Victoria. Melbourne, CSIRO.

Davison A. D. and Deere D. A. (1999). Safety on tap. *Microbiology Australia*, **20**, 28–31.

Deere D. A. and Davison A. D. (1998). Safe drinking water. are food guidelines the answer? *Water: Journal of the Australian Water Association*. November/December, pp. 21–24.

Davison A., Davis S., and Deere D. (1999). Quality assurance and due diligence for water – can HACCP deliver? In: Proc. of the 17th Federal Convention of the Australian Water and Wastewater Association. Adelaide, Australia.

Dewettinck T., Van Houtte E., Geenens D., Van Hege K. and Verstraete W. (2001). HACCP (Hazard Analysis and Critical Control Points) to guarantee safe water reuse and drinking water production – a case study, *Water Science and Technology*, **43**(12), 31–38.

Gikas P. and Tchobanoglous G. (2009). The role of satellite and decentralized strategies in water resources management. *Journal of Environmental Management*, **95**(1), 144–152.

Havelaar A. H. (1994). Application of HACCP to drinking water supply. *Food Control*, **5**(3), 145–152.

Hermanowicz S. W. and Asano T. (1999). 'Abel wolman's "the methabolism of cities" revisited: a case for water recycling and reuse'. *Water Science and Technology*, **40**(4–5), 29–36.

Heist J. A. and Davey A. (2002). A new high rate sewage clarification process. *Proceedings of WEFTEC 2002 – The Water Quality Event*. Water Environment Federation. **17**, 420–436.

Howard G. and Bartram J. (2010). Vision 2030: The Resilience of Water Supply and Sanitation in the Face of Climate Change. Technical report WHO/HSE/WSH/10.01. WHO Press, Geneva, Switzerland.

Holt P. and James E. (2006). Wastewater Reuse in the Urban Environment: Selection of Technologies. A report prepared by Ecological Engineering for Landcom. www.landcom.com.au/downloads/uploaded/Wastewater%20reuse%20technology%20report_links2_d960_de33.pdf (accessed 28 August 2009).

Khan S. (2007). Water Recycling in Australia – A Blog site. http://waterrecycling.blogspot.com/2007/03/sewer-mining.html (accessed 4 March 2007).

Kogarah Municipal Council (2007). HACCP Beverley Park Water Reclamation Project Plan. November 2007, Kogarah (Unpublished).

Levine A. D., Tchobanoglous G. and Asano T., (1985). Characterisation of the size distribution of contaminants in wastewater: treatment and reuse implications. *Journal of Water Pollution*, **57**, 805–816.

Okun D. A. (2000). Water reclamation and unrestricted non-potable reuse: a new tool in urban water management. *Annual Review of Public Health*, **21**, 223–245.

Phillips S. (2004). Is there gold in sewer mining? *Waste Management & Environment*, September 8, 2004.

Stevens D. P., Smolenaars S. and Kelly J. (2008). Irrigation of Amenity Horticulture with Recycled Water: A Handbook for Parks, Gardens, Lawns, Landscapes, Playing Fields, Golf Courses and Other Public Open Spaces. Arris Pty Ltd., Melbourne.

US EPA (2004). Guidelines for Water Reuse. US EPA Municipal Support Division, Washington DC, September 2004.

Westrell T., Schönning C., Stenström T. A. and Ashbolt N. J. (2004). QMRA (quantitative microbial risk assessment) and HACCP (hazard analysis and critical control points) for management of pathogens in wastewater and sewage sludge treatment and reuse. *Water Science and Technology*, (**50**)2, 23–30.

White S. (1998). Wise Water Management: A Demand Management Manual for Water Utilities. Water Services Association of Australia.

WHO (2003). Guidelines for Drinking Water Quality, Third Edition. World Health Organization, Geneva, Switzerland. http://www.who.int/water_sanitation_health/dwq/gdwq3rev/en/ (accessed 12 November 2011).

Yule V. (2008). Who Owns Your Sewage? ON LINE Opinion Australia's E-journal of Social and Political Debate. Posted on Thursday the 3rd July 2008. http://www.onlineopinion.com.au/view.asp?article=7572 (accessed 28 August 2009).

Chapter 15

The Queen Elizabeth Olympic Park water recycling system, London

Siân Hills and Christopher James

15.1 INTRODUCTION AND PROJECT OVERVIEW

The 2012 Olympic and Paralympic Games in London had the aspiration of being the 'Greenest Games' ever. To achieve this, a number of initiatives were undertaken by the Olympic Delivery Authority (ODA), including the implementation of a Sustainable Water Strategy. This strategy had a target to reduce potable water by 40 percent across the 2.4 km^2 site that formed the Queen Elizabeth Olympic Park (the Park). To help achieve this target, the ODA worked with Thames Water Utilities Ltd (Thames Water) to provide a water recycling system to supply 574 m^3/day of reclaimed water for non potable uses on the Park (Knight *et al.* 2012).

The recycling system 'mines' raw sewage from a large sewer in East London at a location known as Old Ford, which is adjacent to the Park. The sewage is treated in a unique and innovative treatment train consisting of a membrane bioreactor (MBR) followed by granular activated carbon (GAC) filtration. The reclaimed water is chlorinated before being supplied to the Park via a dedicated 3.6 km reclaimed water network. It is used for a variety of purposes such as toilet flushing, topping up rainwater harvesting systems at venues, and parkland irrigation. It is the UK's largest wastewater recycling scheme at this community-equivalent scale. As well as helping to achieve the ODA's sustainability targets, the system also provides a focus for Thames Water's research into water recycling and reuse (including Indirect Potable Reuse) and the project has a research programme associated with it. Figure 15.1 shows an overview of the system, illustrating source to end user.

This chapter will cover both the technical and sociological aspects of the project. It will give details of the system's performance, the reclaimed water usage to date and outputs of the on-going research programme, including Park visitors' attitudes to the reclaimed water. It will explain the uses of the reclaimed water and the management practices and procedures adopted to ensure all risks were considered and mitigated to satisfy the relevant UK regulatory bodies. It will also give details of the interactions with venue operators and reclaimed water users with respect to UK Water Regulations compliance and appropriate use. Finally, the chapter will discuss the challenges and lessons learnt to inform future water recycling initiatives and the wider reuse debate within the UK.

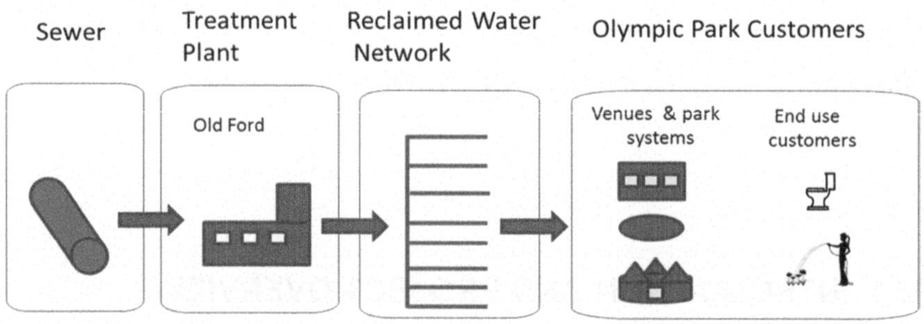

Figure 15.1 Overview of the Queen Elizabeth Olympic Park water recycling system.

15.2 THE OLD FORD WATER RECYCLING PLANT AND RECLAIMED WATER NETWORK

15.2.1 The source influent from the northern outfall sewer and the site

The Northern Outfall Sewer (NOS) is configured as a bundle of five large (each 3 m diameter) brick-built sewer barrels, which convey sewage and rainfall runoff from a substantial area of north London (population of 2.5 million) to be treated at Beckton sewage treatment works on the northern bank of the River Thames Estuary. The relevant portion of the NOS mined for this project has a daily average flow of 116,000 m^3/day, serving a population of approximately 363,000. The NOS is elevated in the section that runs adjacent to the Park and effectively forms the Park's southern boundary. It was ideally situated to provide a constant source for recycling. Additionally, Thames Water owned a convenient portion of land at Old Ford, near the Olympic Stadium, where a treatment plant could be located. The sewer was tapped into and an inlet pumping station constructed. At the point of extraction at the Old Ford Water Recycling Plant (OFWRP), the sewage flows under gravity towards Beckton sewage treatment works. The flow predominantly comprises domestic and light commercial sewage plus surface runoff, as it originates from a combined sewerage system. The representative quality of the sewage is provided in Table 15.1.

Table 15.1 Typical influent quality to the OFWRP.

Parameter	BOD_5 mg/l	COD mg/l	TSS mg/l	TKN mg/l	NH_3-N mg/l	Temp °C
Mean	202	504	296	51	38	Range 11–24
St. deviation (+/–)	66	170	126	15	11	3

The Old Ford site itself is a Site of Importance for Nature Conservation (SINC) and the treatment plant building was architecturally designed to be sympathetic to its surroundings (Figure 15.2), by including natural materials, timber and gabion basket cladding, and a bio-diverse green roof.

Figure 15.2 The OFWRP building and MBR activated sludge tank.

15.2.2 Pre-treatment

The crude wastewater (raw sewage) is first treated by passing through two, underground septic tanks. The septic tanks have been designed to remove any rags and gross solids through a settlement process, as well as up to 30% of the organic loading to reduce the impact on subsequent treatment stages. The resulting settled sewage from the septic tanks is pumped through two 1 mm rotating screen units to remove any particulate matter (hair and fibres) that could cause damage to the membrane process. The captured screenings fall by gravity into a screenings bagging unit so they can be easily removed. Odour control was an important consideration in the design and the plant includes a dedicated odour treatment unit. A full process schematic is shown in Figure 15.3.

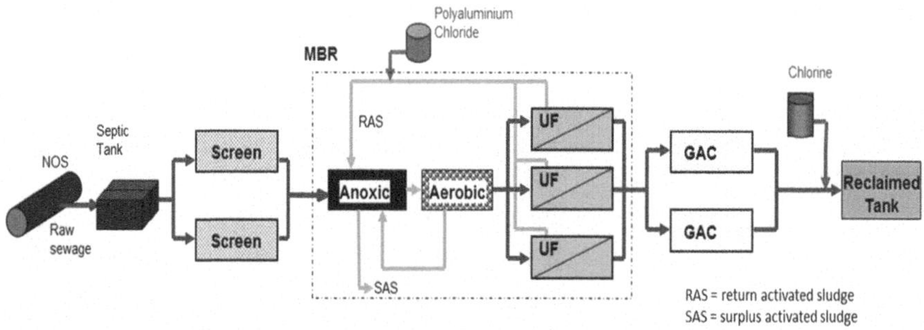

Figure 15.3 The Old Ford Water Recycling Plant process schematic.

15.2.3 Membrane bioreactor

The membrane bioreactor (MBR) consists of two process units. The first, located externally, is a 339 m^3 above-ground tank and contains the activated sludge with a segregated anoxic and aerobic zone. The ultra-filtration (UF) membranes (arranged in three racks), with a nominal pore size of 0.04 microns, are located inside the building. The membrane tank is aerated and the membranes are periodically cleaned in place (CIP) (see Table 15.2 for further details). The building has been designed with a higher roof space in this area so the membranes can be winched out for visual inspections or external cleaning, which avoids disrupting the biological system and plant operation. In addition, the membranes have the ability to automatically adjust filtration rates to compensate for membrane cell downtime or being offline for long periods (e.g., due to CIP or membrane failure).

15.2.4 Post-treatment

There is also a post-treatment step to remove any remaining colour and provide a residual of chlorine in the reclaimed water. This acts as an additional barrier to microbiological contamination and helps to keep the distribution system free of biofilms. First the water is passed through Granular Activated Carbon (GAC) filters. There are two GAC vessels that are operated in parallel, enabling operations to be maintained during a backwash or air scour. The GAC vessels are backwashed with reclaimed water. The final step is disinfection with sodium hypochlorite to achieve a chlorine residual of between 0.3 & 1.5 mg/l in the reclaimed water leaving the plant.

15.2.5 The reclaimed water distribution network

The reclaimed water distribution network delivering the reclaimed water from the plant to the end users is 3.65 km long. The trunk main section runs from Old Ford in the south, to Eton Manor (the northern most located venue on the Olympic Park), with branches connecting to other venues and take-off points in between. The customer meters are located in Venue Termination Pits (VTPs) and there are also adjacent, bespoke-designed, network sample points. The network pipework comprises Polyethylene (PE) pipe, colour-coded to UK British Standard specifications for reclaimed water (black with 4 green stripes) in sizes ranging from 180 mm to 63 mm. The network conveys reclaimed water to a variety of venues and park facilities for non-potable uses such as WC flushing, irrigation and supplementing rainwater harvesting systems. The network has been designed to minimise the risk of accidental misconnection and maximise connection opportunities; all valves being coloured red (not blue) and left-hand closing, all covers labelled 'non-potable' and connection points that are non-standard specification. The reclaimed water currently supplies seven uses as shown in Figure 15.4.

Table 15.2 MBR specifications.

Parameter	Value
Membrane material:	PVDF (polyvinylidene difluoride), 0.04 μm pore size
Area/module	38 m^2
Configuration	3 units of 16 submerged hollow fibre modules
Filtration cycle/mode	11 minute filtration, 1 minute relaxation/ Out-in
Average design flux	20 l/(m^2h) (Peak design flux 40 l/(m^2h)
Aeration	Constant aeration: 330 mbar, 64 Nm^3/hr
SAD_m	0.105 Nm^3/m^2/hr
SAD_p	7.35 Nm^3/m^3
Maintenance wash (MW)	Interval: 7 days, 300 mg/l NaOCl (3.8l per MW)
Standard Clean in Place (CIP)	Duration 6–7 hrs; Hypochlorite Interval: 90 days; Hypochlorite (10%w/w) strength: 1500 mg/l NaOCl (86.3l per CIP); Citric acid interval: 180 days; Citric acid (50%w/w) strength: 2%w/w (230 l per CIP)
Operating temperature	>5–35°C (design range)
Operating pH	6–9 (design range)
Trans-membrane pressure (TMP)	<0.1 bar (Design TMP shutdown 0.5 bar)
Hydraulic retention time (HRT)	14.91 h (septic tank to MBR permeate)
Sludge retention time (SRT)	27 days
Mixed liquor suspended solids (MLSS)	7–8 g/l

SAD = Specific Aeration Demand (SAD_m based on membrane area, SAD_p based on permeate volume).

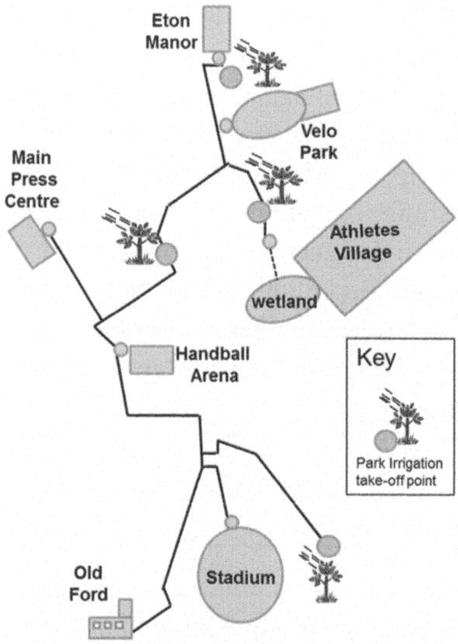

Reclaimed water end uses
IRRIGATION
(1) Olympic Parklands – 4 main areas
(2) Stadium landscaping
WC FLUSHING
(3) Eton Manor Sports Hall (4) Main Press Centre Building
SUPPLIMENTING RAINWATER HARVESTING SYSTEMS (consequently used primarily for WC flushing or irrigation)
(5) Velopark (6) Multi-Use Arena (formally Handball) (7) East Village (formally Athletes Village) – via pond

Figure 15.4 Reclaimed water distribution network and uses.

15.3 RECLAIMED WATER QUALITY

The OFWRP is contractually required to achieve a percentile-based water quality standard, informed by the USEPA guidelines for 'unrestricted urban reuse' (US EPA, 2004) and other water quality criteria specified by end user applications, particularly for irrigation. Consideration was also given to comparison with UK drinking water standards, as discussed later (DETR, 2000). Monitoring is undertaken at prescribed frequencies (daily to monthly) for different parameters and reported on a quarterly basis.

Table 15.3 provides a list of the reclaimed water quality standards and performance data over the first year of operations in which full compliance with the agreed standards was consistently achieved. It should be noted that both 90 and 95 percentile figures are provided in the table for information, but only four parameters (BOD, Colour, TDS and Turbidity, as shown in bold and underlined) are required to meet 95 percentile compliance. Total phosphorus and chlorine were shown to occasionally be outside of the 95 percentile window during the year. These peak concentrations were ascribed to operational issues or system failure, however they were within the required 90 percentile standard. Conductivity, iron and molybdenum were also detected at concentrations above the water quality standards, but did not fail under the percentile arrangements. Although these cited excursions were infrequent, this demonstrates the limited ability of the process to remove certain compounds, the final concentrations of which are dependent upon their occurrence and concentration in the sewerage catchment.

Table 15.3 OFWRP reclaimed water quality, February 2012 to January 2013.

Parameters	Units	Water quality standard	No. of samples	Min.	Mean	Max	90%ile	95%ile
Alkalinity	mg/l HCO_3	<500	237	167	230	298	252	263
Aluminium	µg/l	<5000	237	6.3	14.8	267.5	18.9	31.6
Ammonium	mg/l N	–	237	<0.03	0.04	0.21	0.03	0.16
Arsenic	µg/l	<100	237	<0.7	1.3	6.4	1.7	1.8
Berylium	µg/l	<100	237	<0.5	0.7	53	0.5	0.5
BOD (5 day ATU)	mg/l	<10	226	<1.9	<1.9	<1.9	1.90	**1.90**
Boron	mg/l	<1	237	0.017	0.097	0.117	0.110	0.111
Cadmium	µg/l	<10	237	<0.2	<0.2	<0.2	0.2	0.2
Calcium	mg/l	<250	237	80	102	137	111	115
Choride	mg/l	<250	237	38	95	135	107	111
Chlorine into supply	mg/l	>0.3 to <1.5	250	0.11	0.53	1.00	0.69	0.78
Chromium	µg/l	<100	233	<1.2	1.4	53.8	1.2	1.2
Cobalt	µg/l	<50	237	<0.7	0.9	19.8	0.7	2.3
Total Coliform	No./100 ml	<10/100 ml	234	0	0	0	0	0
Colour	mg/l Pt/Co	<20	238	1	2	10	3	**4**
Conductivity	µS/cm	<1000	238	661	863	1051	935	953
Copper	µg/l	<2000	237	<10	<10	<10	10	10
Cryptosporidium	No./100 ml	–	7	0	0	0	0	0
E. Coli	No./100 ml	0	234	0	0	0	0	0
Fluoride	µg/l	<15,000	230	66	452.6	<600	600	600
Hardness	mg/l	<450	236	232	281	333	302	309

(Continued)

Table 15.3 OFWRP reclaimed water quality, February 2012 to January 2013 (*Continued*).

Parameters	Units	Water quality standard	No. of samples	Min.	Mean	Max	90%ile	95%ile
Iron	µg/l as Fe	<200	237	<1	14.1	365.3	17.4	23.3
Lead	µg/l	<5000	237	<0.3	0.4	13.3	0.3	0.4
Lithium	mg/l	<2.5	237	0.006	0.012	0.017	0.014	0.015
Magnesium	mg/l	<50	237	5.3	8.2	11.3	9.7	9.9
Manganese	µg/l	<200	236	<0.7	0.82	23.30	0.70	0.90
Molybdenum	µg/l	<10	236	<1.6	2.5	10.8	3.4	3.6
Nickel	µg/l	<200	236	<1.6	1.9	13.4	2.1	2.3
Odour	OU/m^3	0	6	0	0	0	0	0
pH	pH Unit	6.5 to 9	238	7.3	7.6	8.1	7.8	7.9
Phosphorus	mg/l	<2.5	233	0.9	1.8	3.9	2.3	2.6
RSC***	mEq/l	<1.25	236	–4.1	–2.7	–1.3	–2.3	–2.2
SAR***	mEq/l	<3	237	0.7	1.9	2.5	2.2	2.3
Selenium	µg/l	<20	237	<0.8	1.1	10.4	1.3	1.4
Reactive Silica	mg/l SiO_2	<25	236	4	11	14	14	14
Sodium	mg/l	<200	237	31	75	100	87	90
Sulphate	mg/l SO_4	<250	237	57	91	121	104	107
Sulphide	mg/l S	<0.05	237	<0.01	0.012	0.16	0.02	0.02
Temperature	°C	–	39,136	3.9	18.7	30.1	23.5	25.2
Total suspended	mg/l	<5	226	<2	2.0	3.5	2.0	2.0
Total dissolved	mg/l	<1,000	238	423	552	673	598	**610**
Total nitrogen***	mg/l N	<30	237	6.8	13.2	23.8	18	20
Turbidity	NTU	<2	236	<0.07	0.10	0.41	0.14	**0.16**
Vanadium	µg/l	<100	237	<1.6	1.8	10.5	2	2.34
Zinc	µg/l	<5,000	237	<7	19	1663	18	22.2

***Calculated Value. RSC = Residual Sodium Carbonate. SAR = Sodium Adsorption Ratio.

15.4 RECLAIMED WATER CONSUMPTION

Reclaimed water was contractually delivered into the Olympic Park reclaimed water distribution network from the OFWRP in early April 2012. This permitted the irrigators to use a reclaimed water source for landscape irrigation across the Park and hence avoid breaching a temporary use ban on certain uses of potable water, imposed by Thames Water due to drought conditions at that time. From 5th April 2012 to 31st January 2013, more than 60,000 m^3 of reclaimed water was supplied into the network for non-potable uses, to support the challenging ODA target of reducing potable consumption by 40%.

Irrigation of the Olympic Park public landscape and parkland areas has been the predominate use of reclaimed water, consuming 78% of the total amount provided over this period. This is followed by venues using the reclaimed water for a variety of applications such as their specific landscape irrigation and top-up of rainwater harvesting systems (including pond filling) for onward use for WC flushing and irrigation (20%). Only 2% of the reclaimed water provided has been used directly for toilet flushing. However, it should be noted that the accurate monitoring of reclaimed water usage has been a challenge for this project due to issues with both the meters and with location and upkeep of elements of the automatic meter reading system. Unfortunately the meters were damaged either directly due to construction activities, or indirectly due to debris entering the reclaimed water system or becoming inaccessible for manual reading or maintenance.

A network demand profile from the initial supply date until the end of January 2013 is provided in Figure 15.5. This shows that demand was at its highest during the transition period between the Olympic and Paralympic Games at 670 m^3/d on average, when it was used for irrigation. Peak demands were also observed during the pre- and post-Games periods, although there was a significant variation in the daily consumption. Daily variations were attributed to the intermittent nature of landscape preservation/rejuvenation activities, seasonal influences (i.e., rainfall negating the necessity for irrigation) and different watering requirements for various types of vegetation (turf, shrubs and wild meadow flowers etc.). Consumption on the Park has significantly diminished moving into the post-Games Legacy period with an average usage of only 40 m^3/d. This is because irrigation systems have been decommissioned (winterised) and also modifications to expand the irrigation network are in progress. In addition, venues are unoccupied during the Park transformation works.

15.5 OPERATIONAL EXPERIENCES

The OFWRP has been operating with consistent performance since start-up in 2011, maintaining a non-potable supply availability of greater than 95% for the first full year of operation. Equally, the general performance and condition of the membranes has been stable, showing low Trans-Membrane Pressures (TMPs

10 kPa at 20l/m^2 · h @ 20°C) and producing excellent filtrate quality (turbidity <0.1 NTU), with no detection of microbial pathogens (coliforms and *E. Coli*) to date. Initial membrane autopsy results, after one year of service, revealed the fibres to be intact, although some minor abrasion was apparent. Minimal evidence of screenings present within the membrane modules has demonstrated that the pre-treatment is very effective. In particular, the use of septic tanks which allows for considerable settlement has resulted in the subsequent 1 mm screens being very lightly challenged to date. Technical specifications and configuration of the MBR including cleaning regimes can be found in Table 15.2.

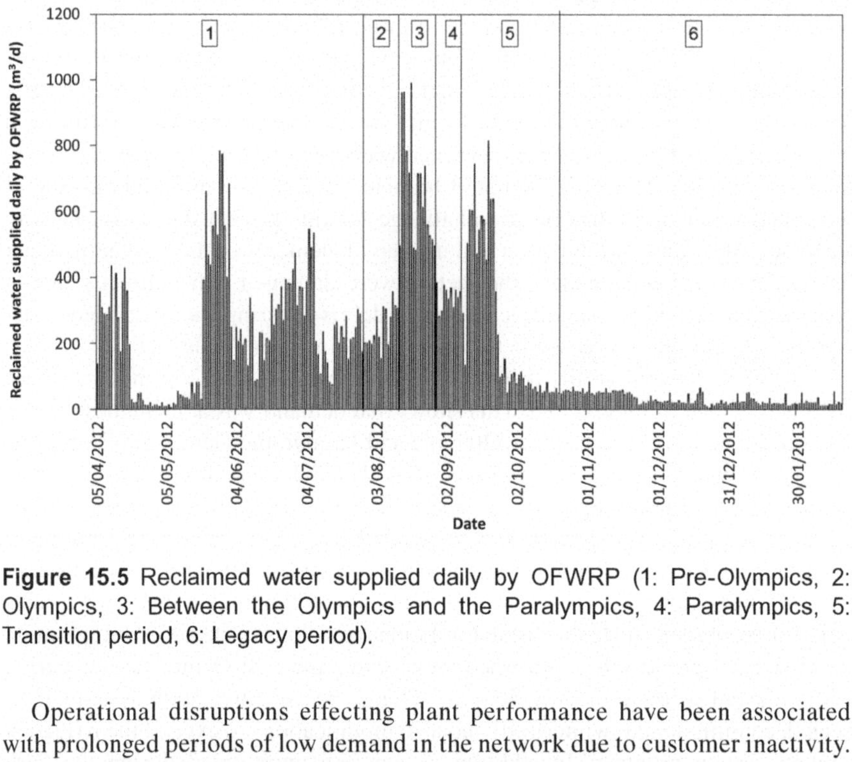

Figure 15.5 Reclaimed water supplied daily by OFWRP (1: Pre-Olympics, 2: Olympics, 3: Between the Olympics and the Paralympics, 4: Paralympics, 5: Transition period, 6: Legacy period).

Operational disruptions effecting plant performance have been associated with prolonged periods of low demand in the network due to customer inactivity. This has caused the biological system in the MBR to be devoid of incoming raw sewage, thereby elevating dissolved oxygen concentrations and inhibiting the de-nitrification process. Consequential nitrate peaks (NO_3-N) observed in the MBR effluent, threaten the ability of the reclaimed water to meet the required total nitrogen consent of 30 mg/l. However, fail-safe mechanisms are in place via an online nitrate analyser to shut-down the treatment works in this instance. To overcome this issue temporarily, a proportion of the reclaimed water produced is diverted to waste in order to maintain incoming sewage to the biological process.

In addition, low network turnover/usage means that chlorine residuals in the reclaimed water network are difficult to maintain. This has prompted a regular flushing procedure to be implemented at the dead-end points of the network, which also aids the suppression of biological re-growth within the pipes. Long term, the option of chloraminating the reclaimed water (i.e., using chloramine instead of chlorine for disinfection) is being considered, particularly as chloramine is a much more stable disinfectant than free chlorine. This also overcomes the issue of chlorine residuals potentially being reduced when the chlorinated reclaimed water is topped-up with chloraminated back-up potable supply, as a result of breakpoint reactions (Cl_2:N ratio increased) forming N_2, NO_3 and NCl_3. In the overall design of a reclaimed water system, a relatively constant demand is preferable to seasonal fluctuations as the latter can lead to water quality issues due to stagnation. In addition the treatment works operate more efficiently, particularly from an energy perspective, with a constant throughput.

With regard to energy use, the OFWRP used 954 kWh per day on average during the first year of operations, equating to approximately 3 kWh for every m^3 of water treated (not necessarily supplied). In terms of carbon footprint, the expected operational carbon footprint of the OFWRP was comparable to the net carbon footprint for processing an equivalent amount of drinking water and wastewater from local sources. For example, in order to treat the 574 m^3/day output a carbon footprint in the range of 243 to 1184 $kgCO_{2-e}$/d was estimated (taking into account both MBR and granulated activated carbon technologies). This compared to 369 $kgCO_{2-e}$/d, which was the total to treat the same volume of potable water at the nearest water treatment works (Coppermills) plus the equivalent wastewater treated at the local sewage treatment works (Beckton). As an additional comparison, the expected operational CO_{2-e} for 574 m^3 for desalination (treating brackish water) is 501–681 $kgCO_{2-e}$/day (Pearce, 2007).

15.6 RECLAIMED WATER SAFETY PLAN

As a non-potable system of this type and scale was new to the UK, close working with regulators, particularly the Environment Agency and the Health Protection Agency, was necessary to agree an implementation procedure. To address this, a Reclaimed Water Safety Plan (RWSP) was produced, based on a model used by the UK Drinking Water Inspectorate (DWI, 2010). The reclaimed water was considered from source (i.e., the sewerage system) through treatment and distribution to the customer (i.e., venue operator or end-user) in four, discrete asset groups, as shown in Figure 15.1. For each asset group the specific hazards that might be encountered were identified with commentary on the risk and the mitigation measures, including control and contingency plans. Information on the monitoring and validation measures was also captured. Potential hazards across the reclaimed water system (51 in number) were ranked using the standard Thames Water Risk Matrix, with each risk being allocated a score (relating to a Likelihood score multiplied by a Consequence score), from 1 (lowest) to 25 (highest). The risks were considered in relation to their effect on the subsequent asset group. To allow easy identification, the risks scores were also colour-coded as high (red), medium (amber), low (green) according to the matrix, following a traffic light warning protocol. Of the 51 hazards identified, only 3 were ranked as red risk. Figure 15.6 shows the distribution of risk across the four asset groups.

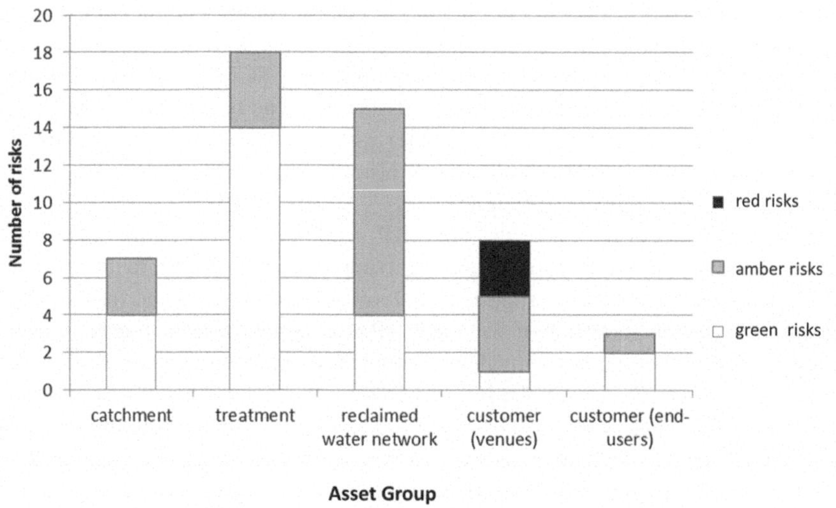

Figure 15.6 Olympic Park risk profile on first iteration of Reclaimed Water Safety Plan (red, amber and green risks represented by black, grey and white, respectively).

The three red risks (shown black on Figure 15.6) that were identified, related to risks at the Customer Venues or Park Systems asset group. This was due to potential contamination as a result of unmarked underground pipework, un-lagged pipework (Legionella risk due to heating) or from the reclaimed water being mixed with polluted rainwater. It was not unexpected that this might be the area of highest risk, as the implementation of a reclaimed water supply was a first for the majority of UK contractors. Following discussions via the ODA these risks were mitigated by (i) a series of initiatives including retro-fitting identification marker tape to un-marked pipework; (ii) dye testing; (iii) additional specific Legionella risk assessments; and (iv) enhanced liaison with venue and irrigation system operators to advise on appropriate management practices.

The amber risks (shown grey on Figure 15.6) again related to the fact that the provision of reclaimed water was a new initiative for the UK and that it was necessary to instigate a range of appropriate new management procedures. Briefing material was produced and new liaison protocols and audit procedures were set up with the Venue operators. Thames Water was also responsible for ensuring that the UK Water Fittings Regulations (DETR, 1999) were complied with, and implemented appropriate inspection regimes, accordingly.

For the sewerage catchment, and particularly the newly-built OFWRP, the majority of the risks were of the lowest category – green (shown white on Figure 15.6). This reflected that the Northern Outfall Sewer is a well understood asset that has been managed by Thames Water for many years with a good record of trade effluent control. The volumes of flow are great and substantially diluted with rainfall runoff (being a combined sewerage system), mitigating risk from any potential pollution incident. Similarly the treatment plant has been newly designed and will be operated by highly-skilled research staff that have significant experience of operating advanced technology and this type of reuse plant at pilot scale.

Following audit of the RWSP by the Health Protection Agency, the Environment Agency issued a Regulatory Position Statement sanctioning the use of the reclaimed water on the Olympic Park. The outputs were also shared with the UK Drinking Water Inspectorate, where a key concern was prevention of any cross-connection to drinking water supplies. Furthermore, Thames Water took advice from an independent public health expert who confirmed that the risks to health from accidental ingestion, splashing or breathing of reclaimed water were low, apart from in immune-compromised or vulnerable groups, who should always be considered at marginally heightened risk from drinking any water that did not strictly comply with drinking water quality guidelines. Overall, all parties agreed that a safety plan approach was useful and appropriate. The safety plan is considered a 'live' document and is audited and updated on a regular basis.

15.7 RECIPIENT COLLABORATION

Due to the unique (to the UK) nature of the water recycling system in the Park, Thames Water engaged extensively with the recipients of the reclaimed water to ensure that Water Regulations (DETR, 1999) were complied with and guidance on system management was disseminated and implemented. It should be noted that although Thames Water has a statutory duty to ensure systems comply with Water Regulations (e.g., with respect to pipe-labelling and prevention of cross contamination to potable water systems), the actual management of customer-side reclaimed water systems is the responsibility of the customer. However, all parties involved in the project recognised their Duty of Care obligations and, to this end, Thames Water undertook a series of reclaimed water briefings with a variety of end-users including venue operators, facilities managers, landscapers and irrigation personnel.

An added challenge to achieving continuity in reclaimed water provision across the Park was that, due to the extensive amount of venue and infrastructure construction, a large number of different contractors were employed across the site, each legitimately tackling their tasks within their own procedures. Despite high-level co-ordination through the ODA, throughout the entire venue project lifespan (from concept through to design, construction and into supply), at venue-level, a variety of approaches to implementing the reclaimed water systems were observed. This necessitated additional communication activities and highlighted the importance of Water Regulations inspections.

In order to assist the contractors, Thames Water and the ODA compiled and issued a guidance note on reclaimed water signage and labelling to be used across the Park. This set out all the UK sources of legislation and guidance information (WRAS, 2011) with examples of what would be acceptable on site. In areas the UK Water Regulations did not cover, recommendations were made based on overseas examples. For example, signage was considered necessary to inform Park visitors that reclaimed water was being used to irrigate the landscape and the wording and location of the signs was jointly agreed between Thames Water and the ODA. An example of the signage and labelling is shown in Figure 15.7.

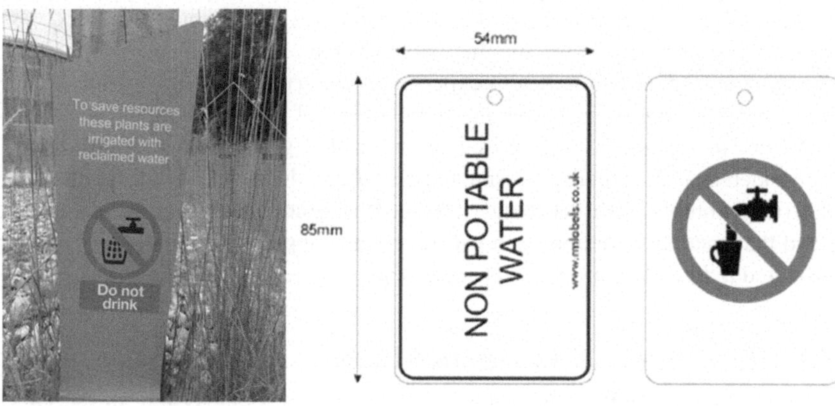

Figure 15.7 Reclaimed water signage and labelling used on the Olympic Park.

In a number of locations, it transpired that guidance had not been followed. For example, longer lead-times on the procurement of the correctly-coloured reclaimed water pipe (black with green stripe) meant that for expediency some contractors incorrectly installed black (only) pipework. This was subsequently identified through Water Regulations inspections and required the contractor to undertake a lengthy (and costly) procedure of re-excavating the buried pipework and retrospectively labelling it as reclaimed water. As an additional mitigation measure, a series of dye and pressure tests were insisted upon by Thames Water before connection to the reclaimed water network was allowed. Additionally, at a number of locations, despite information and briefing to the contrary, illegal cross-connections (no air gap) between the potable and non-potable networks were made in some plant rooms. These were identified and rectified through the close liaison between Thames Water and the customers.

15.8 PUBLIC PERCEPTION

As part of the research programme associated with the project, Thames Water commissioned a survey of visitors to the London 2012 Olympic and Paralympic Games to gauge their opinions and responses towards the recycled water system. The survey was administered just outside the Olympic Park site by a team of researchers from Cranfield University. It was conducted during both the Olympic Games (August 2012) and the Paralympic Games (September 2012). A total of 309 surveys from British residents were collected during the two Games, all of which were useable in the subsequent analysis.

The results show that the overall, respondents were extremely supportive of using non-potable, recycled blackwater (sewage), assuming a dual supply system; 96% indicated that they were supportive of using it in public venues such as the Olympic Park (Figure 15.8) and 90% indicated that they were supportive of using it in homes. When compared with previous studies by Thames Water, these results seemed to show a higher level of support for using non-potable recycled water via dual supply systems. Additionally, the results from this study show that, when purchasing a home, the vast majority of respondents (95%) claimed that they would not be put off by, and may even welcome, the presence of a dual supply system supplying recycled blackwater. The findings are encouraging, in that they may indicate a growing maturity in the UK's general public dialogue around water reuse and particularly its use for residential water supplies. It is increasingly recognised that on-going and meaningful public engagement can be critical to the success of reuse projects.

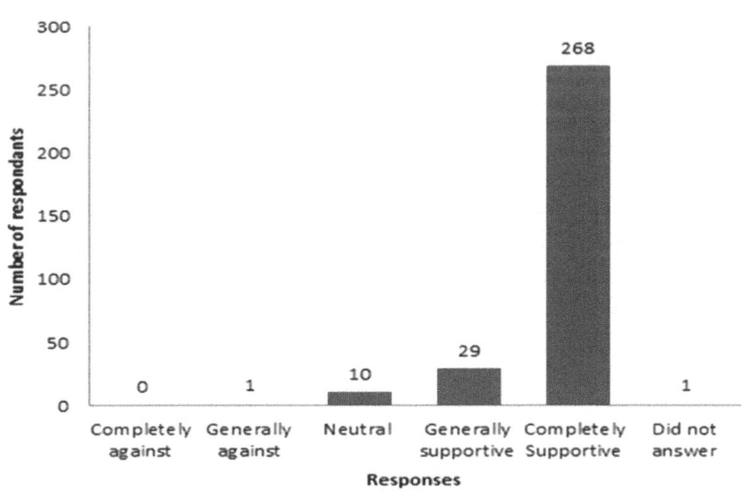

Figure 15.8 Olympic Park, UK, visitors' level of support for using recycled blackwater in public venues via dual supply (Total number of respondents = 309).

15.9 COST-BENEFIT AND COMPARISON WITH OTHER STUDIES

The cost-benefit of reclaimed water systems is difficult to quantify, considering that the environmental and social implications are not easily monetised. Strictly in terms of cost per m^3 to produce, the reclaimed water from Old Ford is more expensive if compared to Thames Water's potable water, where economies of scale can be achieved in large water treatment works. However, conventional water treatment works are not designed to treat crude sewage, so as with the carbon footprint calculation, if sewage treatment costs are included, the production costs may be more comparable. In addition, the provision of reclaimed water supplies has wider environmental and social benefits in a region that has been classified as water stressed.

It was understood from the outset that this project was primarily a research activity with the objective of learning and understanding more about the provision of reclaimed water in the UK context. To that end, it was not undertaken as a truly commercial venture and it was appreciated that the commercial implications would also be part of the learning. To encourage take-up, Thames Water provided the reclaimed water at a price that was 10% lower than potable water. Although the current annual operating costs (OPEX) of £585,000 are acknowledged as being high, they reflect the research nature of the project, including enhanced staff activities, water quality analytical costs and customer/network liaison and audits. The aim is to reduce this OPEX significantly over time through optimisation, experience and other innovative initiatives, to a target of £350,000/year. This would equate to a reclaimed water production cost of £1.75/m^3 at maximum output, which should be covered by the income from reclaimed water sales. However, it should be noted that no consideration is given to the initial capital cost of the installation in these comparisons.

There are a number of other international examples where reclaimed water has been provided by MBR treatment, but few with sewer-mining on the scale of Old Ford. It is not always easy to obtain accurate information for comparison. A 2006 review (Verrecht, 2010) identified 61 MBRs for urban reuse purposes (restricted to toilet flushing and urban irrigation), the majority of which were situated in the USA and Japan. The most numerous examples at that time, were in the East Coast of the USA where environmental regulations and the absence of sewers dictated that building development could only proceed if recycling were undertaken. These specific circumstances provided a favourable commercial environment for reuse. More recently MBRs for reuse purposes are gaining momentum in Australia, Europe and China.

The 2008 Beijing Olympics (Evoqua, 2011) implemented large scale recycling by expanding the Beixiaohe Wastewater Treatment Plant using MBR technology. The MBR treats 60,000 m^3/day of water for reuse including irrigation of the Beijing Olympic Park, with 10,000 m^3/day being further treated with a reverse osmosis system for decorative fountains and a themed lake in the Olympic Village central area.

In 2010, the Sydney Olympic Park Authority carried out a '10 years on' review of the dual reticulation system they provided for the 2000 Olympic Games (Mustor, 2010). The reclaimed water was produced using a sequencing batch reactor process followed by MF and RO, as MBR technology was not really established at that time. This review also highlighted the difficultly of obtaining comparative information from other schemes. However, in terms of expansion of the system, there was a very positive message in that the additional commercial developments in the Sydney Park since the Olympics, including hotels, offices, shops and restaurants, have continued to take-up the reclaimed water offered. The price of reclaimed water is set lower than potable. It may be assumed that the later expansion of the reclaimed water take-up may have helped the economics, as a report on the pricing regime in 2006 indicated that the costs were not being recovered at that time (DTI, 2006).

15.10 LESSONS LEARNT

The experiences of implementing a recycling system for the 2012 Olympic Park, and saving 60 million litres of potable water (to the end of January 2013) have resulted in substantial learning for future non-potable reuse schemes in the UK. As well as the technical knowledge gained, as already described, the learning also spans other aspects of the project. It has relevance not only for Thames Water, but also for contractors, consultants, operators, regulators and others involved in the provision of sustainable water.

15.10.1 Advanced preparation, awareness and guidance

On the Olympic Park, the final decision to implement sewer-mining as a recycling solution was relatively late in the Park build programme, which was not ideal. The importance of early engagement for the provision of a dual reticulation supply, at the concept design stage, cannot be over-emphasised. This will help ensure that correct infrastructure and procedures are in place from the outset and that the supply chain is equipped with the required materials. For example, difficulties in procurement of the appropriately coloured reclaimed water pipe was cited as an issue by some contractors. Internal pipework labels and reclaimed water signage were also not always readily available.

In the UK, dual reticulation schemes of this type are currently rare and the guidance available is not very specific. In addition to water quality, which is discussed in detail in a subsequent section, the current UK guidance on a whole range of related topics, such as pipework labelling, could benefit from being streamlined and consolidated. Classifications of types of reclaimed water can be confusing, with a variety of terms such as greywater, rainwater and non-potable water being used for water supplies, with no indication of the level of treatment (if any) to which the water has been subjected and hence no reference to the actual water quality. Overall, it was felt that at all stages of the project (design, build and operate), the reclaimed water supply was not as visible or considered robustly, as the more orthodox potable water and sewage provision. This was assumed to be because it was a more innovative concept and standard procedures for its implementation had not yet been adopted.

15.10.2 Reclaimed water quality

Water quality was an area that required much attention and effort in this project. As there are no regulations for reclaimed water quality from sewage, Thames Water opted to be informed by the USEPA guidelines for unrestricted urban reuse. However, for certain determinands, such as boron, copper and zinc, the UK drinking water limits were less stringent, so the drinking water standards took precedent. In addition, certain end-users had very particular requirements concerning water quality; for example, the irrigators were concerned about the potential for iron to deposit within the irrigation equipment and cause blockages. Due to this, an initial iron level of 1 mg/l was agreed which was subsequently revised down to 0.2 mg/l. This is equivalent to the UK drinking water standard and well below both the long term and short term standards of 5 mg/l and 20 mg/l in the USEPA recommended guidelines.

Varying quality requirements from different end-users can also be a challenge to accommodate. Irrigators' iron requirements have already been mentioned, but they also requested nutrient-poor reclaimed water as this was more beneficial to promote flowering of wild meadows. Initial aspirations were to provide the Energy Centre with a supply of reclaimed water for cooling purposes and this reinforced the need for some inorganic dissolved solid standards. This use would have taken a substantial volume of the reclaimed water supplied from the OFWRP. However, it was ultimately decided by the operators of the Energy Centre to continue with UK practice of using high quality potable water for their application. The water quality requirements for use during play on artificial hockey pitches (when they are flooded to a depth of several millimetres) were extensively discussed, but again a decision was made to use potable supplies for this purpose, as it was for watering of the grass, field of play grass within the Olympic Stadium. Discussions reflected the topic of 'appropriate' reclaimed water quality, which was often returned to in this project. There is an argument that the provision of very high quality reclaimed water, as provided on the Olympic Park, for uses such as toilet flushing and irrigation may be unnecessarily high for such end uses. This has a negative impact on the cost-benefit and hence may stifle wider uptake of reclaimed water. However, it is appreciated that high quality reclaimed water is appropriate for some applications and can also be an important risk-mitigation factor in the case of illegal or accidental cross connections to the drinking water (potable) supply.

Hence, derivation of and specific guidance on water quality standards for non-potable reuse from sewage effluent by UK Regulators was highlighted as an area that would be useful for future reuse schemes. One example being that the regulatory mechanism for permitting irrigation of landscape in the UK with treated effluent could benefit from being better defined. Treated sewage effluent could be classed as a 'regulated waste' under the current regime and therefore subject to regulations concerned with the concept of 'disposal'. However, it was felt by all parties that this was not an appropriate way to manage or promote reclaimed water, so in this instance the Environment Agency used the issue of a 'Regulatory Position Statement' as the authorising instrument.

15.10.3 Communication and liaison

An additional challenge in this project was posed by the variety of contractors involved at the different venues, which added a level of complexity to the liaison, both at the construction and operational phases. This is further complicated going into the Legacy Phase, where yet a different workforce may be used. An important lesson learnt was the benefit to all parties of Thames Water being proactive in liaison. This included delivering briefings and information packs and providing an active site presence to quickly respond and give guidance on water regulations. Liaison, communication and education were identified as very important. This is particularly relevant where a known quality of reclaimed water (in this case supplied by Thames Water) is being used to supplement other non-potable sources, such as via rainwater harvesting systems. The resultant water mixture can then be taken for a variety of end-uses that are a number of stages removed from the original supply. The areas of responsibility can become more blurred and risk can increase, which is where a Reclaimed Water Safety Plan can help to identify and address where additional procedures could be implemented to benefit all parties.

Within Thames Water itself, as with most UK Water Utilities, the functions of treating sewage and the provision of potable water are generally undertaken by different operating divisions. This is often for historic reasons reflecting the fact that different entities have been responsible for the provision of these services since their inception. This has understandably resulted in some different practices and approaches. Indeed, this is also reflected in the regulatory authorities, with the Drinking Water Inspectorate being responsible for potable water quality and the Environment Agency for sewage effluent quality. This separation is an added challenge for the implementation of a reclaimed water system such as the one described here, which bridges both clean and dirty water disciplines.

15.11 CONCLUSIONS

To meet the aspiration of being the 'Greenest Games' ever, the Olympic Delivery Authority (ODA) had a target to reduce potable water by 40 percent, which led to the implementation of a water recycling system to supply 574 m^3/d of reclaimed water for non-potable uses on the Olympic Park.

Overall, the project has been considered a great success and a key milestone in the provision of reclaimed water at this scale via dual reticulation in the UK. The plant has operated well, both in terms of quantity and quality, and customers have benefited from a supply of reclaimed water, particularly the irrigators who were able to continue to water the parkland despite the summer drought conditions. Much has been learnt that can benefit other schemes going forward, such as the advantages of considering systems at the design stage of developments and in ensuring all parties concerned are involved from early stages.

ACKNOWLEDGEMENTS

The authors would like to thank all parties who assisted on this project including all relevant Thames Water departments, Black and Veatch, Cranfield University, ODA, HPA, EA, DWI and many other contractors, expert advisors and consultants who were integral to the project's success.

REFERENCES

DTI (2006). Global Watch Mission Report: Water Recycling and Reuse in Singapore and Australia. DTI and British Water, London.

Knight H., Maybank R., Hannon P., King D., Wright S., Rigley R. and Hills S. (2012). Lessons Learnt from the London 2012 Games Construction Project. The Old Ford Water Recycling Plant and non-potable water distribution network. Ref: ODA 2012/031. http://learninglegacy.independent.gov.uk/documents/pdfs/sustainability/old-ford-case-study.pdf (accessed 7 June 2013).

Mustor M. (2010). Sydney Olympic Park Reuse Scheme 10 Years On – An interview with Andrzej Listowski. IWA Specialist Group on Reuse – Newsletter. http://www.iwahq.org/1nb/home.html (accessed 7 June 2013).

Pearce G. K. (2007). UF/MF pre-treatment to RO in seawater and wastewater reuse applications: a comparison of energy costs. *Desalination*, **222**, 66–73.

Evoqua Water Technologies (2011). MBR Technology Helps the 2008 Games Go "Green". http://www.water.siemens.com/en/applications/wastewater_treatment/Pages/mbr_technology_green_games.aspx (accessed 7 June 2013).

Drinking Water Inspectorate (2010). Water Safety Plans. http://dwi.defra.gov.uk/stakeholders/water-safety-plans/ (accessed 7 June 2013).

DETR (1999). The Water Supply (Water Fittings) Regulations 1999. http://www.legislation.gov.uk/uksi/1999/1148/pdfs/uksi_19991148_en.pdf (accessed 7 June 2013).

DETR (2000). The Water Supply (Water Quality) Regulations 2000. http://www.legislation.gov.uk/uksi/2000/3184/pdfs/uksi_20003184_en.pdf (accessed 7 June 2013).

US EPA (2004). Guidelines for Water Reuse. EPA/625/R-04/108. http://water.epa.gov/aboutow/owm/upload/Water-Reuse-Guidelines-625r04108.pdf (accessed 7 June 2013).

Verrecht B. (2010). Optimisation of a Hollow Fibre Membrane Bioreactor for Water Reuse. PhD thesis, Canfield University, UK.

WRAS (2011). Marking & Identification of Pipework for Reclaimed (Greywater) Systems. Water Regulations Advisory Service Information & Guidance Note No. 9-02-05 Issue 2, April 2011. http://www.wras.co.uk/Publications_DEFAULT.HTM (Issue 3, accessed on 02 June 2014).

Chapter 16

Decentralised wastewater treatment and reuse plants: Understanding their fugitive greenhouse gas emissions and environmental footprint

Peter W. Schouten, Ashok Sharma, Stewart Burn, Nigel Goodman and Shiv Umapathi

16.1 INTRODUCTION

Centralised urban water and wastewater systems serving large urban areas are rapidly being subjected to a growing number of operational challenges, primarily due to increasing urbanisation and ongoing population growth (Sharma *et al.* 2010a). To accommodate this population growth, urban densities are increasing and urban boundaries are expanding. Centralised systems have provided considerable benefits to the general public, via the provision of reliable water treatment services alongside increased health and sanitation. However, due to the increased demand on water resources, the aging and refurbishment needs of our infrastructure, as well as the need to minimise contaminant loads to receiving environments, questions have been raised regarding the long term viability of conventional centralised solutions for providing ongoing water treatment services. Moreover, some new urban areas are developing in close proximity to environmentally sensitive areas, where wastewater cannot be discharged through conventional approaches. Also, not all of the new developments can be connected to conventional centralised systems due to long distance transportation needs for bringing freshwater in and taking wastewater out, which makes these approaches economically difficult (Sharma *et al.* 2010b).

In order to meet these challenges, decentralised water and wastewater reuse systems are being implemented either in combination with centralised systems or as stand-alone systems (Sharma *et al.* 2010b; Sharma *et al.* 2013a). Decentralised reuse systems are planned using Integrated Urban Water Management (IUWM) and Water Sensitive Urban Design (WSUD) approaches, which provide the potential to implement local water systems by considering local requirements (Sharma *et al.* 2013b). These approaches can also offer the opportunity to use local water sources and close the loop on waste streams through taking a 'fit for purpose' approach, matching the quality of source water to the quality requirements of their end-use. Separate collection and treatment of various waste streams and recovery of valuable water, nutrients and energy is also possible through these systems (Wilderer, 2001).

As decentralised reuse systems are comparatively novel compared to conventional centralised approaches, the understanding and knowledge of these systems is still being developed in regards to planning, design, implementation, operation and maintenance, health impacts and environmental impacts (Sharma *et al.* 2012). Therefore knowledge gaps exist in selecting suitable servicing options. For decentralised wastewater treatment and reuse systems, limited information is available on the total environmental footprint related to their day to day operation, in particular the amount of non carbon dioxide (CO_2) fugitive greenhouse gas (GHG) emissions that they release (Sharma *et al.* 2009). This is very important to quantify, as it is necessary to determine if the increasing installation of decentralised reuse systems will be environmentally sustainable over an extended time period. This chapter details a short-term pilot study that attempts to begin to address these knowledge gaps related to the emission of fugitive GHGs, by analysing the emissions from a cross-section of decentralised wastewater reuse systems. It is believed that the outcomes from this research will assist water professionals in better understanding the potential environmental footprint of decentralised wastewater reuse systems and consequently may also help to guide water professionals in selecting appropriate decentralised wastewater reuse systems in the future.

16.2 EMISSION MECHANICS OF N_2O AND CH_4 FROM WASTEWATER TREATMENT SYSTEMS

Fugitive emissions of the GHGs, mainly methane (CH_4) and nitrous oxide (N_2O) are known to be produced and released by various anaerobic and aerobic wastewater treatment processes. Once released into the atmosphere, N_2O and CH_4 are much more effective at trapping heat in comparison to CO_2. More precisely, N_2O has 298 times greater atmospheric heating potential over 100 years in comparison to CO_2 and CH_4 has 25 times greater atmospheric heating potential over 100 years in comparison to CO_2 (Intergovernmental Panel on Climate Change, 2007). Additionally, N_2O can breakdown and eliminate ozone (O_3) in the stratosphere (Ravishankara *et al.* 2009). This has a negative impact upon the Earth's ecosystem, as the reduction of O_3 results in greater levels of biologically damaging downwelling solar UV radiation (UV_{BE}) being able to penetrate the atmosphere and reach the surface of the Earth.

In the activated sludge wastewater treatment process, N_2O is generated as a by-product of the autotrophic nitrification and heterotrophic denitrification processes (Law *et al.* 2012). Specifically, nitrification takes place under aerobic conditions when ammonium-oxidizing bacteria (AOB) and ammonium-oxidizing archaea (AOA) convert ammonia into nitrite, and nitrite-oxidizing bacteria (NOB) converts nitrite into nitrate (Kampschreur *et al.* 2009). In wastewater treatment processes, nitrification is most likely to be initiated by autotrophic AOB and NOB using either ammonia or nitrite as a source of energy and CO_2 as a source of carbon (Kampschreur *et al.* 2009). AOB have been shown to produce N_2O, as they contain the enzymes to breakdown NO_2^--N and NO leaving N_2O as a remainder (Global Water Research Commission, 2011). Denitrification occurs in anaerobic conditions and is facilitated metabolically by a large range of bacteria, archaea and micro-organisms that couple the oxidation of organic or inorganic substrates to the reduction of nitrate, nitrite, NO and N_2O (Global Water Research Commission, 2011).

Since N_2O is generated as an intermediate product during denitrification, incomplete denitrification may result in the release of N_2O into the surrounding environment. The factors most closely associated to the generation of N_2O during nitrification and denitrification are still not completely understood. However, N_2O production may be correlated to a number of wastewater treatment plant (WWTP) operating parameters such as dissolved oxygen concentration and various mass transfer/solids retention conditions. Recent studies performed at centralised WWTPs indicate that levels of dissolved oxygen, ammonia and nitrite can be used as predictors of the extent of N_2O emission taking place in treatment reactors and that nitrification is generally a higher contributor to N_2O output in comparison to denitrification (Rassamee *et al.* 2011; Ho Ahn *et al.* 2010). In addition, N_2O emissions measured from aerated treatment processes are generally much higher than those measured from non-aerated treatment processes (Rassamee *et al.* 2011; Ho Ahn *et al.* 2010). Sampling at centralised WWTPs is showing that the N to N_2O conversion percentage can vary extensively from WWTP to WWTP depending on a wide variety of process parameters (Kampschreur *et al.* 2009; Foley *et al.* 2010; Townsend-Small *et al.* 2011; Global Water Research Commission, 2011) and could be greatly dependent upon the nutrient loading existent in the influent, which can vary extensively throughout a short time period. Recent diurnal N_2O measurements are showing that a peak in N_2O generation and emissions occurs during the interval over which the maximum daily N loading arrives into a WWTP (Lotito *et al.* 2012).

In a typical wastewater treatment system, CH_4 is produced predominantly via the anaerobic decomposition of organic matter by methanogenic bacteria. This process is generally referred to as anaerobic digestion. In some cases, aerobic wastewater treatment processes can require more oxygen than is delivered via diffusion, and as a result, when surplus mechanical aeration is not available, methanogenic bacteria begin the anaerobic digestion process from which CH_4 is generated and released (Czepiel *et al.* 1993). The anaerobic digestion process occurs over four stages, with each stage requiring a specific group of micro-organisms to initialise: (i) *Hydrolysis*: the breakdown of non-soluble biopolymers to soluble organic compounds; (ii) *Acidogenesis*: the breakdown of soluble organic compounds to CO_2 and volatile fatty acids; (iii) *Acetogenesis*: the breakdown of volatile fatty acids to H_2 and acetate; and (iv) *Methanogenesis*: the breakdown of acetate, CO_2 and H_2 to CH_4 (Mes *et al.* 2004). The rate and extent of anaerobic digestion and CH_4 production are both positively correlated to temperature, toxicity, pH and chemical oxygen demand (COD). COD is used as a direct indictor to predict the potential of biogas emission to occur from a wastewater sample (Mes *et al.* 2004). CH_4 can still be emitted from aeration tanks, even when a high oxygen concentration is present. This is possible as most CH_4 is produced earlier in the sewer pipeline system adjoining the WWTP or it is delivered via rejection water released from sludge handling processes (Global Water Research Commission, 2011). The CH_4 that arrives into aeration tanks is released into the atmosphere via gas stripping.

16.2.1 Study specification and objectives

Both CH_4 and N_2O emissions from centralised WWTPs have been well documented and quantified by the application of a variety of online gas analysis instrumentation and grab sampling techniques. Some examples of these studies include Czepiel *et al.* (1993), Czepiel *et al.* (1995), Ho Ahn *et al.* (2010), Global Water Research Commission (2011), Townsend-Small *et al.* (2011) and Winter *et al.* (2012). Conversely, there is limited information available on the temporal and spatial distribution of the fugitive emissions released at decentralised reuse systems treating smaller daily volumes of sewage. As such, real-world CH_4 and N_2O emission studies carried out at small-scale decentralised reuse systems are necessary in order to resolve this gap in the knowledge base. In order to better ascertain the amount of fugitive CH_4 and N_2O emissions produced by decentralised reuse systems, a series of CH_4 and N_2O gas flux measurements were performed at three different Australian decentralised reuse systems over the months of spring in 2012 using an online non-dispersive infrared (NDIR) gas analyser combined with a gas capture flux hood. This chapter details the extent of the CH_4 and N_2O emissions measured throughout this research campaign and directly compares these results to emissions estimates calculated from a wastewater GHG emission model that is currently employed by Australian wastewater treatment operators to evaluate the annual environmental footprint of their systems. The model is a series of analytical equations that has been developed by experts working in consultation with the Australian Government, but is not commercially available. A complete description of the CH_4 and N_2O components of the model is included in Section 16.3.3.

16.3 MEASUREMENT CAMPAIGN SPECIFICATION AND ANALYSIS METHODOLOGIES

16.3.1 Reuse systems specifications

For the purposes of this research, a decentralised reuse system is defined as a wastewater treatment system managing influent from a population with no greater than 75,000 people with a daily flow rate of no more than 5×10^6 m^3/year. To acquire an inventory of fugitive N_2O and CH_4 emissions data, a series of field investigations were carried out over the months of spring at three decentralised reuse systems situated over the greater metropolitan area of a large Australian city (Melbourne, Victoria, 37°48′49″S 144°47′57″ E, Altitude: 31 m). Off-gases from one reuse system with a large catchment population (Site A), one reuse system with a small catchment population (Site B) and one sewer mining facility used for irrigation over an 18-hole golf course (Site C) were analysed. All three of these reuse systems use the activated sludge treatment process. The reuse systems were chosen due to their varying age, spatial footprint, catchment area, treatment regime and daily average inflow and organic loading, in order to provide a generally representative cross-section of the types of decentralised reuse systems currently working in countries with modern sewage treatment infrastructure. Table 16.1 provides the operational data and calculated influent water quality parameters for each of the three reuse systems evaluated in this study. Generalised treatment regime schematics for the three reuse systems have been presented previously in Schouten *et al.* (2013a) and Schouten *et al.* (2013b).

Table 16.1 Decentralised reuse system operational metadata and water quality parameters.

	Site A	**Site B**	**Site C**
Function	Wastewater treatment for local community	Wastewater treatment for local community	Sewer mining for golf course irrigation
Effluent reuse application	Irrigation of local vineyards, tree plantations and on-site lawns	Redistribution into river system for reuse by local farmers and businesses	Irrigation of greens, fairways, lawns and gardens around the golf course
Predominant influent type	Domestic	Domestic	Domestic
Catchment Population (approximate)	75,000	400	3000
Yearly COD mass load (metric tonnes)	2660	25	88
Yearly BOD mass load (metric tonnes)	1662	9.4	46.4
Yearly total N mass load (metric tonnes)	335	1.7	11
Yearly inflow (m^3) (Daily flow rate (m^3/ day))	4.86×10^6 (13.3×10^3)	21×10^3 (57.5)	1×10^5 (274)

Each site extensively reuses its treated effluent. Site A reuses a sizeable volume of its treated effluent to irrigate local vineyards and tree plantations. The treated effluent is also used to water lawns around the treatment systems and the on-site control office. In addition, Site A collects a stockpile of dried digested sludge for use as fertiliser on nearby lawns and gardens. Site B sends its treated effluent back into a local river system, from which it can be readily collected and used by local farmers, horticulturalists and businesses for irrigation purposes. Site C recycles all of its treated effluent to continuously irrigate the fairways and greens on each hole at the golf course. The treated effluent produced at Site C is also used to water the gardens and lawns around the greens and the clubhouse.

Fugitive N_2O and CH_4 fluxes were measured only from the aeration tank systems at Site A, Site B and Site C, as at these sites aeration tanks had the most sizeable spatial footprint and were completely atmospherically exposed, and consequently had the most potential to release the highest cumulative emissions. N_2O and CH_4 fluxes were measured simultaneously at evenly spaced positions across the length of each aeration tank in order to evaluate and quantify the spatial distribution of the gas emissions. The measurements were performed over a single day during spring (September 2012 to November 2012 in the Southern Hemisphere) in order to obtain a general 'snapshot' of the fugitive emissions produced at each site. Table 16.2 displays the average values and standard deviations (1σ) for water quality parameters measured at the reuse systems during the measurement campaign along with the averaged N_2O and CH_4 fluxes recorded during the sampling interval.

Table 16.2 Reuse system peripheral water quality data and averaged CH_4 and N_2O fugitive emissions for Site A, Site B and Site C.

	Dissolved oxygen (mg/l) (Mean, ±Std. Dev)	**pH (Mean, ±Std. Dev)**	**Conductiv. (μS/cm) (Mean, ±Std. Dev)**	**Sewage temperature (°C) (Mean, ±Std. Dev)**	**CH_4 (g CH_4/ m^2/d) (Mean, ±Std. Dev)**	**N_2O (g N_2O/ m^2/d) (Mean, ±Std. Dev)**	**Number of Measurements (N)**
Site A Aeration tank *Spring*	0.94 (±0.28)	7.14 (±0.17)	994.8 (±36.03)	18.4 (±0.05)	2.4 (±3.4)	1.09 (±0.8)	18
Site B Aeration tank *Spring*	0.099 (±0.05)	6.7 (±0.02)	470.6 (±6.06)	18.3 (±0)	0.07 (±0.02)	0 (NA)	18
Site C Aeration tank *Spring*	1.29 (±0.39)	6.4 (±0.08)	705.2 (±11.1)	22.5 (±0.14)	0.22 (±0.16)	0.71 (±0.25)	18

16.3.2 Gas analysis instrumentation and sampling technique

The gas sampling carried out through the measurement campaign followed a methodology similar to the gas sampling procedures used by Tremblay *et al.* (2004) and Carignan (1998). In-situ gas capture was made on the sewage surface with a commercially available buoyant airtight flux hood (St Croix Sensory Inc., United States) connected to a primary standard calibrated (traceable to the National Institute of Standards and Technology) NDIR gas analyser (Horiba Ltd., Japan). Emitted gases were trapped in the flux hood and sent to a gas conditioning unit (Horiba Ltd., Japan) via a pump operating at a constant flow rate. The gas conditioning system removed water vapour, acids and other pollutants from the gas flow before it entered the NDIR gas analyser. After the CH_4 and N_2O gas concentrations had been calculated by the NDIR gas analyser, the sampled gases were returned into the flux hood so they could mix continuously, enabling a more accurate estimate of gas concentration to be obtained over time (Lambert & Frechette, 2005). Generally, for gas measurements in remote or isolated field locations (such as decentralised WWTPs), in-situ gas collection with a gas capture hood combined with a NDIR gas analyser is the most appropriate. This is due to its relative ease of use, portability and rapid flux calculation capability.

Gas concentrations were recorded over 10 (±5) minute intervals to reduce the influence of the chamber effect (Venterea *et al.* 2009). The flux hood was held in place with a tightened rope to minimise the effect of wave action and turbulence occurring on the sewage surface and to keep the flux hood in the same position over the measurement interval. Following each measurement, the flux hood was removed from the sewage surface so that gas concentrations could return to ambient levels before the next measurement. The gas concentration data from each reuse system was logged continuously to a laptop computer beginning at 10:00 AM (±2 hours) until 2:30 PM (±2 hours) during each day of sampling. From this sampling regime, trends in gas flux could be readily quantified over the time of day at which the influent flow rate and nutrient loading was most likely to be at its highest value. Figure 16.1a depicts the deployment of the flux hood on top of the Site C aeration tank and Figure 16.1b displays the operation of the NDIR gas analyser, the gas conditioner and the laptop computer over the Site B aeration tank. The method described by Tremblay *et al.* (2004) was used to calculate CH_4 and N_2O gas flux. For this, a linear regression was applied separately to the CH_4 and N_2O gas concentration data recorded over the sampling time during each particular measurement interval. From the gas concentration data, CH_4 and N_2O gas flux could be calculated directly using Equation 16.1 (Tremblay *et al.* 2004):

$$\text{Flux} = \frac{m \times V \times \alpha \times \beta}{A \times \gamma} \qquad (16.1)$$

where m is the slope from the linear regression set to the gas concentration data over the sampling time (ppm/second); V is the volume under the flux hood (m^3); α is a gas concentration conversion factor (for CH_4: 655.47 μg/m^3/ppm; for N_2O: 1798.56 μg/m^3/ppm); β is a temporal conversion factor (86,400 seconds/day); A is the area under the flux hood (m^2) and γ is a magnitude conversion factor (1×10^6 μg/g). Flux is given in g/m^2/day.

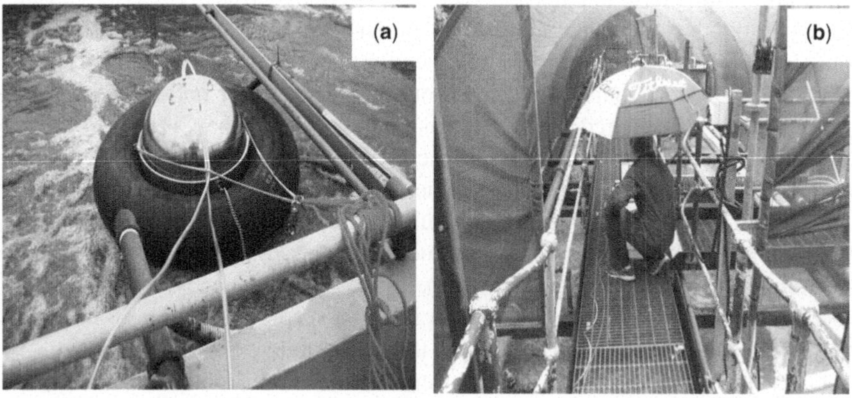

Figure 16.1 Example deployment of the gas analysis system in the field: (a) The flux hood on top of the Site C aeration tank; (b) The gas analysis and data collection workstation being worked on over the Site B aeration tank.

Continuous measurement of various water quality parameters were made simultaneously to the gas flux measurements using a water quality sonde (Aquaread Ltd., United Kingdom), with the sonde positioned in close proximity to the flux hood. The sonde was lowered to a depth of approximately 0.5 to 1 m under the sewage surface during sampling. The water quality parameters that were measured were temperature, pH, dissolved oxygen and conductivity. At the start of each site visit, the water quality sonde was calibrated to a standard calibration solution formulated by the sonde manufacturers (Aquaread Ltd., United Kingdom). More water quality and daily operational data such as influent flow rate, effluent flow rate, influent total N mass load, influent total COD and BOD mass load, oxygen delivery profiles and inflow pump timings were obtained from the site engineers when necessary.

To calculate a basic estimate of the total percentage conversion of influent nutrient loading to CH_4 and N_2O gas taking place at each decentralised system, basic emission factor calculations were performed. To do this, the mean fugitive N_2O emission measured over spring was normalised to the total annual $N_{INFLUENT}$, and the mean CH_4 emission recorded over spring was normalised to the total annual $COD_{INFLUENT}$ for the aeration tanks at Site A, Site B and Site C. These normalisations were made using Equation 16.2 and 16.3:

$$N_2OEF = \left[\frac{(N_2O)_{MEAN}}{(N_{INFLUENT})_{ANNUALTOTAL}} \right] \times 100\% \tag{16.2}$$

$$CH_4EF = \left[\frac{(CH_4)_{MEAN}}{(COD_{INFLUENT})_{ANNUALTOTAL}} \right] \times 100\% \tag{16.3}$$

where: N_2OEF and CH_4EF are the annual emission factors for N_2O and CH_4 respectively; $(N_2O)_{MEAN}$ is the mean of the N_2O flux measurements (tonnes) integrated over the tank surface area made over spring; $(N_{INFLUENT})_{ANNUALTOTAL}$ is the annual total N (tonnes) arriving in the influent at each system; $(CH_4)_{MEAN}$ is the mean of the CH_4 flux measurements (tonnes) integrated over the tank surface area made over spring and $(COD_{INFLUENT})_{ANNUALTOTAL}$ (tonnes) is the annual total COD arriving in the influent at each system.

16.3.3 Wastewater GHG emissions modelling

The cost and time required to set up and maintain online gas measurement instrumentation at both centralised and decentralised WWTPs is often highly prohibitive. Consequently, semi-empirical modelling techniques are predominantly used by wastewater treatment operators to estimate the amount of fugitive N_2O and CH_4 emissions being released from their wastewater treatment systems each year. These models generally employ input data obtained from real-world continuous or intermittent measurements made using online or handheld water quality instrumentation, from wastewater grab samples analysed using chemical or optical methods in a laboratory, or from inferences or extrapolations taken from previously published investigations. Currently, in Australia a large number of wastewater treatment operators are required to report on the total N_2O and CH_4 emissions produced annually by their WWTPs as part of the National Greenhouse Energy Reporting Scheme (NGERS). NGERS currently uses two separate models to estimate direct N_2O and CH_4 emissions released during wastewater treatment. A simplified version of the NGERS N_2O emission model applicable to this research is as follows (Global Water Research Commission, 2011):

$$N_2O_{WWT} = \left[(N_{IN} - N_{OUT}) \times EF_{N_2O_{WWT}} \right] + \left[N_{OUT} \times EF_{N_2OD} \right] \tag{16.4}$$

where N_2O_{WWT} is the amount of N_2O gas emitted from wastewater (tonnes $CO_{2EQUIVALENT}$), N_{IN} is the amount of nitrogen entering the WWTP/system (tonnes), N_{OUT} is the amount of nitrogen leaving the WWTP/system to re-enter the local environment (tonnes), $EF_{N_2O_{WWT}}$ is the default N_2O emission factor for domestic wastewater treatment (4.9 tonnes $CO_{2EQUIVALENT}$/tonnes $N_{REMOVED}$) and EF_{N_2OD} is the default N_2O emission factor for treated wastewater discharge (4.9 tonnes $CO_{2EQUIVALENT}$/tonnes $N_{REMOVED}$). $EF_{N_2O_{WWT}}$ is always set to zero for WWTPs that do not have secondary nitrification-denitrification treatment in place.

The basic NGERS CH_4 emission model used to predict fugitive CH_4 emissions from wastewater treatment is given below (Global Water Research Commission, 2011):

$$CH_{4WWT} = \left[\left(COD_{IN} - COD_{SLUDGE} - COD_{OUT}\right) \times EF_{CH4WWT} \times F_{ANAEROBIC}\right] \quad (16.5)$$

where CH_{4WWT} is the quantity of CH_4 gas emitted from wastewater (tonnes $CO_{2EQUIVALENT}$), COD_{IN} is the amount of COD coming into the WWTP/system (tonnes), COD_{SLUDGE} is the amount of COD removed in sludge and treated in the WWTP/system (tonnes), COD_{OUT} is the quantity of COD exiting the WWTP/system, EF_{CH4WWT} is the default CH_4 emission factor for domestic wastewater treatment (5.3 tonnes $CO_{2EQUIVALENT}$/tonnes $COD_{REMOVED}$) and $F_{ANAEROBIC}$ is the fraction of COD treated anaerobically within the WWTP/system each year. For all WWTPs running a managed aerobic treatment processes, $F_{ANAEROBIC}$ is automatically set to zero.

Emissions estimations made by the NGERS models (Equations 16.4 and 16.5) were directly evaluated against the measured N_2O and CH_4 emissions data collected at Site A, Site B and Site C over the March 2012 to April 2013 measurement campaign period. These comparisons are discussed in further detail in Section 16.4.1.

16.4 MEASUREMENT CAMPAIGN RESULTS AND DISCUSSION

16.4.1 Fugitive emissions

Figures 16.2 (a, b and c) display the N_2O and CH_4 emissions measurements taken from the aeration tanks over a single day of sampling at the three sites. In the Site A data displayed in Figure 16.2a, six emissions measurements were obtained in one section of the aeration tank from 9:49 AM to 10:45 AM. Following this, the flux hood was moved approximately five metres lengthways across the tank where a further six emissions measurements were performed from 11:02 AM to 11:55 AM. After this, the flux hood was moved ten metres lengthways once again where a final series of six emissions measurements were taken from 12:09 PM to 1:07 PM. In the Site C data shown in Figure 16.2c emissions measurements were made in six separate sections across the aeration tank, each spaced approximately one metre apart. Each of these sections was evaluated over the following intervals: 10:41 AM to 10:57 AM (section 1), 11:04 AM to 11:25 AM (Section 2), 11:37 AM to 11:56 AM (section 3), 12:05 PM to 12:22 PM (Section 4), 12:35 PM to 12:52 PM (Section 5) and 1:03 PM to 1:24 PM (Section 6). Due to limited available space at Site B, flux measurements were recorded in same position on the aeration tank surface over the single day sampling period.

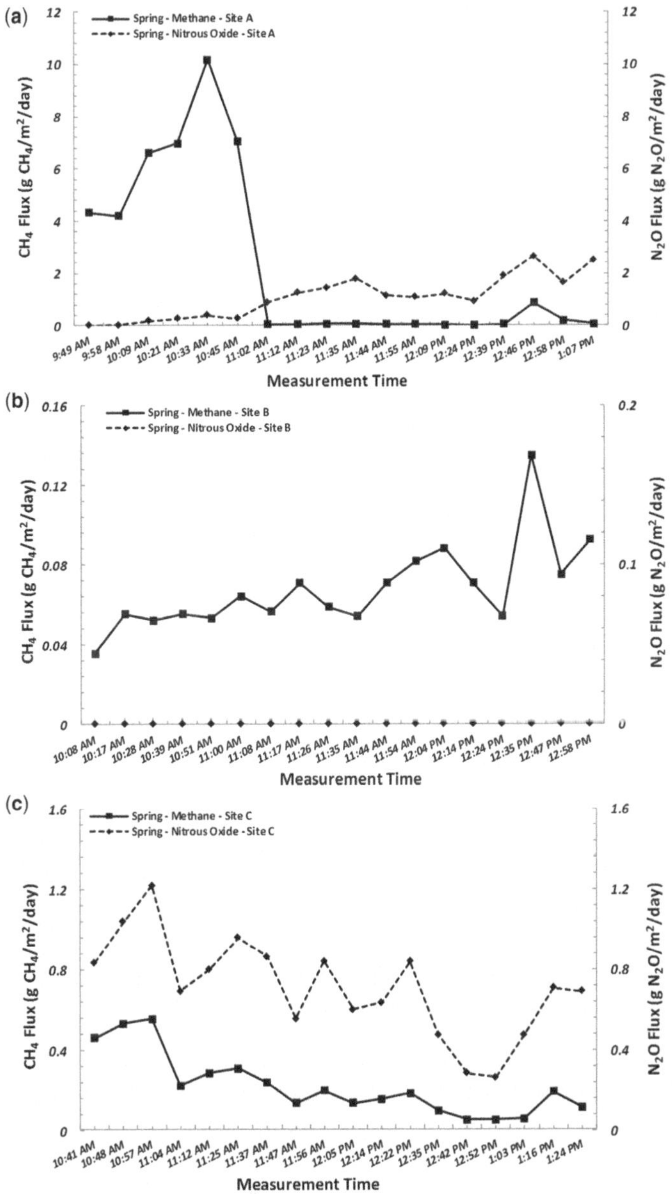

Figure 16.2 (a) N_2O and CH_4 emissions measurements taken from the aeration tanks made over a single day of sampling at Site A. (b) N_2O and CH_4 emissions measurements taken from the aeration tanks made over a single day of sampling at Site B. (c) N_2O and CH_4 emissions measurements taken from the aeration tanks made over a single day of sampling at Site C.

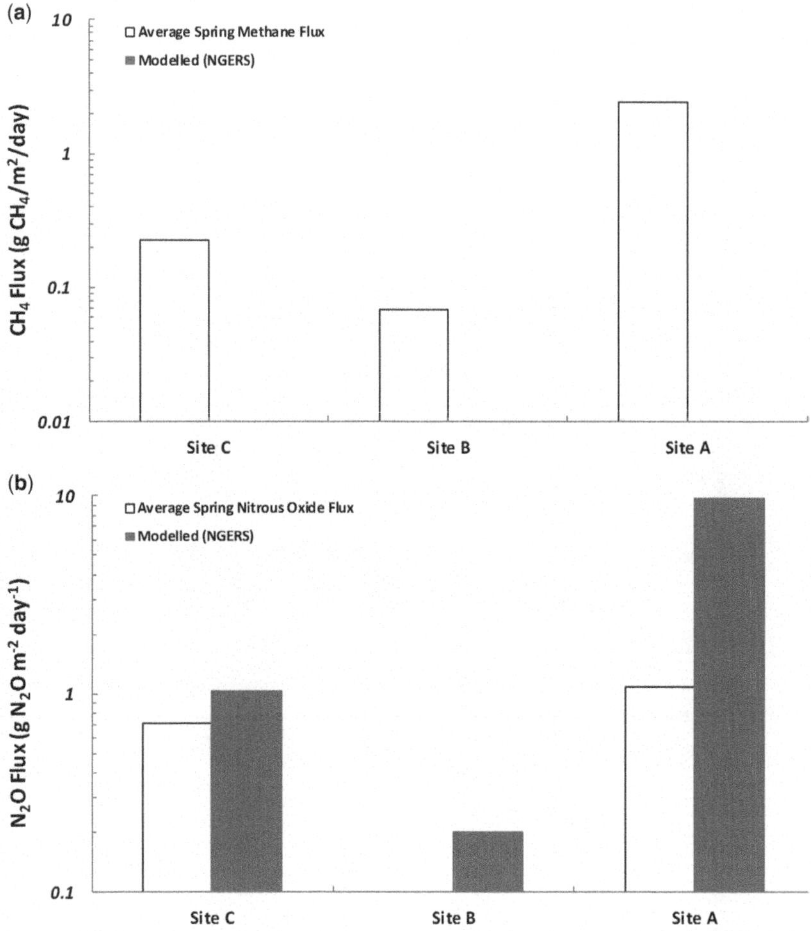

Figure 16.3 (a) Daily averaged CH_4 emissions data measured from the aeration tanks located at Site A, Site B and Site C over spring 2012. The measured CH_4 data is compared to CH_4 emissions estimates calculated by the NGERS models. (b) Daily averaged N_2O emissions data measured at the Site A, Site B and Site C aeration tanks over spring 2012. The measured N_2O data is compared to N_2O emissions estimates calculated by the NGERS models.

From these staggered emissions measurements, significant variations in the spatial (lengthways across the tank) and short-term temporal (with measurements being made in the same position) distributions of CH_4 and N_2O emissions were found to occur at Site A, Site B and Site C. Some of the most extensive temporal variations in CH_4 flux found were: (i) a 32% increase (7 g CH_4/m^2/day to 10.2 g CH_4/m^2/day) over a time interval of twelve minutes at Site A in section 1; (ii) an increase of gas output by a factor of close to 2½ (0.054 g CH_4/m^2/day to 0.13 g CH_4/m^2/day) over ten minutes at Site B; and (iii) a tripling of flux (0.061 g CH_4/m^2/day to 0.19 g CH_4/m^2/day) over thirteen minutes at Site C in section 6. N_2O flux was also found to fluctuate by substantial amounts over short time periods, with the most prevalent instances being a doubling of flux (0.94 g N_2O/m^2/day to 1.89 g N_2O/m^2/day) taking place over a time interval of 15 minutes at Site A in section 3, and a decrease of 36% (0.87 g N_2O/m^2/day to 0.56 g N_2O/m^2/day) over ten minutes at Site C in section 3. Additionally, N_2O fluxes measured in quick succession (<10 minutes apart) over a distance of no greater than five metres were found to increase by as much as a factor of three at Site A (0.27 g N_2O/m^2/day to 0.89 g N_2O/m^2/day between section 1 and section 2) and could decrease by as much as 43% at Site C (1.2 g N_2O/m^2/day to 0.7 g N_2O/m^2/day between section 1 and section 2). Similar levels of high variability were also measured for CH_4 fluxes measured over the same distances at Site A and Site C.

As mentioned in the previous paragraph, spatial distributions of CH_4 and N_2O flux were not obtained at Site B due to the small aeration tank surface area over which there was minimal space to reposition the flux hood. However, it could be assumed that similar substantial spatial variations in both CH_4 and N_2O flux may also take place there. It is difficult to pinpoint the exact cause of these considerable variations in the spatial and short-term temporal distribution of CH_4 and N_2O fluxes at each of the decentralised reuse systems. The spatial variations may be linked to oscillations in the amount of organic content and dissolved gas entering the aeration tank during the measurement interval. As a result of variability in the inflow organic and dissolved gas loading, the amount of biological activity and gas available for stripping within a given volume at one section of the tank may be completely different to another section of the tank. Fluctuations in flux output measured over time and across the surface area of an aeration tank could also be due to gradual changes occurring to the chemistry of the mixed liquor located within different positions. These changes are most likely to be facilitated by changes in total aeration delivery and the positioning of the bubblers/oxygen supply in the tank aeration system.

Figure 16.3a and Figure 16.3b show respectively the daily averaged CH_4 and N_2O emissions data recorded from the aeration tanks located at Site A, Site B and Site C over spring. In these two figures, the measured CH_4 and N_2O emissions data is compared directly to CH_4 and N_2O emissions estimates calculated by the NGERS modelling framework specified in Section 16.3.3. From Figure 16.3a it can be seen that Site A emitted the highest amount of CH_4 on average in comparison to Site B and Site C, with Site A producing peak CH_4 emissions of 2.4 g CH_4/m^2/day. This was expected as Site A treats a far greater daily loading of COD in comparison to Site C and Site B and as such has the highest potential for anaerobic digestion to occur throughout its treatment regime. Site C released the next highest amount of CH_4, that being 0.22 g CH_4/m^2/day and Site B emitted the lowest CH_4 level of 0.07 g CH_4/m^2/day. In Figure 16.3b, it is clear that Site A also emitted a higher average amount of N_2O gas in comparison to Site B and Site C. Specifically, Site A emitted an average N_2O emission output of 1.09 g N_2O/m^2/day, in comparison to 0.71 g N_2O/m^2/day from Site C and 0 g N_2O/m^2/day from Site B. Once again, this outcome was anticipated to occur as Site A takes in a much larger volume of N in its influent flow every day compared to Site C and Site B. Subsequently, of all the three sites, Site A has by far the greatest potential for larger amounts of N_2O gas conversion to take place during nitrification and denitrification processes occurring within its treatment system. Despite its much lower yearly inflow and total N mass load, Site C still managed to emit close to 65% of the total average N_2O gas output delivered by Site A. It is unclear why Site C was emitting this substantial amount of N_2O. However, this disproportionate output may be occurring due to periods of inadequate aeration (as a result of an inefficient aeration system) or could be due to large peaks in ammonium entering the system, which can lead to an increase in N_2O generation and its subsequent release (Global Water Research Commission, 2011).

It is clear from Figure 16.3a that the NGERS CH_4 emissions model (Equation 16.5) predicts that no CH_4 emissions should occur from the aeration tanks at Site A, Site B and Site C. This is due to the initial assumption that a managed aerobic treatment process does not treat any organic loading anaerobically and as such does not have the potential to emit any fugitive CH_4 gas. This assumption effectively sets the $F_{ANAEROBIC}$ term in the NGERS CH_4 emissions model to zero. However, dissolved CH_4 gas can still enter aerobic treatment processes, after being generated in upstream treatment systems and sewers, and can be readily stripped out and released. Consequently, the current NGERS CH_4 model may need to be modified to factor in the possibility that CH_4 gas stripping can take place on a constant basis within an aerobic treatment system.

Figure 16.3b show that the NGERS N_2O emissions model (Equation 16.4) can make approximate estimations of the actual total N_2O output from an aeration tank operating within a decentralised system. However, the accuracy of these emissions calculations was found to vary significantly from site to site. When compared directly to the measured N_2O emissions value, the NGERS N_2O emissions model had an overall percentage error ($|N_2O_{MODELLED} - N_2O_{MEASURED}|/N_2O_{MEASURED}) \times 100\%$) of 799% for Site A and a percentage error of 45% for Site C. No percentage error estimate could be calculated for Site B, as N_2O emissions were not released there. However, the NGERS model did estimate that a relatively small amount of N_2O gas (0.2 g N_2O/m^2/day) could emit from Site B. These results indicate that the input variables employed by the NGERS N_2O emissions model may not provide enough detail to reliably estimate real-world N_2O fugitive emissions within an acceptable error limit. As such it is appears that the NGERS N_2O emissions model could be redeveloped to factor in site-specific information relating to other important process parameters and operational conditions influencing N_2O generation. These parameters and conditions could include dissolved oxygen level, aeration regime type (intermittent or continuous) and solids retention time. In addition to this, the default N_2O emission factor for domestic wastewater treatment may need to be recalculated and defined specifically for different types of treatment systems running dissimilar treatment processes.

The respective CH_4 and N_2O emission factors calculated over the spring measurement campaign for the Site A, Site B and Site C aeration tanks are provided in Tables 16.3 and 16.4. All of these emission factors were inside the expected range of emission factors that have previously been reported for large-scale centralised activated sludge WWTPs using in-situ gas flux measurement methods (Global Water Research Commission, 2011). The N_2O emission factors calculated for Site A, Site B and Site C are within the expected range of 0% to 4% conversion of influent total N to N_2O emission (Kampschreur *et al.* 2009). Also, the CH_4 emission factors calculated for Site A, Site B and Site C are all within range of CH_4 emission factors estimated for large-scale WWTPs located in the Netherlands and France (0.005 to 0.04) (Global Water Research Commission, 2011). It is possible that these emission factors may increase at each of the sites if CH_4 and N_2O emissions occurring from treatment processes adjoining the aeration tanks are taken into account. However, the overall impact of these adjoining treatment processes may be regarded as being low or negligible, as at each of the decentralised sites the aeration tank systems had the largest spatial footprint and operational volume in comparison to the rest of treatment processes at each particular site and were completely open to the atmosphere. They therefore had the most potential to release the largest amounts of fugitive emissions.

Table 16.3 CH_4 emission factors and the daily average CH_4 emission per m^3 of influent calculated for Site A, Site B and Site C.

Site Name	CH_4 Emission Factor (CH_4EF)	Daily average CH_4 emission per m^3 (g $CH_4/m^2/m^3$)
Site C	0.0088	8.3×10^{-4}
Site B	0.00596	1.1×10^{-3}
Site A	0.0365	1.8×10^{-4}

The daily average CH_4 and N_2O emissions per m^3 of influent calculated over the measurement campaign for Site A, Site B and Site C are given in Tables 16.3 and 16.4. Site B measured the largest daily average CH_4 emission per m^3 of 1.1×10^{-3} g $CH_4/m^2/m^3$, which was an order of magnitude greater than the next highest measurement made for Site C (8.3×10^{-4} g $CH_4/m^2/m^3$). For N_2O, Site C delivered the greatest daily average emission per m^3 of 2.6×10^{-3}g $N_2O/m^2/m^3$, with Site A providing the second highest daily average emission per m^3 of 8.2×10^{-5} g $N_2O/m^2/m^3$. These results indicate that the CH_4 and N_2O output per m^3 of infuent can be relatively high and may be comparable to centralised systems, even for very small decentralised systems like Site B treating small catchment areas.

Table 16.4 N_2O emission factors and the daily average N_2O emission per m^3 of influent calculated for Site A, Site B and Site C.

Site Name	N_2O Emission Factor (N_2OEF)	Daily Average N_2O Emission per m^3 (g $N_2O/m^2/m^3$)
Site C	0.23	2.6×10^{-3}
Site B	0	0
Site A	0.13	8.2×10^{-5}

16.4.2 Total carbon footprint for each reuse system

To evaluate the total daily carbon footprint for each reuse system, the average daily CH_4 and N_2O fluxes measured over the entirety of the measurement campaign reported in Section 16.4.1 were converted to their equivalent CO_2 (CO_{2EQ}) values by multiplying them by their respective 100 year global warming potential conversion factors (25 for CH_4 and 298 for N_2O) given by the Intergovernmental Panel on Climate Change (2007). The average daily CH_4 flux after conversion to the equivalent CO_2 value was calculated to be 50.4 ± 71.4 g CO_{2EQ}/m^2/day for Site A, 1.43 ± 0.46 g CO_{2EQ}/m^2/day for Site B and 4.7 ± 3.3 g CO_{2EQ}/m^2/day for Site C. The daily average N_2O flux following conversion to its equivalent CO_2 value was 336.5 ± 249.9 g CO_{2EQ}/m^2/day for Site A and 220.7 ± 76.6 g CO_{2EQ}/m^2/day for Site C. The total annual carbon footprint due to CH_4 and N_2O emissions for each reuse system was estimated by integrating the average daily CO_{2EQ} CH_4 and N_2O fluxes over the atmospherically exposed surface area of each of the specific treatment processes under analysis and by multiplying this value by the number of days in a common year (365). As a result, the total annual carbon footprint due to CH_4 emissions was 6.8 tonnes (Site A), 0.031 tonnes (Site B) and 0.16 tonnes (Site C). The total annual carbon footprint due to N_2O emissions for each reuse system was calculated to be 45.4 tonnes (Site A) and 7.7 tonnes (Site C).

From these results it is clear that each of the reuse systems emitted relatively significant amounts of CH_4 and N_2O gas (in particular Site A and Site C), which may have an eventual long-term impact upon the atmospheric infrared radiation budget and as a consequence, global climate change. Subsequently, it is important to evaluate various strategies to mitigate and reduce these emissions before they can enter the atmosphere, or to capture and recycle them for other applications, such as providing a localised energy supply to power daily treatment systems operations. This is of critical importance, as the installation of decentralised reuse systems is increasing rapidly in both developed and developing countries, and as such, the cumulative fugitive emissions released from of all these decentralised sites may become greater than the cumulative fugitive emissions output of established centralised treatment systems. Various mitigation and recycling schemes are further detailed in Section 16.4.3.

16.4.3 Emissions mitigation and gas reuse strategies

Strategies for reducing emissions from wastewater treatment processes can have direct and indirect benefits in terms of energy usage and environmental footprint mitigation. The immediate cost/emission avoidance is realised from the generation and exploitation of energy from the capture and reuse of biogas, as well as an indirect reduction of GHG emissions through improved energy and resource efficiency, and a decrease in the use of fossil fuels (Bogner *et al.* 2008). There are many approaches to capturing biogas including covering anaerobic lagoons, membrane capture systems and up flow anaerobic digesters. According to the Australian Water Recycling Centre of Excellence (2010), the anaerobic digester is the most important component of a wastewater treatment process as it generates a constant energy source in the form of biogas, readily mobilises nutrients and it is usually fully enclosed. Biogas capture approaches from anaerobic digesters are usually applied to those treating high strength wastewaters, such as those derived from industrial and agricultural applications.

Biological nitrogen removal is one of the main drivers for treating wastewater and inefficiency in treatment processes and reaction kinctics can lead to the production of high levels of N_2O and other unwanted compounds. Ammonia removal in traditional wastewater treatment processes requires an oxidation phase and a reduction phase and includes multiple process steps. In the denitrification step a carbon source is required if the reaction is to proceed. This is rate limiting and needs careful management. Approaches that utilise anaerobic ammonia oxidation techniques, such as DEMON (DEAmMONification) and ANAMMOX (ANaerobic AMMmonium OXidation), have been developed and provide significant energy and cost saving benefits. An example of a wastewater treatment plant that has utilised anaerobic ammonium oxidation techniques (DEMON) at full scale is the Strass WWTP in Austria. This plant serves a population which varies from 60,000 in summer to 250,000 in winter. In 2004 the plant implemented the DEMON process providing deammonification without the need for a supplemental carbon supply (Wett, 2007). The two specific advantages of employing this process on site were that the energy requirements for nitrification of the side stream ammonia were reduced and the organic sludge previously needed for denitrification of the side stream was now available for conversion to biogas in the anaerobic digesters (WERF, 2010). Following the commissioning of the DEMON process, the onsite specific energy demand of the side-stream process was reduced to 1.16 kWh/kg ammonia nitrogen removed, compared to approximately 6.5 kWh/kg ammonia nitrogen removed using traditional biological nitrogen removal methods at the same plant (Wett, 2007).

Real-world studies, such as the ones described in this chapter, provide a better understanding of the seasonal and operational variability of wastewater treatment processes and associated fugitive GHG emissions. Further and more detailed understanding could be gained via continuous ongoing online real-time monitoring of N_2O and CH_4 gas concentrations as well as process gas concentrations (DO and NH_3), which can be correlated to process conditions and variability in treatment loading. The incorporation of biogas recovery systems, such as a fixed film anaerobic digester or membrane based gas recovery system, to decentralised systems will be site specific and depend on available side streams and existing processes. As additional carbon sources may be required to maximise energy recovery from waste streams, co-location of decentralised wastewater treatment water reuse facilities near other industries that produce high strength wastewaters is highly desirable. A detailed analysis also needs to be performed to determine the long-term cost benefit of deploying such technology at decentralised reuse systems. Additionally, to make a significant impact on energy reduction and resource recovery a step change in current practice is required. Innovative solutions (e.g., the incorporation of technologies such as DEMON or ANAMMOX) for side stream treatment to efficiently remove nitrogen and more effectively utilise the carbon in waste streams for energy recovery, offer considerable potential.

16.5 CONCLUSION

This chapter has described the results of a measurement campaign designed to ascertain levels of CH_4 and N_2O from water reuse facilities at three sites in Melbourne, Australia. It was found that typical decentralised reuse systems running an activated sludge treatment regime can emit greatly variable, but still measurably high levels of CH_4 and N_2O per unit area from their aeration tanks. As decentralised reuse systems are being installed in growing numbers around the world, the cumulative fugitive emissions released from separate decentralised reuse systems serving different communities may become greater than the total fugitive emissions output of established centralised treatment systems. In addition, the emissions data measured at each of the decentralised reuse systems was generally not well correlated to the emissions data calculated by the current NGERS semi-empirical models. Therefore, it is recommended that further revisions be made to these models to make them more applicable to a wider variety of wastewater treatment systems. This will help to greatly improve the accuracy of emissions reporting performed by wastewater treatment operators.

The measured data also showed that both N_2O and CH_4 fluxes can vary dramatically in magnitude over small distances across the surface area of a treatment reactor. Accordingly, neglecting the fact that the magnitude of both N_2O and CH_4 fluxes can vary by measurable amounts over small distances across an aeration tank may result in inaccuracies in recorded N_2O and CH_4 emissions data, and consequently in the overall environmental footprint calculated for the reuse system under investigation. Therefore, in order to gain a more accurate determination of both temporal and spatial flux output, it is recommended that flux measurements are made in rapid succession at multiple evenly spaced positions across the entire measurable surface of an aeration tank.

The emissions dataset presented in this investigation was obtained over a single season. However, modifications to physical and chemical parameters brought on by seasonal change may influence the amount of both CH_4 and N_2O emissions from wastewater treatment processes, particularly variations in sewage temperature, which can have a direct impact upon the production and release of CH_4. Consequently, further field studies spanning across an entire calendar year are required to measure long-term CH_4 and N_2O emissions, in order to better ascertain the overall effect of seasonal change and to assimilate a larger emissions dataset to compare directly to emissions data estimated by the NGERS models. In addition, a direct comparison between the annual emissions output from the decentralised reuse systems and nearby centralised systems treating similar influent is required to quantitatively determine if the decentralised reuse systems produce emissions at a similar magnitude (per litre of influent) to the centralised systems. At this stage, it is not possible to directly compare the emissions data measured from the decentralised reuse systems to any emissions data from centralised systems, as there are no long-term studies available detailing online measurements of CH_4 and/or N_2O gases from centralised wastewater treatment processes in Australia. It is anticipated that this deficiency in the knowledge base will soon be resolved, as a large number of Australian research groups and system operators are beginning to actively measure and record gas emissions from a wide variety of treatment systems.

Although a number of technologies have been developed and implemented, their long term performance in reducing GHG emissions, suitability for different contexts and overall sustainability requires further detailed research.

REFERENCES

Australian Water Recycling Centre of Excellence (AWRCoE) (2010). Sustainability – Life Cycle Assessment and Integrated Resource Management Strategies. Australian Water Recycling Centre of Excellence Technical Report.

Bogner J., Pipatti R., Hashimoto S., Diaz C., Mareckova K., Diaz L., Kjeldsen P., Monni S., Faaij A., Gao Q., Zhang T., Abdelrafie M. A., Sutamihardja R. T. M. and Gregory R. (2008). Mitigation of global greenhouse gas emissions from waste: Conclusions and strategies from the Intergovernmental Panel on Climate Change (IPCC) fourth assessment report. Working Group III (Mitigation). *Waste Management and Research*, **26**(1), 11–32.

Carignan R. (1998). Automated determination of carbon dioxide, oxygen, and nitrogen partial pressures in surface waters. *Limnology and Oceanography*, **43**, 969–975.

Czepiel P. M., Crill P. M. and Harriss R. C. (1993). Methane emissions for municipal wastewater treatment processes. *Environmental Science and Technology*, **27**, 2472–2477.

Czepiel P., Crill P. and Harriss R. (1995). Nitrous oxide emissions from municipal wastewater treatment. *Environmental Science and Technology*, **29**, 2352–2356.

Foley J., de Haas D., Yuan Z. and Lant P. (2010). Nitrous oxide generation in full-scale biological nutrient removal wastewater treatment plants. *Water Research*, **44**, 831–844.

Global Water Research Commission (2011). N_2O and CH_4 Emission from Wastewater Collection and Treatment Systems. Technical Report. Report of the GWRC research strategy workshop. Global Water Research Coalition. London, United Kingdom.

Ho Ahn J., Kim S., Park H., Rahm B., Pagilla K. and Chandran K. (2010). N_2O emissions from activated sludge processes, 2008–2009: results of a national monitoring survey in the United States. *Environmental Science and Technology*, **44**, 4505–4511.

IPCC (Intergovernmental Panel on Climate Change) (2007). Climate Change 2007: The Physical Science Basis, Contribution of Working Group I to the Fourth Assessment Report of the Intergovernmental Panel on Climate Change. Cambridge University Press, Cambridge United Kingdom and New York, USA.

Kampschreur M. J., Temmink H., Kleerebezem R., Jetten M. S. and van Loosdrecht M. C. (2009). Nitrous oxide emission during wastewater treatment. *Water Research*, **43**(17), 4093–4103.

Lambert M. and Frechette, J.-.L. (2005). Analytical techniques for measuring fluxes of CO_2 and CH_4 from hydroelectric reservoirs and natural water bodies. In: Greenhouse Gas Emissions – Fluxes and Processes: Hydroelectric Reservoirs And Natural Environments, A. Tremblay, L. Varfalvy, C. Roehm and M. Garneau (eds), Springer Publishing, Germany, pp. 37–60.

Law Y., Ye L., Pan Y. and Yuan Z. (2012). Nitrous oxide emissions from wastewater treatment processes. *Philosophical Transactions of the Royal Society of London Series B-Biological Sciences*, **367**, 1265–1277.

Lotito A. M., Wunderlin P., Joss A., Kipf A. and Siegrist H. (2012). Nitrous oxide emissions from the oxidation tank of a pilot activated sludge plant. *Water Researcg*, **46**(11), 3563–3673.

Mes T. Z. D., de Stams A. J. M., Reith J. H. and Zeeman G. (2003). Chapter 4. Methane production by anaerobic digestion of wastewater and solid wastes. In: Biomethane and Biohydrogen. Status and Perspectives of Biological Methane and Hydrogen Production. J. H. Reith, R. H. Wijffels and H. Barten (eds), Dutch Biological Hydrogen Foundation, Netherlands, pp. 58–102.

Rassamee V., Sattayatewa C., Pagilla K. and Chandran K. (2011). Effect of oxic and anoxic conditions on nitrous oxide emissions from nitrification and denitrification processes. *Biotechnology and Bioengineering*, **108**(9), 2036–2045.

Ravishankara A. R., Daniel J. S. and Portmann R. W. (2009). Nitrous oxide (N_2O): The dominant ozone-depleting substance emitted in the 21st century. *Science*, **326**(5949), 123–125.

Schouten P., Sharma A., Burn S., Goodman N. and Umapathi S. (2013a). Evaluation of fugitive greenhouse gas emissions from decentralized wastewater treatment plants. In *Proceedings of the World Environmental and Water Resources Congress 2013*, Cincinnati, 19–23 May.

Schouten P. W., Sharma A., Burn S. and Goodman N. (2013b). Spatial and short-term temporal distribution of fugitive methane and nitrous oxide emission from a decentralised sewage mining plant: a pilot study. *Journal of Water Climate Change*, **5**(1), 1–12.

Sharma A. K., Grant A. L., Grant T., Pamminger F. and Opray L. (2009). Environmental and economic assessment of urban water services for a Greenfield development. *Environ. Eng. Sci.*, **25**(5), 921–934.

Sharma A., Burn S., Gardner T. and Gregory A. (2010a). Role of decentralised systems in the transition of urban water systems. *Water Science Technology*, **10**(4), 577–583.

Sharma A. K., Tjandraatmadja G., Grant A. L., Grant T. and Pamminger F. (2010b). Sustainable sewerage servicing options for peri-urban areas with failing septic systems. *Water Science Technology*, **62**(3), 570–585.

Sharma A., Cook S., Tjandraatmadja and Gregory A. (2012). Impediments and constraints in the uptake of water sensitive urban design measures in greenfield and infill developments *Water Science Technology*, **65**(2), 340–352.

Sharma A. K., Tjandraatmadja G., Cook S. and Gardiner T. (2013a). Decentralised systems – Definitions and drivers in the current context. *Water Science Technology*, **67**(9), 2091–2101.

Sharma A., Chong M. N., Schouten P., Cook S., Ho A., Gardner T., Umapathi S., Sullivan T., Palmer A. and Carlin G. (2013b). Decentralised Wastewater Treatment Systems: System Monitoring and Validation. Urban Water Security Research Alliance Technical Report No. 70.

Townsend-Small A., Pataki D. E., Tseng L. Y., Tsai C.-Y. and Rosso D. (2011). Nitrous oxide emissions from wastewater treatment and water reclamation plants in southern California. *Journal of Environment Quality*, **40**, 1542–1550.

Tremblay A., Lambert M. and Gagnon L. (2004). Do hydroelectric reservoirs emit greenhouse gases? *Environ. Manage.*, **33**(1), S509–S517.

Venterea R. T., Spokas K. A. and Baker J. M. (2009). Accuracy and precision analysis of chamber-based nitrous oxide gas flux estimates. *Soil Science Society of America Journal*, **73**, 1087–1093.

Water Environment Research Foundation (WERF) (2010). Sustainable Treatment: Best Practices from The Strass im Zillertal Wastewater Treatment Plant. Water Environment Research Foundation Online Technical Report OWSO4R07b. http://www.werf.org/c/_FinalReportPDFs/OWSO/OWSO4R07b.aspx (accessed 07 July 2014).

Wett B. (2007). Development and implementation of a robust deammonification process. *Water Science Technology*, **56**(7), 81–88.

Wilderer P. A. (2001). Decentralised versus centralised wastewater management. In: Decentralised Sanitation and Reuse, Concepts, Systems and Implementation, Lens P., Zeeman G. and Lettinga L. (eds), IWA Publishing, London, United Kingdom, pp. 39–54.

Winter P., Pearce P. and Colquhoun K. (2012). Contribution of nitrous oxide emissions from wastewater treatment to carbon accounting. *J. Water Clim. Change*, **3**(2), 95–109.

Venterea R. T., Spokas K. A. and Baker J. M. (2009). Accuracy and precision analysis of chamber-based nitrous oxide gas flux estimates. *Soil Sci. Soc. Am. J.*, **73**, 1087–1093.

Chapter 17

Large-scale water reuse systems and energy

Valentina Lazarova

17.1 INTRODUCTION

The ambitious goals of sustainable development and achieving a zero net carbon and pollution emission footprint call for a new holistic approach to the management of the water cycle with an increased role for water reuse (Daigger, 2009; Novotny *et al.* 2010; Lazarova *et al.* 2012). With the further growth of megacities and increasing efforts to optimise energy efficiency, water recycling is of growing interest and will take a leading role in the future of sustainable urban water cycle management.

Sustainability in water resource planning requires the consideration of the embodied energy in the water cycle and in particular in water recycling (Cornel *et al.* 2011; Wilson, 2012). Wastewater treatment and reclamation plants have the potential to become environmental platforms, as well as energy sources for tomorrow's eco-cities as part of a system characterised by the smallest possible ecological footprint (GWRC, 2011). Water recycling is enabling the optimisation of energy intensity within the water cycle, especially within decentralised and semi-autonomous urban systems, with a treatment level that adheres to the 'fit for use' principle (Asano *et al.* 2007; Lazarova *et al.* 2012). Recycled water needs to be delivered at a cost justifiable for its purpose; therefore, energy-intensive processes should be limited when exceptionally high water quality is required.

The energy demand of wastewater or greywater treatment depends on the water quality requirements, the selected treatment process and the plant capacity. More intensive water treatment has a higher environmental impact in terms of carbon footprint. For example, the state of the art advanced water reclamation based on high-tech, energy-intensive technologies has a carbon footprint which is five times higher than the conventional water reclamation processes. Nevertheless, the energy consumption of water reclamation is significantly lower than other alternative resources such as desalination, or water transportation over long distances and represents only a fraction of the energy demand (e.g., specific energy use) for water supply, treatment and distribution (Lazarova *et al.* 2012).

This chapter summarises sources of energy consumption and carbon emissions from water reuse systems and explores methods to minimise the energy and carbon footprints of such systems.

17.2 ENERGY FOOTPRINT OF THE URBAN WATER CYCLE

17.2.1 Typical components of energy consumption in the urban water cycle

The typical specific energy consumption of the major components of the urban water cycle and treatment processes is shown in Figure 17.1 (Lazarova *et al.* 2012). The energy required to convey and treat water to acceptable levels is in the range of 0.05–5 kWh/m^3, depending on the water source (freshwater, seawater or wastewater) and on specific regional factors such as climate, water availability, water use and population density. The energy consumption of water and wastewater treatment facilities varies across a similar range, from 0.2 to 1.4–1.5 kWh/m^3, depending on the pumping head, level of treatment and plant capacity. Energy optimised nutrient removal with anaerobic digestion, as found for example, at the Strass wastewater treatment plant in Austria (220,000 p.e.), not only has a relatively low-energy footprint of 0.35 kWh/m^3, but also produces electricity, achieving energy self-sufficiency (Wett *et al.* 2007). Water conveyance can reach values of 1.1 kWh/m^3 and may have a significantly higher energy footprint in specific cases, such as long-distance transportation. For example, the State Water Project in California requires 2.5 kWh/m^3 for water delivery.

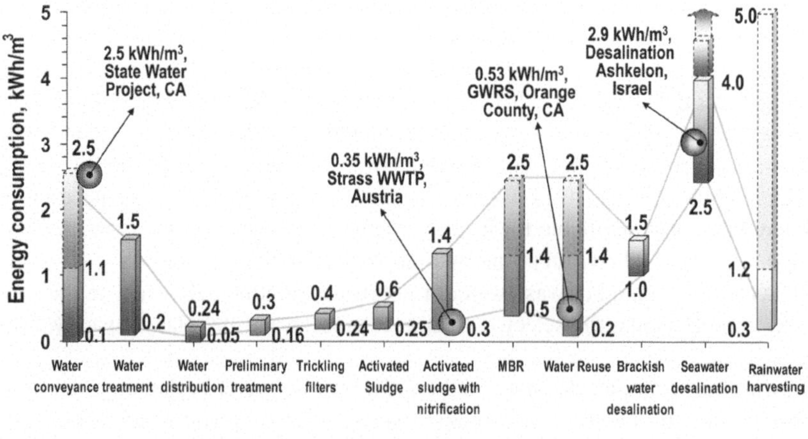

Figure 17.1 Typical energy footprint of the major components and processes in urban water cycle management (Lazarova *et al.* 2012).

Globally, rainwater harvesting technologies are being more readily adopted, as the desire increases for buildings to become more adaptable and resilient to climate change and population growth. The majority of these systems use pumping equipment and schemes, characterised by high-energy consumption, typically in the range of 0.3–1.2 kWh/m^3. However, rainwater harvesting systems are generally applied at the small scale in individual or high-rise buildings and thus a direct comparison with large facilities is not appropriate. Innovative gravity-driven rainwater harvesting systems, recovering the kinetic energy of the water flow or using solar energy, may help to reduce the energy consumption and thereby to improve the sustainability of rainwater harvesting.

17.2.2 Energy consumption of wastewater treatment and reuse

Advanced wastewater treatment for nutrient removal and water reuse requires more energy compared to conventional wastewater treatment. The type of sewer system (combined or separated) and the degree of plant utilisation greatly influence the specific energy consumption, especially in tourist areas, where the ratio between hydraulic loading in the summer and winter can vary by a factor of 7–10. For this reason, several performance indicators are used in different countries and for various purposes, such as kWh/p.e. (population equivalent, which is calculated on the basis of the specific daily pollution of one person as COD, BOD_5 or m^3, determined as an average value for a given period of time), kWh/m^3 or kWh/kg BOD_5 removed (on a yearly basis).

Typical energy consumption levels of activated sludge plants in Europe are in the range of 2.0 ± 0.5 kWh/kg BOD_5 for carbon removal and 3.0 ± 0.5 kWh/kg BOD_5 for activated sludge with nitrification/denitrification. Specific energy consumption per unit of water volume is a useful benchmark when comparing selected processes. For wastewater treatment plants, pumping head has a great importance and typical values vary widely from 0.4 to 1.2 kWh/m^3. The benchmark used in several European countries such as Austria, Germany, Sweden and Switzerland is expressed as population equivalent (p.e.) corresponding to 110 g COD/d or 60 g BOD_5/d for a one-year period (pe110.yr). Typical energy consumption of large treatment plants over 100,000 p.e. in Austria and Germany is 23 kWh/pe110.yr, while a higher value of 42 kWh/pe110.yr has been reported in Sweden, probably due to more diluted wastewater (Jonasson, 2007). Energy input rises with the increasing level of treatment and nutrient removal, as well as with decreasing plant size. In addition to plant specificity, the difference in the applied methodology and wastewater characteristics for the estimation of energy consumption makes comparison of the results very difficult.

Figure 17.2 illustrates such an example, comparing the results of a survey using questionnaires (Lazarova & Kamisoulis, 2005) and on-site energy audits of several full-scale WWTPs in Europe, with measurement of the absorbed power and power factor for all major electromechanical equipment such as motors and blowers. A major part of the facilities is an activated sludge plant with combined sewers, designed for carbon and nitrogen removal, with few tertiary treatment facilities and membrane bioreactors (MBR). The majority of the wastewater treatment plants in Spain included in the survey is also equipped with tertiary treatment such as UV disinfection and/or chlorination.

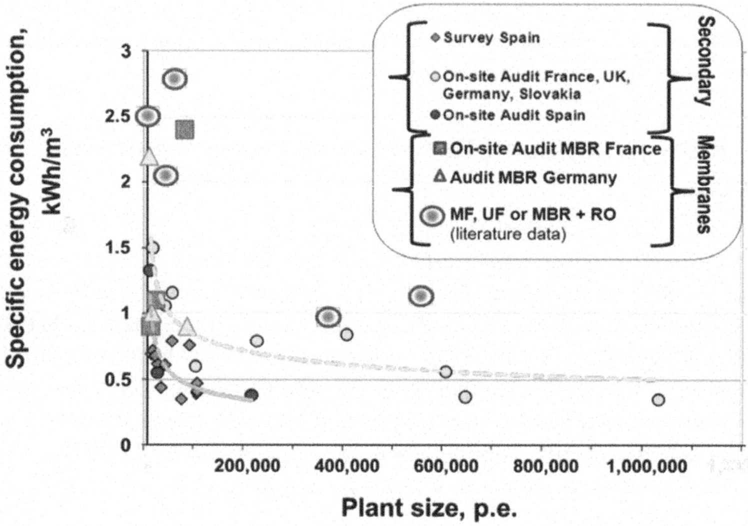

Figure 17.2 Distribution of specific energy consumption of wastewater treatment plants in Europe, predominantly using activated sludge, as a function of plant capacity.

A relatively good correlation was observed between plant size and specific energy consumption from 0.38 to 1.5 kWh/m^3 with lower values for large facilities in Spain, 0.56 ± 0.16 kWh/m^3. It should be noted that on-site energy audits typically show higher energy use as compared to information received from plant responses to questionnaires (in range of 0.94 ± 0.53 kWh/m^3). The highest value of 2.5 kWh/m^3 was reported for a new MBR facility with a nominal capacity of 75,000 p.e., operated at high hydraulic loads (73% of design capacity) and low BOD loads (26% of design capacity). The addition of a conventional tertiary treatment or odour removal accounted for an increase in energy consumption by 10–15%. Significant influence of sludge treatment on energy use was not observed (anaerobic digestion was implemented in 25% of the surveyed wastewater treatment facilities). Nevertheless, it is important to underline that some sludge treatment processes (e.g., thermal drying or incineration), implemented for beneficial treatment of biosolids, are characterised by high-energy demand.

To produce high-quality recycled water and close the urban water cycle, energy demanding water reclamation processes are needed (Figure 17.3), such as highly efficient tertiary treatment with chemical addition, MBRs, advanced disinfection and/or membrane filtration such as microfiltration (MF), ultrafiltration (UF), nanofiltration (NF) and reverse osmosis (RO). Tertiary disinfection by UV or ozone leads to an increased energy demand of 7–40%, with higher values typical for high ozone doses. Compared to conventional activated sludge, the implementation of MBR technology for the production of disinfected and free of suspended matter effluents, requires 30–50% more energy at optimal operating conditions, for example 0.83 kWh/m^3 compared to 0.55 kWh/m^3 (an average annual value). Large variations of 0.7 to 2.5 kWh/m^3 for MBR plants have been reported as a function of plant size, hydraulic loading and wastewater quality (Lazarova *et al.* 2011; Barillon *et al.* 2013). The production of high-quality recycled water by combined membrane treatment, such as MF/RO or UF/RO, requires 100–150% more energy than conventional wastewater treatment, which is however, significantly lower compared to desalination.

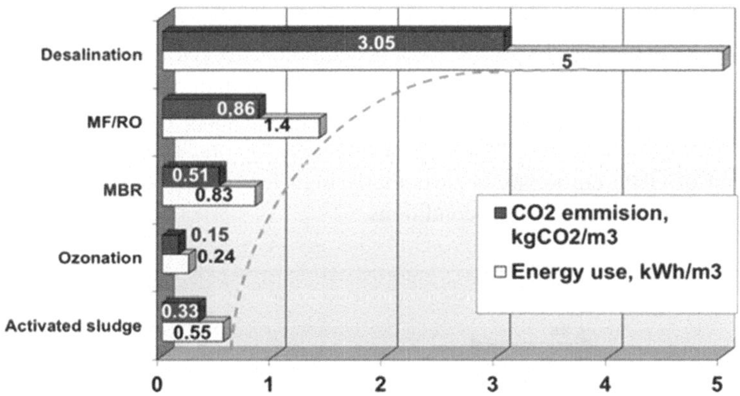

Figure 17.3 Energy footprint of wastewater treatment processes compared to desalination for a typical water reuse plant with a size of 10,000 m^3/d (adapted from Novotny, 2010; Lazarova *et al.* 2012).

Despite the requirement for a higher level of wastewater treatment, water reuse is one of the most cost and energy effective alternative water resources compared to desalination and long distance water transportation. Energy-efficient advanced water recycling plants such as the Groundwater Replenishment System (GWRS) project in Orange County, California (shown in Figure 17.1), are producing recycled water of drinking water quality with a relatively low-energy footprint of 0.53 kWh/m^3 for the RO process. This consumption is almost 50% of the total energy demand of 1.1 kWh/m^3 of the GWRS plant (Mehul, 2012).

17.2.3 Carbon footprint of wastewater treatment and reuse

The carbon footprint of the elements of the urban water cycle, including wastewater treatment and reuse, is proportional to energy use, as shown in Figure 17.3. For the calculation of these values, an emission factor of 0.61 kg CO_2eq/kWh was used based on US EPA data for the average emission in the United States from a mix of fuel sources (WNA, 2013). It is important to underline that the emission factors vary from year to year and from country to country. For this reason, the carbon footprint of a given treatment process could be very different depending on the energy source.

There are many different electricity generation methods, each having advantages and disadvantages with respect to operational cost, environmental impact and other factors. In relation to greenhouse gas (GHG) emissions, each generation method produces emission fluxes in varying quantities through construction, operation and decommissioning (including fuel supply activities). As demonstrated by numerous recent studies (World Nuclear Association, 2013), some generation methods such as coal- and oil-fired power plants release the highest GHG emissions (Figure 17.4). GHG emissions of nuclear power plants, hydroelectric plants and wind farms are among the lowest: on a lifecycle basis they are about 30 times lower than coal generation. Lifecycle emissions for power generation using natural gas are about 15 times lower than power generated from coal. Within each of the power generation methods, there is a great variability of GHG emissions due to region- and site-specific factors. The reported range of variation of GHG emissions is particularly high for solar photovoltaic systems, mainly due to specific country conditions.

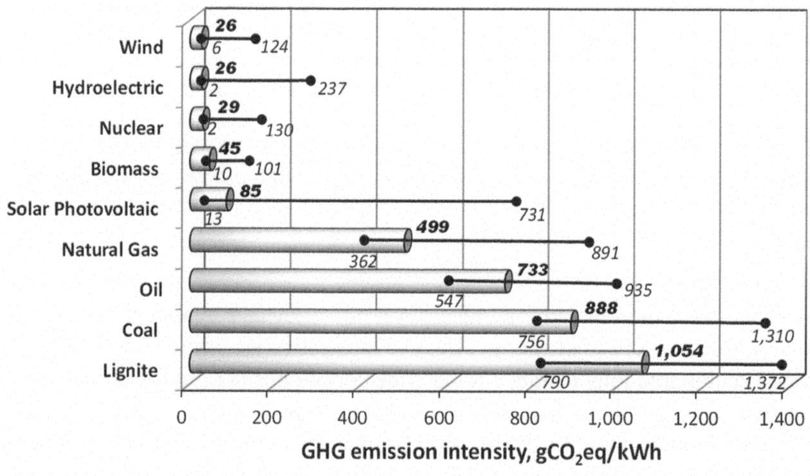

Figure 17.4 Influence of energy source on GHG emission factors (adapted from World Nuclear Association, 2013).

A good example of the influence of energy source on the carbon footprint of advanced water recycling is provided by Delgado *et al.* (2012). The operational carbon footprint was assessed for the Advanced Water Recycling Plant (AWRP) of Bundamba, Australia, with a treatment capacity of 66,000 m^3/d and a treatment train including pre-treatment, MF, 3 stage RO and post-treatment. The specific energy consumption was estimated at 1.14 kWh/m^3, of which membrane pretreatment by MF accounts for 21% and RO system for 53%. As the energy supply in Australia originates mainly from coal thermal power plants (emission factor (EF) of 0.921 kg CO_2eq/kWh), the calculated carbon footprint is high; being 10,200 ton CO_2eq per year (82% of the GHG emission is due to energy consumption). If the same AWRP was constructed and operated in Spain (EF of 0.35 kg CO_2eq/kWh, mainly due to use of natural gas) or France (EF of 0.085 kg CO_2eq/kWh, mainly because of use of nuclear energy), the carbon footprint would be two times lower (5000 ton CO_2eq) or 4 times lower (2580 ton CO_2eq), respectively. In this case, the impact of energy consumption is 60% and 30% lower, respectively (Figure 17.5).

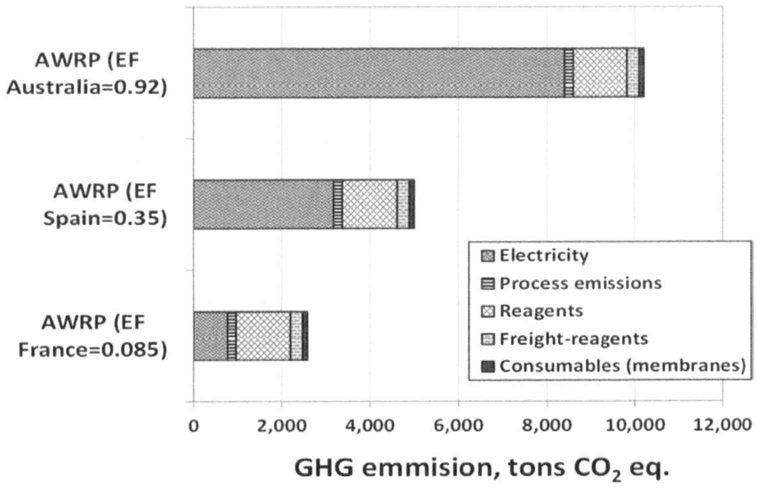

Figure 17.5 Influence of the electricity source mix on the carbon footprint of a 66,000 m^3/d advanced water recycling plant (AWRP) with MF/RO treatment.

The requirement for zero liquid discharge associated with treatment of MF and RO brines may lead to a significantly higher footprint (up to +300%) compared to conventional treatment processes. As compared to advanced MF/RO treatment, the carbon footprint of conventional tertiary treatment (sand filtration, GAC filtration, UV disinfection, chlorination) is typically 20–25% lower.

17.3 KEY ENERGY USE COMPONENTS OF WASTEWATER TREATMENT AND REUSE

17.3.1 Typical distribution of energy consumption

Typically, the most energy consuming processes in conventional wastewater treatment plants (activated sludge) are aeration, pumping, mechanical treatment and ventilation for odour control (Figure 17.6a). Aeration of activated sludge is characterised with the highest energy consumption for conventional wastewater treatment facilities, accounting typically for 45 ± 15% of total energy needs depending on plant size, design and additional treatment processes (Lazarova *et al.* 2012). In more advanced MBR systems, where final clarification is replaced by UF membranes, the overall aeration demand is increased by membrane air scouring, which typically accounts for 65 ± 10% of WWTP energy consumption for hollow fibre submerged membranes (Figure 17.6b). The contribution of air scouring is higher for flat sheet MBRs, accounting for up to 58% of energy consumption (Barillon *et al.* 2013), typically in the range of 0.45–0.55 kWh/m^3 compared to 0.2–0.3 kWh/m^3 for MBRs equipped with hollow fibre membranes. Tertiary treatment, in particular membrane processes, require more energy for the production of recycled water. Energy consumption increases with an increase in the level of treatment for the removal of pathogens, trace organics and salts.

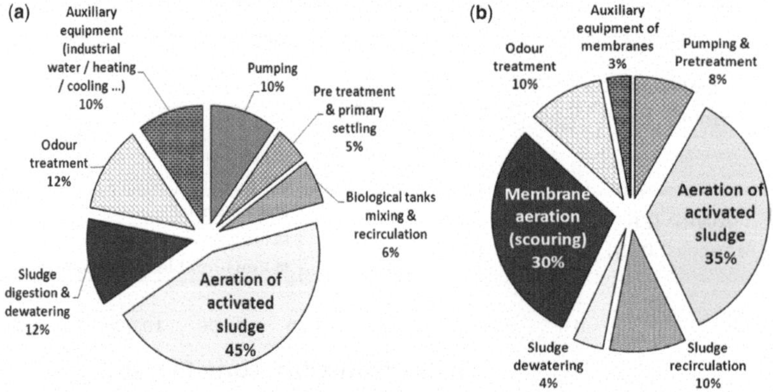

Figure 17.6 Energy consumption breakdown for conventional activated sludge plants (a), and MBR hollow fibre systems (b).

17.3.2 Energy consumption of large water recycling facilities

The energy demand of tertiary treatment for water reclamation and advanced water recycling is also greatly influenced by the plant size and treatment process.

Energy demand decreases with an increase of plant size and increases with the level of treatment. The highest specific energy consumption is typical for advanced oxidation and membrane treatment. The typical range of specific energy demand for the most common tertiary treatment processes used in water reuse plants is shown in Figure 17.7. Rapid sand filtration and vacuum-driven submerged UF or MF membrane are characterised by the lowest energy consumption values. Pressure-driven sidestream UF and MF and more complex treatment schemes with UV or ozone disinfection have higher energy use, which however, is lower than the energy demand of RO. The analysis of the energy consumption breakdown of large advanced water recycling facilities (advanced treatment by MF/RO/UV + H_2O_2) confirms that the highest energy consumption is required for the RO operation, accounting for about 50% of the total energy demand.

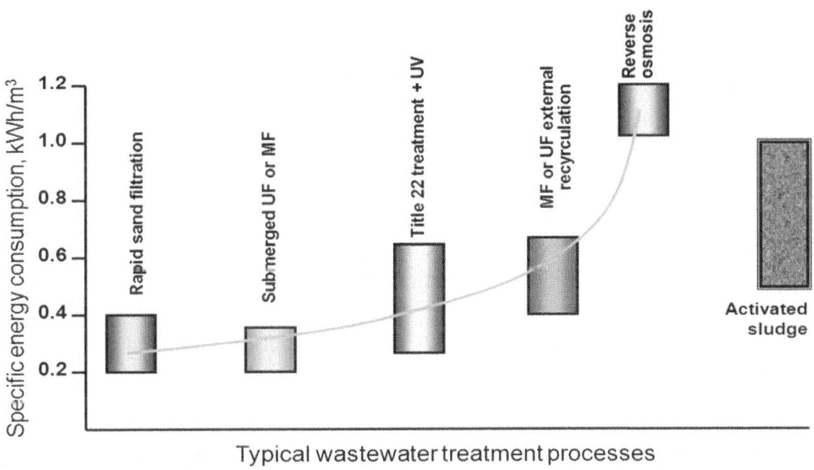

Figure 17.7 Typical energy consumption breakdown of tertiary treatment processes in large water reuse plants (>30,000 m^3/d) compared to conventional activated sludge with nitrogen removal.

The distribution of energy consumption for two large advanced water recycling plants (AWRP) is shown in Figure 17.8 for Bundamba AWRP, Australia (66,000 m^3/d) and GWRS AWRP in Orange County, California (265,000 m^3/d). The specific energy consumption is very similar, 1.14 and 1.1 kWh/m^3, respectively. The major differences in the energy footprint of these two plants is in the relatively high pumping requirement for the GWRS to convey recycled water to the recharge basins, accounting for 18% of the total energy consumption, about 0.2 kWh/m^3. MF typically requires 20–25% of the energy, which represents 0.275 kWh/m^3 for the GWRS (Mehul, 2012).

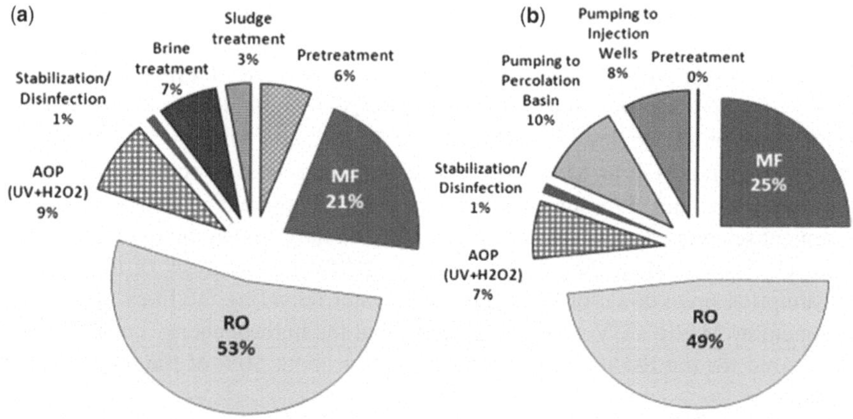

Figure 17.8 Energy consumption breakdown of advanced MF/RO/UV + H_2O_2 recycling plants in (a) Bundamba AWRP, Australia, 66,000 m³/d and (b) GWRS AWRP in Orange County, California, 265,000 m³/d.

Conventional tertiary treatment for non-potable reuse requires significantly less energy. For example, one of the largest recycling facilities in the world is in Barcelona, the El Prat de Llobregat Water Reclamation Plant, with a design flow of 300,000 m³/d. The plant has a specific energy consumption of 0.28 kWh/m³, the major part of which is for sludge treatment (Delgado *et al.* 2012). The treatment train includes a high-rate clarification with sand ballast and lamella settling, followed by MF through a 10 μm mesh, UV disinfection and final chlorination with sodium hypochlorite.

17.4 METHODS FOR ENERGY AND CARBON FOOTPRINT MINIMIZATION

Energy conservation is being widely pointed out as the key element in a more sustainable energy future, because it can be implemented immediately using existing technologies. The most efficient and proven methods for energy optimisation are as follows (Lazarova *et al.* 2012):

- System design and process optimisation;
- Pumping optimisation: pumping consumes the most energy and it is recommended to adapt pump design and selection to the specific operating conditions, as well as to use VFD (variable frequency drive) and premium efficiency motors. Pump efficiency increases with increasing pump size and with decreasing delivered pressure;
- Membrane design and optimisation: pilot testing is highly recommended to select the most efficient equipment (MF, UF, RO) and adapt the design to

the raw water quality and other specific requirements. Control of membrane fouling is a critical issue with a strong influence on energy demand;
- Energy recovery for RO systems;
- Use of renewable energy for the reduction of carbon footprint.

Significant energy savings can be achieved in wastewater treatment plants not only from the implementation of technologies and best practices for low-energy consumption, but also by energy recovery from sludge and sewage (Lazarova *et al.* 2012a). Examples of wastewater treatment plants that are energy self-sufficient are the Strass WWTP in Austria, 220,000 p.e. (Wett, 2006) and Al Samra WWTP in Jordan, 2.2 million p.e. (Fievez, 2009). Various energy saving strategies have been implemented in many recent facilities, such as optimisation of aeration system design; efficient aeration control; use of premium efficiency motors and variable frequency drives for large pumps and aeration devices (blowers, mechanical aerators). Moreover, innovative waste stream handling technologies have been implemented, such as treatment of concentrated return flow from anaerobic digestion by deammonification, as well as the next generation of technologies using anaerobic treatment or pre-treatment of the main wastewater flow. Anaerobic digestion of sludge has been and will remain a major source of energy from the production and use of biogas. Biogas production may be improved by innovative technologies such as co-digestion, combined heat and power generation from digester gas (cogeneration, fuel cells, microturbines) or mechanical energy (direct drive or stirling engines) and gasification (La Cour Jansen *et al.* 2004; Bolzonella *et al.* 2006; Deublein & Steinhauser, 2008; Bouchy *et al.* 2012; Lazarova *et al.* 2012a). Similar energy saving strategies are also applied in advanced water recycling plants with the use of innovative low-fouling and low-energy consuming membranes, as well as innovative processes such as forward osmosis or pressure retarded osmosis (Hoek & Tarabara, 2013).

Energy recovery from RO concentrate is a common practice in recent desalination plants (Huehmer, 2013). A common misconception exists that energy recovery is not economical for water reuse RO systems, due to the low feed pressures and high product water recovery. The recent feedback from the MF/RO Ulu Pandan advanced recycling plant in Singapore demonstrated that the use of an energy recovery device (ERD) instead of a booster pump reduces pumping energy by 6.5%. Six energy recovery devices (one per unit) are under implementation in the new extension of the GWRS Orange County to treat 114,000 m^3/d. It is expected that the implementation of ERDs would offer energy savings of 182,000 kWh/year and would result in an estimated payback period for the ERDs of 2–7 years, depending on operating conditions.

The use of renewable energy is becoming another relevant option enabling to decrease carbon footprint of advanced wastewater treatment and reclamation plants. Historically, wastewater treatment plants have been using the thermal part of solar energy in very specific applications namely, sludge drying and water desalination. The most popular use of solar energy is in small reverse osmosis desalination plants, which apply mostly photovoltaic (PV) panels. Because of the need for a large area for the installation of solar panels, the contribution of this green energy to the power supply of large WWTPs is relatively limited.

A good example of using solar energy is the WWTP of the city of Chino, California, which meets 14% of its electricity demand from renewable energy produced by 3047 solar panels installed on a 6475 m^2 area (Crawford and Sandino, 2010). This plant is an activated sludge plant with tertiary filtration and disinfection with a treatment capacity of 43,200 m^3/d. The capital cost for a 0.7 MW photovoltaic system was US$3500/kW. Another good example of the use of solar energy in an advanced MF/RO recycling facility is the Edward C. Little Water Recycling in West Basin, California. The implementation of 2848 solar panels with a total area of 5574 m^2 installed on the roofs of the recycled water storage reservoirs covers 10% of peak energy needs (Walters *et al.* 2013). In addition to the requirement of high surface area, another disadvantage of PV technologies is the relatively low energy net efficiency, typically 12–15% for crystalline silicon modules and below 10% for flat-module thin film technologies. An ambitious project recently implemented in France at the new WWTP of Cannes, is satisfying 22% of total plant electrical demand by an innovative photovoltaic-thermal system (PVT) offering improved energy efficiency (Lazarova *et al.* 2012).

Wind generators with vertical or horizontal axes are considered a mature technology. Onshore wind generators produce electricity at relatively low cost and are already largely deployed within the zones with high wind potential. Offshore wind parks are considered to become booming markets of the years ahead because of their higher energy yield and easier installation. Nevertheless, they still present numerous challenges. The overall impact of wind power, for the improvement of energy efficiency of wastewater treatment plants, still remains relatively low. One of the largest WWTPs that meets about 67% of its average annual electrical demand with wind power, is located near the Atlantic Ocean in Atlantic City, New Jersey in the United States (Crawford and Sandino, 2010). The capacity of this activated sludge plant with disinfection is 1.75 m^3/s and the daily power demand is 2.5 MW. The capital cost for the wind farm was relatively high at US$2000/kW and was funded by means of a public–private partnership. When the prevailing wind speed exceeds 19 km/h, each of the 5 installed wind turbines produce 1.5 MW of electricity. Several challenges had to be overcome with the installation, such as turbine design in a hurricane zone, community acceptance and environmental impacts (e.g., periodic monitoring of the impact of the wind turbines on birds).

Wind power is used mainly for small capacity desalination plants with increasing potential for application for large plants. The largest wind-powered desalination plant has been operating since 2006 in Perth, Australia (Crisp & Rhodes, 2007; Sanz & Stover, 2007). The capacity of the Kwinana Desalination Plant is 143,000 m^3/d with a design to expand to 250,000 m^3/d. To minimise GHG emissions, the plant electrical requirement of 180 GWh/year is satisfied by renewable energy generated by an 83 MW wind farm located 200 km north of Perth. This seawater reverse osmosis (SWRO) desalination plant has also implemented a number of measures to reduce energy consumption, including energy recovery devices, variable frequency drives on supply pumps, booster pumps and second pass pumps and state-of-the art low-energy membrane elements. At the nominal capacity and with an overall water recovery rate of 42%, the plant consumes less than 4.2 kWh/m^3 including intake, pre-treatment, both RO passes, post-treatment, potable water pumping and all electrical losses. This is a low energy level compared to the best performing plants treating seawater of salinity of 35,000–37,000 mg/l.

17.5 CONCLUSIONS

The ambitious goals of sustainable development and achieving a zero net carbon and pollution emission footprint call for a new holistic approach to the management of the urban water cycle with an increased role for water reuse. It is important to stress, however, that advanced water treatment required for urban, industrial and potable reuse is, at present, characterised by a high-energy demand, in particular for membrane treatment processes. This chapter has summarised sources of energy consumption and carbon emissions from water reuse systems and explored methods to minimise the energy and carbon footprints of such systems. For example, reverse osmosis usually is the most expensive component of wastewater reclamation, accounting for approximately half of the total energy demand of large advanced water reuse systems. Compared to conventional tertiary treatment for non-potable reuse purposes, the energy footprint of advanced MF/RO/UV treatment schemes is up to 3–4 times higher. More advanced wastewater treatment has a higher environmental impact in terms of carbon footprint. Nevertheless, for the same advanced water recycling scheme, the carbon footprint can vary within a large range up to 20 times, depending on the electricity source mix.

Several proven strategies and innovative techniques have been developed and implemented to improve energy efficiency and to reduce the carbon footprint of wastewater treatment and reuse. The successful experience of several large water recycling schemes, for example, the Groundwater Replenishment System (GWRS) project in Orange County, California, indicates that the energy footprint of advanced treatment could be more efficient compared to other water supply alternatives, such as water transportation over long distances or desalination.

REFERENCES

Asano T., Burton F. L., Leverenz H. L., Tsuchihashi R. and Tchobanoglous G. (2007). Water Reuse Issues, Technologies, and Applications. McGraw-Hill, New York.

Barillon B., Martin Ruel S., Langlais C., Lazarova V. (2013). Energy efficiency in membrane bioreactors. *Water Science and Technology*, **67**(12), 2685–2691.

Bolzonella D., Pavan, P., Battistoni, P. and Cecchi, F. (2006). Anaerobic co-digestion of sludge with orther organic wastes and phosphorous reclamation in BNR wastewater treatment plants. *Water Science Technology*, **53**(12), 177–186.

Bouchy L., Pérez A., Camacho P., Rubio P., Silvestre G., Fernández B., Cano R., Polanco M. and Díaz N. (2012). Optimization of municipal sludge and grease co-digestion using disintegration technologies. *Water Science and Technology*, **65**(2), 214–220.

Cornel, P., Meda, A. and Bieker, S. (2011). Wastewater as a source of energy, nutrients and service water. In: Treatise on Water Science, P. Wilderer, (ed.), Elsevier-Verlag, Oxford, UK, pp. 337–375.

Crawford, G. and Sandino, J. (2010). *Energy Efficiency in Wastewater Treatment in North America: A Compendium of Best Practices and Case Studies of Novel Approaches*. WERF Report OWSO4R07e, Water Environmental Research Foundation and IWA Publishing, p. 80.

Crisp, G. and Rhodes, M. (2007). Perth seawater desalination plant – blazing a sustainability trail. *IDA World Congress*, Maspalomas, Gran Canaria, Spain, October 21–26.

Daigger, G. T. (2009). Evolving urban water and residuals management paradigms: Water reclamation and reuse, decentralization, resource recovery. *Water Environment Research*, **81**(8), 809–823.

Delgado, L., Poussade, Y. and Aguiló, P. (2012). Comparative study of carbon footprint of water reuse. In: Water-Energy Interactions in Water Reuse, V. Lazarova, K. H. Choo, and P. Cornel, (eds), IWA Publishing, London, UK, pp. 193–202.

Deublein, D. and Steinhauser, A. (2008). Biogas from Waste and Renewable Resources – an Introduction. Wiley-VCH Verlag GmbH, Weinheim, pp. 389–400.

Fievez E. (2009). AS SAMRA WWTP: towards the energy self-sufficient wastewater treatment plant. *IWA Conference Water and Energy*, Copenhagen, Denmark, 29–31 October 2009.

GWRC (2011). Water & Energy in the Urban Water Cycle: Improving Energy Efficiency in Municipal Wastewater Treatment. Report prepared by Nanyang Environment & Water Research Institute (NEWRI), PUB Singapore and Global Water Research Coalition (GWRC), Singapore.

Hoek, E. M. V. and Tarabara, V. V. (eds.) (2013). *Encyclopedia of Membrane Science and Technology*, 3 Volume Set., Wiley, New York, USA.

Huehmer, R. (2013). Evaluation and Optimization of Emerging and Existing Energy Recovery Devices for Desalination and Wastewater Membrane Treatment Plants. WateReuse Research Foundation Report. https://www.watereuse.org/product/08-14-1 (accessed 02 June 2014).

Jonasson, M. (2007). Energy Benchmark for Wastewater Treatment Processes – a comparison between Sweden and Austria. Dept. of Industrial Electrical Engineering and Automation Lund University, http://www.iea.lth.se/publications/MS-Theses/Full%20document/5247_full_document.pdf (accessed 15 January 2012).

La Cour Jansen J., Gruvberger C., Hanner N., Aspegren H. and Svärd Å. (2004). Digestion of sludge and organic waste in the sustainability concept for Malmö, Sweden. *Water Science Technology*, **49**(10), 163–169.

Lazarova, V., Choo, K. H. and Cornel, P. (eds.) (2012). *Water-Energy Interactions in Water Reuse*. IWA Publishing, London, UK.

Lazarova, V. and Kamisoulis, G. (2005). Financial aspects of the operation of sewage treatment plants. Programme for the Assessment and Control of Pollution in the Mediterranean Region, WHO, MED POL Phase III.

Lazarova, V., Martin Ruel, S., Barillon, B. and Dauthuille, P. (2011). The role of MBR technology for the improvement of environmental footprint of wastewater treatment. *Water Science and Technology*, **66**(10), 2056–2064.

Lazarova, V., Peregrina, C. and Dauthuille, P. (2012a). Toward energy self-sufficiency of wastewater treatment. In: Lazarova, V., Choo, K. H. and Cornel, P. (eds) *Water-Energy Interactions in Water Reuse*. IWA Publishing, London, UK, pp. 87–125.

Mehul, V. P. (2012). Groundwater replenishment system – energy usage implications. In: Water-Energy Interactions in Water Reuse, Lazarova, V., Choo, K. H. and Cornel, P. (eds), IWA Publishing, London, UK, pp. 183–192.

Novotny, V., Ahern, J. F. and Brown, P. R. (2010). *Water Centric Sustainable Communities: Planning, Retrofitting and Constructing the Next Urban Environments*. J. Wiley & Sons, Hoboken, NJ.

Sanz, M. A. and Stover, R. L. (2007). Low energy consumption in the Perth Seawater Desalination Plant. *IDA World Congress*, Maspalomas, Gran Canaria, Spain, October 21–26.

Walters, J., Oelker, G. and Lazarova, V. (2013). Producing designer recycled water tailored to customer needs. In: Milestones in Water Reuse: The Best Success Stories, V. Lazarova, T. Asano, A. Bahri, and J. Anderson, (eds) IWA Publishing, London, UK, pp. 37–52.

Wett, B. (2006). Solved up-scaling problems for implementing deammonification of rejection water. *Water Science and Technology*, **53**(12), 121–128.

Wett, B., Buchauer, K. and Fimml, C. (2007). Energy self-sufficiency as a feasible concept for wastewater treatment systems. *Proceedings of the IWA Leading Edge Technology Conference*, Singapore, Sept. 2007, 21–24.

Wilson, M. (2012). Embodied energy in the water cycle. In: Water-Energy Interactions in Water Reuse, V. Lazarova, K. H. Choo, and P. Cornel, (eds), IWA Publishing, London, UK, pp. 61–73.

World Nuclear Association (2013). Comparison of Lifecycle Greenhouse Gas Emissions of Various Electricity Generation Sources. WNA Report. http://www.world-nuclear.org/uploadedFiles/org/WNA/Publications/Working_Group_Reports/comparison_of_lifecycle.pdf (accessed 02 June 2014).

Chapter 18

Risk mitigation for wastewater irrigation systems in low-income countries: Opportunities and limitations of the WHO guidelines

Bernard Keraita, Javier Mateo-Sagasta Dávila, Pay Drechsel, Mirko Winkler and Kate Medlicott

18.1 INTRODUCTION

With increasing global water scarcity and pollution of water bodies, wastewater irrigation is gaining momentum (Hamilton *et al.* 2007). Existing literature shows that formal and informal wastewater irrigation systems are widespread, regardless of the development level and climatic conditions (Jimenez & Asano, 2008; Raschid-Sally & Jayakody, 2008). The evidence suggests that these systems are beneficial, as they increase agricultural production compared to systems with limited access to water or where available water sources have low nutrient levels. This is particularly true for high-value crops like vegetables, leading to increased income for farmers and others in the supply chain. However, the literature also shows that wastewater can affect the quality of crops and pose public health and environmental risks (WHO, 2006). There are also productivity risks due to salinity, sodicity and ion-specific toxicities (FAO, 1992). This is especially so when partially or untreated wastewater is used; a common practice in most low-income countries (Scott *et al.* 2004; Keraita *et al.* 2008a). Of most concern, are excreta-related pathogens associated with low sanitation coverage in these countries (Blumenthal *et al.* 2000). Such risks jeopardize sustainability of wastewater irrigation systems, calling for a research and development focus on risk mitigation in these irrigation systems.

In this chapter, risk mitigation approaches for wastewater irrigation systems in low-income settings are discussed. The discussion is largely drawn from field experiences in West Africa, which aimed to provide further scientific evidence for the continued updating of the WHO guidelines on safe wastewater use in agriculture (WHO, 2006).

18.2 HEALTH RISKS ASSOCIATED WITH WASTEWATER IRRIGATION SYSTEMS IN LOW-INCOME COUNTRIES

Most wastewater irrigation systems in low-income countries use untreated or partially treated wastewater (Keraita *et al.* 2008a). This is due to low levels of wastewater treatment, with median levels estimated at 35% in Asia, 14% in Latin America and less than 1% in sub-Saharan Africa (WHO-UNICEF, 2000). Although various use scenarios of irrigation systems exist, it is mainly farmers who use wastewater directly for irrigation without it being mixed or diluted by other water bodies (direct use) or much more common after dilution and mixing (indirect use) (Keraita *et al.* 2008a; Mateo-Sagasta & Salian, 2012).

Wastewater contains a variety of pathogens and pollutants. Extensive studies on human health risks posed by wastewater irrigation especially from pathogen contamination have been done (WHO, 2006). Table 18.1 is a simplified presentation of wastewater-related human health risks, affected groups and exposure pathways. Other pollutants include salts, metals, metalloids, pathogens, residual drugs, organic compounds, endocrine disruptor compounds and active residues of personal care products (Tchobanoglous *et al.* 1995). These compounds pose environmental and human health risks. Emphasis in discussions is often given to different types of pollutants depending on the regional risk relevance. For example, in low-income countries, risks from microbiological contaminants receive most attention. This is because people in these countries are most affected by diseases caused by poor sanitation such as diarrhoeal diseases and helminth infections (Prüss-Ustün & Corvalan, 2006). The situation changes significantly in transitional economies and is different in high-income countries, where microbiological risks are largely under control. In this context, chemical pollution (heavy metals, pesticides) and emerging pollutants (such as antibiotics) are a major public health concern.

Humans are mainly exposed to wastewater-irrigation risks by: i) consuming irrigated produce (consumption-related risks); ii) coming into contact with wastewater when working in the farms (occupational risks); and iii) exposure to wastewater and wastewater-irrigated soils when walking by or children playing on the fields (environmental risks). Constituents of most concern in wastewater are excreta-related pathogens and skin irritants (Blumenthal *et al.* 2000; WHO, 2006). For consumption-related health risks, the primary concern is uncooked vegetable dishes such as salad (Harris *et al.* 2003). Several diarrhoeal outbreaks have been associated with wastewater-irrigated vegetables (Shuval *et al.* 1986; WHO, 2006). There is also strong epidemiological evidence for *Ascaris lumbricoides* infections for both adults and children consuming uncooked vegetables irrigated with wastewater (Peasey, 2000). Helminth infections, particularly *A. lumbricoides* and hookworm, have higher importance in relation to occupation-related risks compared to bacterial, viral and protozoan infections (Blumenthal *et al.* 2000). The most affected group is farm workers, owing to the long duration of their contact with wastewater and contaminated soils (Blumenthal & Peasey, 2002;

WHO, 2006). Recent studies from Vietnam, Cambodia, India and Ghana have associated skin diseases such as dermatitis (eczema) to contact with untreated wastewater (Keraita *et al.* 2008a).

Table 18.1 Simplified presentation of the main human health risks from wastewater irrigation.

Type of risk	Health risk	Who is at risk	Exposure pathway
Occupational risks (contact)	– Parasitic worms such as *A. lumbricoides* and hookworm infections – Bacterial and viral infections – Skin irritations caused by infectious and non-infectious agents – itching and blister on the hands and feet – Nail problems such as koilonychias (spoon–formed nails)	– Farmers/ field workers – Marketers of wastewater-grown produce	– Contact with irrigation water and contaminated soils – Contact with irrigation water and contaminated soils – Contact with contaminated soils during harvesting – Exposure through washing vegetables in wastewater
Consumption-related risks (eating)	– Mainly bacterial and viral infections such as cholera, typhoid, ETEC, Hepatitis A, viral enteritis which mainly cause diarrhoeas – Parasitic worms such as ascaris	– Vegetable consumers	– Eating contaminated vegetables, especially those eaten raw
Environmental risks	– Similar risks as those exposed to occupational and consumption risks, but decreasing with distance from farm	– Children playing in wastewater–irrigated fields – People walking on or nearby fields	– Soil particle intake – Aerosols

Source: Adapted from Abaidoo *et al.* (2010).

18.3 RISK MITIGATION PERSPECTIVES FROM THE WHO GUIDELINES

18.3.1 The multiple-barrier approach

Different approaches have been proposed for risk mitigation. For a long time, conventional wastewater treatment was regarded as the ultimate risk mitigation measure (Asano & Levine, 1998). This approach put a strong emphasis on the use of water quality standards in wastewater irrigation systems and strict regulations as used in most high-income countries. This was the primary basis of earlier versions of the WHO guidelines on wastewater irrigation (WHO, 1989). However, the most recent WHO guidelines for wastewater irrigation recommend a shift from water quality standards to health-based targets that can be achieved along a chain of multiple risk reduction measures (WHO, 2006). Such a multiple-barrier approach is based on the Hazard Analysis and Critical Control Points (HACCP) concept as commonly applied in food safety programs (Ropkins & Beck, 2000) and described in more detail in Chapter 14. In contrast to the use of water quality standards, conventional wastewater treatment is regarded as one of the barriers and not the only barrier. Hence, treatment, where possible, is combined with other health protection measures at farmer and consumer levels. Barriers are placed at critical control points along the food chain (from production to consumption), aiming to maximum risk reduction. For example, barriers can be placed at wastewater generation points, on farms, at markets and even at the consumer level. A generic example of these barriers is shown in Figure 18.1. This approach is more applicable in low-income countries where irrigation with untreated wastewater is common and wastewater treatment is limited.

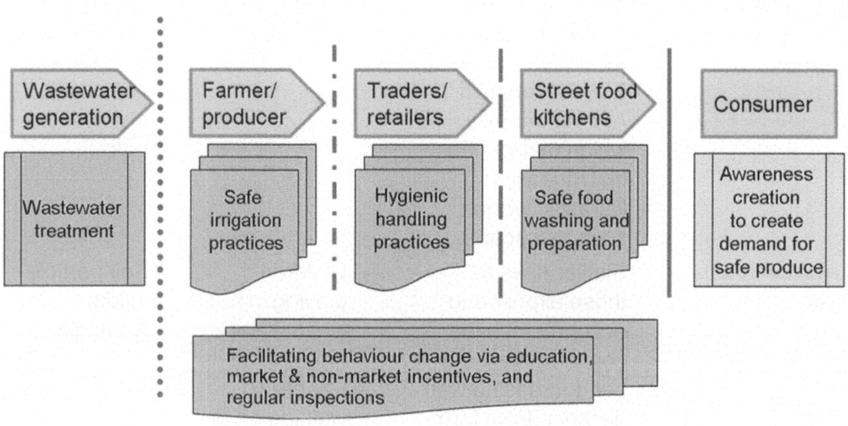

Figure 18.1 The multiple-barrier approach for consumption-related risks along the food chain as applied in wastewater irrigation (Amoah *et al.* 2011).

18.3.2 Evidence of risk mitigation in the WHO guidelines

The WHO (2006) guidelines adopt a health-based target of a tolerable additional disease burden of $\leq 10^{-6}$ disability-adjusted life years (DALYs; a measure of disease burden) per person per year. The guideline translates the health-based target into a performance target of 6–7 log units pathogen reduction at the point of exposure A lower health-based target of $\leq 10^{-4}$ or $\leq 10^{-5}$ may be appropriate as suggested by Mara *et al.* (2010). A set of measures for risk mitigation and their anticipated pathogen reduction levels are shown in Table 18.2. For example, combining: i) a minimal (low-cost) wastewater treatment (1–2 $\log_{10}$ units pathogen reduction); ii) drip irrigation (2–4 $\log_{10}$ units pathogen reduction); and iii) washing vegetables after harvesting (1 $\log_{10}$ units pathogen reduction), can achieve a 4–7 $\log_{10}$ unit pathogen reduction. These $\log_{10}$ pathogen reduction rates proposed by the WHO (2006) guidelines are based on the best available evidence in 2006.

Table 18.2 Pathogen reductions achievable by different health-protection measures.

Level	Control measure	Reduction (log units)	Comments
Pre-farm	Wastewater treatment	1–6	Reduction to be achieved by wastewater treatment depends on type and degree of the treatment process.
Farm-based	Drip irrigation used in: Low-growing crops	2	Root crops and crops such as lettuce that grow just above, but partially in contact with, the soil.
	High-growing crops	4	Crops, such as tomatoes, the harvested parts of which are not in contact with the soil.
	Spray irrigation with; Drift control	1	Use of micro-sprinklers, anemometer-controlled direction-switching sprinklers, inward-throwing sprinklers, and so on.
	Buffer zone	1	Protection of residents near spray irrigation. The buffer zone should be 50–100 m.
	Pathogen die-off	0.5–2 per day	Die-off on crop surfaces that occurs between last irrigation and consumption. The log unit reduction achieved depends on climate (temperature, sunlight intensity, humidity), time, crop type, and so on.
Post-harvest	Produce washing	1	Washing salad crops, vegetables and fruit with clean water.
	Produce disinfection	2	Washing salad crops, vegetables and fruit with a weak disinfectant solution and rinsing with clean water.
	Produce peeling	2	Fruits, root crops.
	Produce cooking	6–7	Immersion in boiling or close-to-boiling water until the food is cooked ensures pathogen destruction.

Source: Amoah *et al.* (2011), WHO (2006).

However, some challenges still remain including: i) the actual verification of the risk response in a low-income context; ii) the field testing and implementation of the suggested measures; iii) how to monitor, at low cost, the acceptance and effectiveness of such practices; and iv) how to translate the guidelines into specific policies to support setting of standards, laws and regulations.

The concept of health-based targets and performance targets expressed as log reductions remains challenging for policy makers and practitioners, especially in low-income countries who are the primary audience for the guidelines. The next revision of the guidelines is likely to include provision for health-based targets expressed at the levels of health outcome, performance, water quality and specified technology to allow flexibility according to capacity.

18.4 EVIDENCE FROM FIELD STUDIES IN WEST AFRICA

18.4.1 Farm-based risk mitigation measures

18.4.1.1 Improving irrigation water quality at farms

Even though conventional wastewater treatment may not be an available option for reducing faecal contamination before or on the farm, an understanding of how it works is helpful in developing appropriate technologies to improve irrigation water quality. Many conventional treatment plants have three stages; primary, secondary and tertiary (Tchobanoglous *et al.* 1995). Primary treatment aims at settling suspended solids. In secondary treatment, soluble biodegradable organics are degraded and removed by bacteria and protozoa through (aerobic or anaerobic) biological processes. Tertiary treatment aims at effluent polishing before the effluent is discharged or reused and can consist of the removal of nutrients (mainly nitrogen and phosphorous), toxic compounds, residual suspended matter, or microorganisms (disinfection with chlorine, ozone, ultraviolet radiation or others). Nevertheless, this third stage is rarely employed in low-income countries. In Sub-Saharan African cities, where pathogens are the main concern, low-rate process technologies such as stabilisation pond systems are well suited for pathogen reduction (Scheierling *et al.* 2010). Field studies conducted in Ghana have largely focussed on two main pathogen removal mechanisms: sedimentation and filtration systems (Keraita *et al.* 2008b, c):

Simple on-farm sedimentation ponds

In Ghana, as in many other countries in West Africa, shallow dugout ponds (usually less than 1 m deep and 2 m wide) are widely used in irrigated urban vegetable farming sites (Figure 18.2). In most cases, shallow dugout ponds are used as storage reservoirs where surface runoff and wastewater effluents are channelled. Other variations include the use of mobile drums and other reservoirs, which are common in areas where irrigation water sources are distant from farm sites. These containers are filled manually or by pumping water from the streams and then used for irrigation when needed. The re-filling frequency of drums and reservoirs depends on their volume and daily water needs.

Figure 18.2 Simple sedimentation pond in an urban vegetable farm in Kumasi, Ghana (Photo: IWMI).

During the storage of water and gradual use in irrigation, sedimentation takes place similar to water storage and treatment reservoirs (WSTRs), although the extent of pathogen removal will be lower depending how long the ponds are not used. Studies conducted in Ghana showed that these ponds are very effective in removing helminths (reduced to less than 1 egg/l) when sedimentation is allowed for 2–3 days (Keraita *et al.* 2007a). Reductions can further be achieved with better pond designs (deeper, wedge-shaped beds) and training farmers on how to collect water (Keraita *et al.* 2007a). In addition, measures that can enhance sedimentation, without disturbing pond beds, such as using natural flocculants in the ponds (e.g., Moringa Oleifera seed extracts), seem to be promising in Ghana (Sengupta *et al.* 2012). Furthermore, use of additional measures that influence pathogen die-off such as sunlight intensity, temperature and crop type, can help in lessening the pathogen load in irrigation water (Keraita *et al.* 2007a).

Filtration systems

There is a wide range of filtration systems, although slow sand filters are probably the most appropriate to treat irrigation water. Sand filters remove pathogenic microorganisms from polluted water by first retaining them in the filtration media before they are eliminated (Stevic *et al.* 2004). The typical pathogen removal range reported by the WHO is 0–3 $\log_{10}$ reduction units and 1–3 $\log_{10}$ reduction units for bacteria and helminths, respectively (WHO, 2006). Research by the authors in Ghana, using column slow sand filters, achieved between 98.2–99.8% of bacteria removal, equivalent to an average of 2 $\log_{10}$ reduction units/100 ml. In addition, 71–96% of helminths were removed (Keraita *et al.* 2008b). This removal was significant but not adequate, as irrigation water had very high levels of fecal contamination.

Farmers in West Africa also use other forms of filtration systems. In Ouagadogou, Burkina Faso, wells are sunk next to wastewater canals creating a hydraulic gradient that enables water to infiltrate into the well. In the course of this filtration process, microorganisms and turbidity are reduced. Wastewater can also be allowed to pass through sand filter trenches, sand embankments, column sand filters and simple sand bags, as farmers channel irrigation water to collection storage ponds. While the reduction of bacteria and virus may be minimal due to their small size, some reduction in protozoa and helminth eggs can be achieved.

In Ghana, it was identified that farmers use different forms of sieves, but mostly use folded mosquito nets over watering cans to prevent particles like algae, gravel and organic particles from entering the watering cans. Studies on this kind of simple filter show about 1 log unit removal for bacteria and 12–62% for helminths when a nylon sieve was used (Keraita *et al.* 2008b). Further modifications could be done to increase removal rates, because these are the systems that many farmers find easier to adopt. Clogging is a limitation when using sand filters, however the proper choice of filtration media (right uniformity coefficient and effective size configurations) can reduce the problem.

18.4.1.2 Drip irrigation

Localized techniques such as drip and trickle irrigation offer farm workers health protection and also have minimal pathogen transfer to crop surfaces because water is directly applied to the root (FAO, 1992). However, when compared to surface and even spray irrigation methods, conventional drip irrigation systems are much more expensive. Low-cost drip irrigation techniques like bucket or sack drip kits (see Figure 18.3) offer more potential for use and adoption in low-income countries, and are now available on the market (Kay, 2001). Nevertheless, in West Africa, the use of drip irrigation in wastewater irrigation systems is rare. For instance, in Ghana the authors purchased the low-cost drip irrigation kits (bucket drip kits) for field assessments with urban vegetable farmers from India, as they were not available in the country.

Unlike the WHO, which had evidence for low- and high-growing crops, the assessment was conducted on lettuce, a low-growing crop. Data obtained in the Ghana study shows a difference of up to 6.1 $\log_{10}$ reduction units and 2.5 $\log_{10}$ reduction units of faecal coliforms/100 g of lettuce for the dry and wet seasons compared to when watering cans are used (Keraita *et al.* 2007a). This comparatively higher reduction than levels recorded in the reviews done by the WHO (2006), (2 $\log_{10}$ reduction units), can be attributed to the much lower pressure of the drippers leading to reduced soil splashing of contaminated soil particles on the leaves. The low pressure kits, however, were very much prone to clogging due to high turbidity levels in polluted water.

A great number of drip lines on the beds also interfered with other farming practices; weeding in particular. In addition, as farmers vary the kinds of vegetables they plant, different emitter spacing was required for adapting the drip kits to the crop spacing specification, which is another challenge (see the low density in Figure 18.3).

Figure 18.3 Low-cost drip irrigation kits on a test station in Ghana (Photo: IWMI).

18.4.1.3 Spray and sprinkler irrigation

Spray and sprinkler are overhead irrigation methods and have the highest potential to transfer pathogens to crop surfaces as water is applied to the edible parts of most crops. Moreover, the spread of pathogens through aerosols is a potential hazard (FAO, 1992). The use of conventional spray and sprinkler irrigation in wastewater irrigation systems is not common in West Africa. However, the watering cans, which are overhead methods applying water directly on to the edible parts of leafy vegetables, are the most commonly used irrigation method (Drechsel *et al.* 2006). Other forms of overhead irrigation observed are the use of buckets and water hoses. In a study done in Ghana, Keraita *et al.* (2007a) show that the watering can method led to high contamination on mature lettuce with mean faecal coliform levels of 6.53 and 8.21 $\log_{10}$ units/100 g for dry and wet seasons, respectively. The method compared poorly to drip irrigation, which recorded mean levels of 0.47 $\log_{10}$ reduction units of faecal coliforms during the dry and 5.65 $\log_{10}$ reduction units during wet season. Similar levels were observed for furrow irrigation where 5.65 $\log_{10}$ units and 7.29 $\log_{10}$ units were recorded during the dry and wet season, respectively.

There is hardly any documented research on modifying indigenous or even conventional irrigation methods to reduce crop contamination. While the aims may be different, it is common to see farmers modifying conventional irrigation systems to suit their farming needs in West Africa. In an attempt to reduce contamination levels while using watering cans, the studies in Ghana tested the influence of using watering cans fitted with caps and lowering heights where watering cans are lifted to. These modifications were based on empirical pathogen transportation models, which showed that detachment and transportation of pathogens on soils are minimized by reducing the size and velocity of water particles striking the soil (Tyrrel & Quinton 2003). Data obtained show that using watering cans with caps reduced faecal coliforms on lettuce by an average of 1.5 $\log_{10}$ and 1.3 helminth egg/100 g of lettuce (Keraita *et al.* 2007a). An average reduction of 2.5 $\log_{10}$ units faecal coliforms and 2.3 helminth eggs/100 g lettuce contamination was achieved through the use of capped watering cans if used from less than 0.5 m height on lettuce, compared with when irrigated with more splashing power from a height over 1 m (Keraita *et al.* 2007a). This example shows that simple changes in the use of watering cans can contribute to an overall reduced crop contamination.

18.4.1.4 Pathogen die-off

In West Africa, the effectiveness of withholding irrigation a few days before harvest to allow pathogen die-off on crop surfaces due to exposure to sunlight and drying-out of surfaces has been tested (Shuval *et al.* 1986). The results from field trials in Ghana (dry season) showed an average daily reduction of 0.65 $\log_{10}$ reduction units of faecal coliforms and 0.4 helminth eggs/100 g of lettuce (Keraita *et al.* 2007b). While the lower coliform counts can be attributed to die-off, lower egg counts could be attributed to fewer additions over the days without irrigation. The studies showed that this measure was not appropriate during the wet season due to lower temperatures and soil splashing from rains. In addition, the studies were only limited to farms, so assessment has not been done on natural die-off after harvesting, as the range of 0.5–2 log units/day in the WHO guidelines also assumes similar die-off patterns between harvesting and consumption (WHO, 2006). Nevertheless, at farm level, the greatest limitation of this measure is the corresponding high loses of crop yield under the hot conditions in Ghana. In the Ghana studies, for the daily pathogen reduction obtained, the corresponding losses were 1.4 tons/ha of fresh weight. This was a major adoption deterrent for many farmers (Keraita *et al.* 2007b). In cooler climates, such as Addis Ababa, irrigation frequency is much lower, favouring natural die-off.

18.4.2 Post-harvest risk mitigation measures

18.4.2.1 Produce peeling at markets

While studies on internalisation of pathogens from wastewater irrigation are very limited, there is a general consensus from a wide range of literature that most pathogenic contamination occurs on the surface of crops (Ilic *et al.* 2010). Only a few crops (e.g., carrots and cucumber) grown in study locations in West Africa could be peeled; usually at kitchens, not at markets. Most of the crops grown were leafy vegetables and 'fruity' vegetables like green pepper and tomatoes. The term 'peeling' was therefore adapted to the removal of outer leaves for vegetables like cabbage and lettuce. Keraita *et al.* (2007b) showed the mean difference in levels of faecal coliforms between the inner and outer lettuce leaves (25% wet weight) to be 1.8 log units/100 g. On cabbage, faecal coliform levels decreased gradually from 5.66 $\log_{10}$ reduction units/100 g in the whole cabbage before any leaves were removed; to 1.24 $\log_{10}$ reduction units after 30% (wet weight) of the outer leaves were removed. No faecal coliforms were recorded after 45% of outer leaves were removed. Thus, while decontamination can be achieved by 'peeling', there could be some yield losses if the 'peeled' parts are edible. Studies on crops that are usually peeled before eating such as onions, carrots and cucumber are encouraged.

18.4.2.2 Produce washing at markets

At markets in warm climates, produce-sellers sprinkle water or wash vegetables periodically to keep them looking fresh, so that they can sell them at a higher price. However, many markets in low-income countries have no running water and produce sellers have to rely on water that they buy from tankers. Due to costs, and in some cases unavailability, the same water (usually in buckets and bowls) is used to wash or refresh vegetables for the whole day. In Kumasi, few studies have been carried out to assess the effects of this practice on crop (de)contamination (Owusu, 2009; Akple, 2009). Owusu (2009) assessed levels of faecal coliforms on spring onions over 5 washing cycles on a bucket of water (1 kg of onions in each washing cycle, as done by vegetable sellers in Kumasi). The study showed a sharp decrease in faecal coliform levels after first washing, from approximately 5 $\log_{10}$ reduction units/100 g to less than 1 $\log_{10}$ reduction unit/100 g. Interestingly, subsequent washing cycles (cycle 2–5) recorded an increase in contamination, with the fifth cycle showing similar faecal coliform levels as those recorded in unwashed spring onions (approximately 5 $\log_{10}$ reduction units). In essence, produce-sellers should change the water more often or stop washing after the first cycle. However, stopping washing will affect the physical quality of the produce, resulting in a lower price. Alternatively, washing using the same water should be done only once, but this is dependant on local water availability and might have cost implications.

18.4.2.3 Produce washing and disinfection at kitchens

A survey involving 210 street restaurants and 950 urban households in major cities across West Africa showed that vegetable washing at household level is common in 56–90% of the households and 80–100% of the restaurants (Amoah *et al.* 2007). In Francophone cities, the most common disinfectants used in restaurants were bleach (Eau de Javel) (55%) and potassium permanganate (31%), followed by salt/lemon or soap (both 7%). Half of the households (50%) reported use of bleach, followed by potassium permanganate (22%), salt (14%), and water only (12%). Every second respondent rinsed the leaves after washing. In contrast, the use of bleach and potassium permanganate was practically unknown as food disinfectants in (Anglophone) Ghana, where various salt and vinegar concentrations were commonly used besides cleaning in water only. Salt was preferred to vinegar because it is cheaper (Rheinlaender, 2006).

Against this background, some of the disinfection practices were simulated in laboratory conditions to assess their effectiveness in reducing faecal coliform levels (Amoah *et al.* 2007). Lettuce, the most commonly grown urban vegetable in West Africa, was used. Results are presented in Table 18.3. The assessment showed that, irrespective of the method used, washing vegetables reduced faecal coliform levels in lettuce, however, the levels varied significantly and common concentrations of salt or vinegar appeared to be of little impact. Pathogen removal through disinfection was largely influenced by contact time, concentration and the type of disinfection. Similar results were obtained for related studies conducted to disinfect cabbage and spring onion using concentrations (Akple, 2009).

Table 18.3 Effect of selected disinfection methods on faecal coliform levels on lettuce in West Africa.

Method	Log reductions	Comments
Dipping in a bowl of water	1.0–1.4	– Increased contact time from a few seconds to 2 minutes improves the efficacy from 1 to 1.4 logs – Not very efficient compared to washing with other sanitizers – Not very effective for helminth eggs if washing has to be done in the same bowl of water – Warming the water did not result in different counts
Running tap water	0.3–2.2	– Comparatively effective compared with washing in a bowl, also for helminth egg removal – Increased efficacy only with increased contact time from few seconds to 2 minutes – Limited application potential due to absence of tap water in poor households
Dipping in a bowl with a salt solution	0.5–2.1	– Salt solution is a better sanitizer compared to dipping in water if the contact time is long enough (1–2 min) – Efficacy improves with increasing temperature and increasing concentration, however, high concentrations have a deteriorating effect on the appearance of some crops like lettuce

(*Continued*)

Table 18.3 Effect of selected disinfection methods on faecal coliform levels on lettuce in West Africa (*Continued*).

Method	Log reductions	Comments
Dipping in a bowl with a vinegar solution	0.2–4.7	– Very effective at high concentration (>20 ml/l) but this could have possible negative effects on taste and palatability of the washed vegetables – To achieve best efficacy and keep the sensory quality of product, the contact time should be increased to 5–10 min – Efficacy is improved even at low concentration if carried out with a temperature over 30°C
Dipping in a bowl with potassium permanganate solution	0.6–3.0	– Most effective at higher concentrations (200 ppm), a temperature of 30°C or higher and a contact time of 5–10 min – Higher concentration colours washed vegetables purple which requires more water for rinsing or may raise questions on a negative health impact
Dipping in a bowl with a solution containing a commercially available washing detergent	1.6–2.6	– Significant reductions could be achieved with 5–10 min contact time – Residual perfumes and soap taste might affect consumer's sensory perception – As the solution contains surfactants which could affect health, thorough rinsing is required
Dipping in a bowl of water with added household bleach	2.2–3.0	– Tested dosages (commercial bleach) resulted in 165–248 μS/cm salinity (= concentration indicator) – Effective with 5–10 min contact time, and widely used in Francophone West Africa – May pose a health risk if dosage is not well explained
Dipping in a bowl of water containing chlorine tablets	2.3–2.7	– Effective at 100 ppm but tablets not commonly available in some West African countries – Effect of higher concentrations on efficacy not tested

Source: Adapted from Amoah *et al.* (2007).

18.5 ADOPTION OF SAFE RE-USE PRACTICES

The adoption of safe practices by key actors means that farmers, produce-sellers and those who prepare foods in households and street restaurants need to change their behaviour and routine practices. However, this change can be slow, dynamic and complicated due to the multiple factors that influence adoption (Karg & Drechsel, 2011). Indeed, for each risk mitigation measure, the key actors will need to make an investment. The investment can be in different forms such as: i) increased labour; ii) monetary, in the form of capital and operational costs; iii) accepting losses of yields; and iv) behaviour change. To support the change and compensate the key actors for investments made, incentive systems are needed (Frewer *et al.* 1998). Nevertheless, this critical area is not addressed in the WHO (2006) guidelines.

Some specific factors that can enhance adoption of the risk mitigation measures are described below. In summary, and based on field studies (Amoah *et al.* 2011), analysis suggested that the key actors could adopt safe practices: i) if safe reuse practices could be translated to higher incomes (economic incentives); ii) if actors know the risks and opportunities of safe reuse (awareness); iii) if it reinforced their social benefits, for example, their status in the community-(social marketing); iv) if actors are provided with appropriate access to land (land tenure); v) if they are trained and qualified to adopt safer practices (training and extension); vi) if they are encouraged by law (laws and regulations); and vii) if all the above are integrated in an effective communication strategy (effective communication).

18.5.1 Economic incentives

Studies show that people are more likely to adopt innovations for direct economic returns on investments (Frewer *et al.* 1998). The adoption of safer practices should then potentially help key actors to sell more or sell at higher prices. However, this will only happen if consumers are willing to pay more for safer products. In low income countries, risk awareness is generally low and even if a consumer gets sick, (s)he might not be able to identify the cause. Farmers and market sellers interviewed about potential risks reported no complaints (Obuobie *et al.* 2006). This situation of low awareness limits the willingness by consumers to pay for safer produce to the better educated minority, therefore, there seem to be two lines of action: i) increase risk perception/awareness to increase demand for safer products and willingness to pay a premium; and ii) serving these consumers through dedicated marketing channels (Boateng *et al.* 2007). Producer groups should be encouraged to sell their products outside the existing marketing channels to avoid mixing-up with unsafe produce. Economic incentives can also come from the public sector, but a quantification of such costs and benefits and the demonstration that benefits are higher than costs will be needed to justify this public support (FAO, 2010).

18.5.2 Raising Awareness: 'making visible the invisible'

To encourage behaviour change, key actors need to be aware of the risks of wastewater irrigation and the benefits of adopting safer practices. This awareness concerns consumers as well as traders and producers. The importance of awareness to increase demand and willingness to pay for safer products has been discussed in the previous section. A particular challenge of pathogenic risks is their invisible nature, which makes it difficult for the key actors to be aware of the risks and to assess the effectiveness and quantify the impacts of the risk mitigation measures. Risk visibility would greatly facilitate risk perceptions and encourage adoption of safer practices.

While many actors such as farmers and produce-sellers in low-income countries use physical indicators such as colour, dirt and odour to assess the cleanliness of the produce, these physical indicators do not always correspond with microbiological indicators. Scientists need to work with farmers to identify physical indicators or combinations of physical indicators that can be used as proxies for microbiological contamination at farm level. Key actors will also like to 'see the impacts' of the risk mitigation measures before changing from their original practices. In this regard, participatory approaches as used in many farming experiments will help key actors to compare new practices with old practices in their own environments before making choices (Doward *et al.* 2003). Participatory water sampling and analysis with low-cost kits and methods such as petrifilms can allow actors to visualize the invisible enemy: pathogens, and be more aware of the risks of unsafe wastewater irrigation.

18.5.3 Social marketing

Education could start in schools, but should be combined with social marketing techniques to address the current generation of key actors, for example in hand-wash campaigns. Social marketing seeks to induce a target audience to voluntarily accept, modify or abandon behaviour for the benefit of individuals, groups or society as a whole (Siegel & Doner-Lotenberg, 2007). This could be an important tool to adopt risk mitigation measures in low-income settings where economic incentives are limited by low risk perceptions among customers (Karg & Drechsel, 2011). Even if health considerations are not valued highly in the target group, social marketing studies can help identify valuable related benefits, including indirect business advantages (e.g., attracting tourists), improved self-esteem, a feeling of comfort or respect for others. Studies must look for positive, core values that trigger the primary target audience to associate with innovative approaches (Siegel & Doner-Lotenberg, 2007). For example, if washing vegetables with vinegar rather than salt is perceived as feeling 'advanced', or using drip kits compared to watering cans, then the social-marketing messages and communication strategies should reinforce this existing positive association (Drechsel & Karg, 2013).

18.5.4 Land tenure security

Concentration of population and economic activities in cities results in very limited land availability and intense competition for its use. Besides, there is often uncertainty regarding the ownership of land. Market forces push up land prices and often make it unaffordable for urban and peri-urban food producers. Land tenure insecurity was often mentioned in Ghana as urban farms are on public or private land and can easily be closed. An incentive such as better tenure security could facilitate farmers' investments in structures that have positive health impacts, such as wastewater treatment ponds. Municipalities may adopt a variety of approaches to securing land for horticulture, including regularization of informal titles, or promoting urban agriculture in public land (such as terraces along urban rivers). Similar incentives are possible for street food restaurants, which are often more informal than formal.

18.5.5 Training and extension

Another key factor for the correct application of safer practices and compliance over time is having trained and qualified actors. Extension services and research-extension linkages will have a significant role to play. Training materials supporting food safety on- and off-farm have been prepared by IWMI and FAO (Box 18.1) and can be used within larger training programs for urban and peri-urban producers (e.g., FAO 2007; see also www.fao.org/fcit).

BOX 18.1 TRAINING AND AWARENESS MATERIALS ON WASTEWATER IRRIGATION AND FOOD SAFETY DEVELOPED BY IWMI AND FAO

FAO (2012). On-farm practices for the safe use of wastewater in urban and peri-urban horticulture. A training handbook for farmer field schools. FAO, Rome http://www.fao.org/docrep/016/i3041e/i3041e.pdf (accessed 02 June 2014).

Drechsel P., Keraita B. and Amoah P. (2012). Safer irrigation practices for reducing vegetable contamination in urban sub-Saharan Africa: An illustrated guide for farmers and extension officers. Accra: IWMI, 30 pp. http://www.iwmi.cgiar.org/Publications/Books/PDF/Farmers_Guide-Low_res-Final2.pdf (accessed 02 June 2014).

IWMI (2007). Improving food safety in Africa where vegetables are irrigated with polluted water. Awareness and training video for staff of street restaurants. http://youtu.be/DXHkQE_hFg4 (accessed 24 August 2014).

IWMI (2008). Good farming practices to reduce vegetable contamination. Awareness and training video for wastewater farmers. http://youtu.be/Aa4u1_RblfM (Accessed 24 August 2014).

18.5.6 Laws and regulations

Karg and Drechsel (2011) identified regulations as an important external factor to institutionalize new food-safety recommendations, to provide the legal framework for both incentives (e.g., certificates) and disincentives (such as fees). To integrate improved food-handling practices into institutional structures, inspection forms can be updated, inspectors and extension officers can be trained and pressure can be applied to farmers and caterers to enhance compliance. However, regulations should not be based on imported (theoretical) standards, but rather on locally feasible standards that are viewed as practical and are not prone to corruption. In this way, regulation and institutionalisation may contribute to ensuring the long-term sustainability of behaviour change, whereas promotional and educational activities are usually limited to a specific time frame. However, educating younger children may be more effective for long term impact.

18.5.7 Effective communication

To be useful, knowledge (whether being farmer's innovations, the latest research findings or pressing policy issues) must be effectively shared amongst people and institutions (FAO, 2011). It is important to understand the knowledge pathways used by key actors (farmers, produce-sellers and those who prepare foods in households and street restaurants) who will adopt risk mitigation measures, so that more effective channels are selected for risk communication. For example, a pilot social marketing study in Ghana showed that it is more likely that innovations spread from farmer to farmer through social networks than through any external facilitation (Keraita *et al.* 2010a; Scott *et al.* 2010). Farmers preferred field demonstrations and/or learning by doing. This also verifies the importance of encouraging the actors' own experimentation, because it promotes knowledge generation as well as self-monitoring and evaluation. However, it is pertinent for the implementation process to recognise the wider system within which key actors operate. The wider system, made up of institutions, regulatory bodies and in- and output markets, can have a significant positive or negative influence on key actors' decision making, but might at least in part be ignored by scientists.

In addition to channels, the materials and media used for knowledge sharing is also critical. For example, research findings from field studies in Ghana were synthesized to make farmer-friendly training and extension materials on the most appropriate risk mitigation measures (Box 18.1). These materials were translated into different local languages and included documentaries (radio and video) as well as illustrated flip charts. In addition, interactive approaches like the Farmer Field School (FFS) can be used for actual training and demonstration of the mitigation measures (Braun & Duveskog, 2008). The training modules were prepared together with various actors such as farmers and marketers' representatives, extension officers from the Ministry of Food and Agriculture and communication experts (FAO, 2012). Modules on risk mitigation measures in Ghana are now being integrated in relevant ministries' formal training curricula, starting with the Urban Agriculture Directorates in Kumasi and Accra. Dissemination of best practices is done by the extension officials from the ministries.

18.6 DISCUSSION AND CONCLUSION

Developing risk mitigation measures in line with the multiple-barrier approach offers a variety of options to achieve a realistic chance of risk mitigation from wastewater irrigation systems in low-income countries. This provides local health risk managers with the flexibility to address wastewater irrigation risks with locally viable means, instead of not taking any action due to unattainable water quality threshold levels. This chapter has demonstrated the potential of the suggested public health protection measures from actual field studies in West Africa. The studies have shown that farm and post-harvest risk mitigation measures provide more direct solutions to preventing contamination and decontamination of vegetables grown in wastewater irrigation systems. However, the effectiveness of individual measures may not be sufficient; they can be used in combination to complement each other in order to achieve the acceptable risk levels. Combination can be achieved within and between operation levels, such as farms, markets and households. The measures presented, although not exhaustive, allow for flexibility to adaptation in different locations.

Additionally, uptake of the WHO (2006) guidelines remains low where this flexibility is most needed, for example, in low-income countries where conventional wastewater treatment coverage is still low. The reason is commonly attributed to the WHO (2006) recommendation to apply an ex-ante quantitative microbial risk assessment (QMRA) to first define locally appropriate health-based targets before mitigation and monitoring options are assessed. The capacity to follow this more complex approach is lacking in many low-income countries, where the easy standards of the previous edition (WHO, 1989) are well known, although hard to apply given the lack of treatment (Keraita *et al.* 2010b). There are also fears that flexibility could result in laxity, especially if the required behaviour change fails or where there is no related compliance monitoring.

Some practitioners, in particular in medium- and high-income countries, where strict water quality standards are possible to implement and control, perceive the WHO (2006) approach as 'a license for using untreated wastewater for irrigation' or as a 'damage control' mechanism, which has shifted focus from comprehensively addressing the problem to short-term, 'piece-meal' solutions. It is seen to have a narrower focus due to its emphasis on food safety, without considering environmental considerations and other benefits of wastewater treatment. The paucity of reliable field assessments and epidemiological studies on the impacts of the suggested health protection measures in low-income country contexts further justifies these assertions. Indeed, unlike wastewater treatment, which has been tested over centuries, many of the proposed barriers have hardly been tested on a wide scale for validation and more has to be done to show that these measures work for different soils, crops and irrigation system.

The even larger limitation is that many of the suggested measures largely rely on behaviour change. However, behaviour change can be complicated, especially in countries where risk perception is low and there is therefore a limited demand for safer food, let alone institutional capacities to prevent unsafe production. Extra effort is therefore needed to raise awareness of health risks and available measures to reduce risks and identify incentives for facilitating behaviour change (Drechsel & Karg, 2013). Consequently, while the 2006 WHO guidelines have shown that risk mitigation in wastewater irrigation systems is also possible in low-income countries where treatment capacities are only slowly emerging, more needs to be done to support wide application by national stakeholders and their potential transposition into legally enforceable national standards.

There are at least three pathways to proceed which are not mutually exclusive and in part already under consideration:

(i) To reduce the complexity and rigour of implementing the 2006 guidelines by expressing health-based targets in terms of health outcome, performance, water quality and specified technology targets that can be adopted according to the capacity of the implementing country to carry out risk assessment. Targets could also reflect the maximum tolerable burden of disease of 10^{-6} DALY and the 10^{-4} DALY per person per year suggested by Mara *et al.* (2010);

(ii) To translate the 2006 guidelines into a Sanitation Safety Plan Manual that provides step by step guidance that will assist countries to institutionalise the implementation of the guidelines in a similar manner to the Water Safety Plans operationalised by the WHO Drinking Water Guidelines (Medlicott *et al.* 2012).

(iii) To consider wastewater-induced food safety risks as one of many food safety challenges in low-income countries where water-borne and food-borne diseases are often interlinked. This would allow definition of health-based or operational targets using a holistic approach where campaigns targeting appropriate washing of vegetables are linked to hand-washing for an integrated behaviour change approach.

REFERENCES

Abaidoo R. C., Keraita B., Drechsel P., Dissanayake P. and Akple M. S. (2010). Soil and crop contamination through wastewater irrigation and options for risk reduction in developing countries. In: Soil Biology and Agriculture in the Tropics, P. Dion (ed.), Springer Verlag, Heidelberg, Germany. pp. 498–535.

Akple M. S. K. (2009). Assessment of Effectiveness of Non-Treatment Interventions in Reducing Health Risks Associated with Consumption of Wastewater – Irrigated Cabbage. Unpublished MSc thesis submitted to the Department of Theoretical and Applied Biology, KNUST, Kumasi.

Amoah P., Drechsel P., Abaidoo R. C. and Klutse A. (2007). Effectiveness of common and improved sanitary washing methods in West Africa for the reduction of coli bacteria and helminth eggs on vegetables. *Tropical Medicine & International Health,* **12**(2), 40–50.

Amoah P., Keraita B., Akple M., Drechsel P., Abaidoo R. C. and Konradsen F. (2011). Low Cost Options for Health Risk Reduction Where Crops are Irrigated with Polluted Water in West Africa. IWMI Research Report Series 141, Colombo http://www.iwmi.cgiar.org/Publications/IWMI_Research_Reports/PDF/PUB141/RR141.pdf (accessed 05 May 2014).

Asano T. and Levine A. (1998). Wastewater reclamation, recycling, and reuse: an introduction. In: Wastewater Reclamation and Reuse, Asano T. (ed.), Vol. **10**, CRC Press, Boca Raton, FL, pp. 1–56.

Blumenthal U. J., Peasey A., Ruiz-Palacios G. and Mara D. D. (2000). Guidelines for wastewater reuse in agriculture and aquaculture: recommended revisions based on new research evidence. WELL study, Task No: 68 Part 1. http://www.lboro.ac.uk/well/resources/well-studies/full-reports-pdf/task0068i.pdf (accessed 12 August 2014).

Blumenthal U. J and Peasey A. (2002). Critical Reviews of Epidemiological Evidence of Health Effects of Wastewater and Excreta in Agriculture. Background paper for WHO guidelines for the safe use of wastewater and excreta in agriculture. World Health Organisation, Geneva.

Boateng O. K., Keraita B. and Akple M. S. K. (2007). Organic vegetable growers combine for market impact. *Appropriate Technology,* **34**(4), 26–28.

Braun A. and Duveskog D. (2008). Farmer Field School Approach – History, Global Assessment and Success stories. Paper commissioned by DFID, UK.

Dorward P., Galpin M. and Shepherd D. (2003). Participatory Farm Management methods for assessing the suitability of potential innovations: A case study on green manuring options for tomato producers in Ghana. *Agricultural Systems*, **75**, 97–117.

Drechsel P., Graefe S., Sonou M. and Cofie O. O. (2006). Informal Irrigation in Urban West Africa: An Overview. IWMI Research Report 102. IWMI, Colombo. http://www.iwmi.cgiar.org/Publications/IWMI_Research_Reports/PDF/pub102/RR102.pdf (accessed 5 May 2014).

Drechsel P. and Karg H. (2013). Motivating behaviour change for safe wastewater irrigation in urban and peri-urban Ghana. *Sustainable Sanitation Practice*, **16**, 10–20.

FAO (1992). Wastewater Treatment and Use in Agriculture. FAO Irrigation and Drainage paper 47. FAO, Rome.

FAO (2007). The Urban Producer's Resource Book. http://www.fao.org/docrep/010/a1177e/a1177e00.htm (accessed 05 May2014).

FAO (2011). Food Security Communication Toolkit. Food and Agriculture Organisation of the United Nations, Rome.

FAO (2010). The Wealth of Waste: The Economics of Wastewater Use in Agriculture. FAO Water Reports 35, FAO, Rome.

FAO (2012). On-Farm Practices for The Safe use of Wastewater in Urban and Peri-Urban Horticulture. A Training Handbook for Farmer Field Schools. FAO, Rome. http://www.fao.org/docrep/016/i3041e/i3041e.pdf (accessed 05 May 2014).

Frewer L. J., Howard C. and Shepherd R. (1998). Understanding public attitudes to technology. *Journal of Risk Research,* **1**(3), 221–235.

Hamilton A. J., Stagnitti F., Xiong X., Kreidl S. L., Benke K. K. and Maher P. (2007). Wastewater irrigation the state of play. *Vedose Zone Journal*, **6**(4), 823–40.

Harris L. J., Farber J. M., Beuchat L. R., Parish M. E., Suslow T. V., Garrett E. H. and Busta F. F. (2003). Outbreaks associated with fresh produce: incidence, growth, and survival of pathogens in fresh and fresh-cut produce. *Compr Rev Food Science and Food Safety*, **2**, 78–141.

Ilic S., Drechsel P., Amoah P. and LeJeune J. T. (2010). Applying the multiple-barrier approach for microbial risk reduction in the post-harvest sector of wastewater irrigated vegetables. In: Wastewater Irrigation and Health: Assessing and Mitigation Risks in Low- Income Countries, P. Drechsel, C. A. Scott, L. Raschid-Sally, M. Redwood and A. Bahri (eds), Earthscan-IDRC-IWMI, UK, pp. 189–207.

Jimenez B. (2008). Water reuse overview for Latin America and Caribbean. Chapter 9. In: Water Reuse: An International Survey of Current Practice, Issues and Needs, B. Jimenez and T. Asano (eds.), IWA Publishing, London, p. 648.

Jimenez B. and Asano T. (2008). Water reclamation and reuse around the world: In: Water Reuse: An International Survey of Current Practice, Issues and Needs, B. Jimenez and T. Asano (eds), IWA Publishing, London, pp. 1–26.

Karg H. and Drechsel P. (2011). Motivating behaviour change to reduce pathogenic risk where unsafe water is used for irrigation. *Water International,* **36**(4), 476–490.

Kay M. (2001). Smallholder Irrigation Technology: Prospects for Sub Sahara Africa. IPRTRID, FAO, Rome.

Keraita B., Blanca J. and Drechsel P. (2008a). Extent and implications of agricultural reuse of untreated, partly treated, and diluted wastewater in developing countries. *CAB Reviews: Perspectives in Agriculture, Veterinary Science, Nutrition and Natural Resources,* **3**(058), 1–15.

Keraita B., Drechsel P. and Konradsen F. (2008b). Using on-farm sedimentation ponds to reduce health risks in wastewater irrigated urban vegetable farming in Ghana. *Water Science and Technology,* **57**(4), 519–25.

Keraita B., Drechsel P. and Konradsen F. (2008c). Potential of simple filters to improve microbial quality of irrigation water used in urban vegetable farming in Ghana. *Journal of Environmental Science and Health, Part A,* **43**(1–7).

Keraita B., Konradsen F. Drechsel P., and Abaidoo R. C. (2007a). Effect of low-cost irrigation methods on microbial contamination of lettuce. *Tropical Medicine & International Health,* **12**, 15–22.

Keraita B., Konradsen F., Drechsel P. and Abaidoo R. C. (2007b). Reducing microbial contamination on lettuce by cessation of irrigation before harvesting. *Tropical Medicine & International Health,* **12**, 8–14.

Keraita B., Drechsel P., Seidu R., Amerasinghe P., Cofie O. and Konradsen F. (2010a). Harnessing farmers' knowledge and perceptions for health-risk reduction in wastewater-irrigated agriculture. In: Wastewater Irrigation and Health: Assessing and Mitigation Risks in Low- Income Countries, P. Drechsel, C. A. Scott, L. Raschid-Sally, M. Redwood and A. Bahri (eds) Earthscan-IDRC-IWMI, UK. pp. 189–207.

Keraita B., Drechsel P. and Konradsen F. (2010b). Up and down the sanitation ladder: Harmonizing the treatment and multiple-barrier perspectives on risk reduction in wastewater irrigated agriculture. *Irrigation and Drainage Systems,* **24**(1–2), 23–35 DOI: 10.1007/s10795-009-9087-5.

Mara D. D., Hamilton A., Sleigh A. and Karavarsamis N. (2010). More Appropriate Tolerable Additional Burden of Disease. Improved Determination of Annual Risks. Norovirus and Ascaris Infection Risks. Extended Health-Protection Control Measures. Treatment and Non-Treatment Options. Discussion paper: options for updating the 2006 WHO Guidelines. WHO-FAO-IDRC-IWMI, Geneva; http://www.who.int/water_sanitation_health/wastewater/guidance_note_20100917.pdf (accessed 05 May 2014).

Mateo-Sagasta J. and Salian P. (2012). Global Database on Municipal Wastewater Production, Collection, Treatment, Discharge and Direct Use in Agriculture. AQUASTAT methodological paper. FAO.

Medlicott K. O., Stenström T. A., Winkler M. and Cissé G. (2012). Draft: Sanitation Safety Plans: A Manual for Stepwise Assessment of and Integrated Health Risk Assessment Management. WHO, Geneva.

Obuobie E., Keraita B., Danso G., Amoah P., Cofie O., Raschid-Sally L. and Drechsel P. (2006). Irrigated Urban Vegetable Production in Ghana: Characteristics, Benefits and Risks. IWMI-RUAF-CPWF, IWMI, Accra, Ghana.

Owusu I. (2009). Evaluation of Non-Treatment Interventions for Reducing Health Risks with Wastewater Irrigated Spring Onions. Unpublished MSc thesis submitted to the Department of Theoretical and Applied Biology, KNUST, Kumasi.

Peasey A. (2000). Human Exposure to Ascaris Infection Through Wastewater Reuse in Irrigation and Its Public Health Significance. Unpublished PhD thesis, University of London, England.

Prüss-Ustün A. and Corvalan C. (2006). Preventing Disease Through Healthy Environments, Towards An Estimate of the Environmental Burden of Disease, WHO, Geneva.

Raschid-Sally L. and Jayakody P. (2008). Drivers and Characteristics of Wastewater Agriculture in Developing Countries: Results from A Global Assessment. IWMI Research Report 127, International Water Management Institute Colombo, Sri Lanka.

Rheinlaender T. (2006). Street Food Quality, A Matter of Neatness and Trust: A Qualitative Study of Local Practices and Perceptions of Food Quality, Food Hygiene and Food Safety in Urban Kumasi, Ghana. Unpublished MSc thesis, Institute of Public Health, University of Copenhagen, Denmark.

Ropkins K. and Beck J. A. (2000). Evaluation of worldwide approaches to the use of HACCP to control food safety. *Trends in Food Science Technology*, **11**, 10–21.

Scheierling S. M., Bartone C., Mara D. D. and Drechsel P. (2010). Improving Wastewater Use in Agriculture: An Emerging Priority. Policy Research Working Paper no. WPS 5412. Washington, DC: World Bank. http://go.worldbank.org/UC28EAWNI0 (accessed 05 May 2014).

Scott C., Drechsel P., Raschid-Sally L., Bahri A., Mara D. D, Redwood M. and Jiménez B. (2010) Wastewater irrigation and health: challenges and outlook for mitigating risks in low-income countries. In: Irrigation and Health: Assessing and Mitigation Risks in Low- Income Countries, P. Drechsel, C. A. Scott, L. Raschid-Sally, M. Redwood, A. Bahri (eds.), *Wastewater* Earthscan-IDRC-IWMI, UK. pp. 189–207.

Scott C. A., Faruqui N. I. and Raschid-Sally L. (2004). Wastewater use in irrigated agriculture: management challenges in developing countries. In: Wastewater Use in Irrigated Agriculture: Confronting the Livelihood and Environmental Realities, C. A. Scott, N. I. Faruqui and L. Raschid-Sally (eds), CABI Publ., Wallingford, UK. pp. 1–10.

Sengupta M. E., Keraita B., Olsen A., Boateng O. K., Thamsborg S. M., Palsdottir G. R. and Dalsgaard A. (2012). Use of Moringa oleifera seed extracts to reduce helminth egg numbers and turbidity in irrigation water. *Water Research,* **46**(11), 3646–3656.

Shuval H. I., Adin A., Fattal B., Rawitz E. and Yekutiel P. (1986). Wastewater Irrigation in Developing Countries: Health Effects and Technical Solutions. World Bank Technical Paper No. 51, Washington, DC.

Siegel M. and Doner-Lotenberg L. (2007). Marketing Public Health: Strategies to Promote Social Change. Jones & Bartlett Publishers, 2nd edn, Boston.

Stevik T. K., Aa K., Ausland G. and Hanssen J. F. (2004). Retention and removal of pathogenic bacteria in wastewater percolating through porous media. *Water Research*, **38**, 1355–1367.

Tchobanoglous G., Burton F. L. and Stensel H. D. (1995). Wastewater Engineering: Treatment, Disposal and Reuse. 3rd Edition, Metcalf and Eddy Inc., Tata McGraw-Hill, New Delhi, India, 1017–1102.

Tyrrel S. F. and Quinton J. N. (2003). Overland flow transport of pathogens from agricultural land receiving faecal wastes. *Journal of Applied Microbiology*, **94**, 87–93.

WHO (1989). Guidelines for The Safe Use of Wastewater and Excreta in Agriculture and Aquaculture. World Health Organisation, Geneva, Switzerland.

WHO (2006). Guidelines for The Safe Use of Wastewater, Greywater and Excreta in Agriculture and Aquaculture. World Health Organisation, Geneva, Switzerland.

WHO-UNICEF (2000) Global Water Supply and Sanitation Assessment 2000 Report. WHO/UNICEF Joint Monitoring Programme for Water Supply and Sanitation, New York.

Section IV

Decision Making and Implementation

Chapter 19

Decision support systems for water reuse in smart building water cycle management

Caryssa Joustra and Daniel H. Yeh

19.1 INTRODUCTION

Buildings focus is often on energy. However, buildings utilise large amounts of potable water, as well as discharge wastewater and contribute to pollutant loadings through stormwater runoff (USEPA, 2009). Buildings in the United States utilise 13 percent of the total water used per day, and 8 percent of the national energy demand is directed to the treating, distribution, and heating of water (USEPA, 2009). Advancements in information technology, in addition to increased demands placed on comfort control within the built environment, led to the pursuit of 'intelligent' or 'smart' buildings (Wong *et al.* 2005). Initial focus was placed on the implementation of technologies that allowed for energy efficiency of heating, ventilation, and air conditioning (HVAC) components; however, smart buildings have grown to incorporate all subsystems housed within the building envelope (Snoonian, 2003). With regards to the water subsystem, the building industry is apt to take a somewhat compartmentalized approach to water management. The use of alternative water sources (e.g., rainwater, municipal reclaimed water, air conditioning condensate, or stormwater) or the reuse of wastewaters (grey or black) significantly complicates the building water cycle. An integrated building water management (IBWM) approach that takes into consideration water from various sources, both inside and outside the building, should be implemented in order to enhance the intelligence of buildings. One way to determine outcomes from possible solutions that aim to alleviate the disparity between supply and demand is the creation and implementation of systems models.

Increased availability of computer systems and decreased technological costs allow information systems to be incorporated by both groups and individual users at all levels of management. Decision support systems (DSSs) are tools, often computerised, used to organise and present information for decision making. Depending on the needs of the user, the complexity of DSSs ranges from simple excel spreadsheets to multi-program complex computer models. The increased complexity inherent in smart buildings with integrated water components supports the need for scalable, adaptable, and flexible DSSs that can track and organise the flow of information, as well as aid decisions regarding water cycle design, operation, and improvements (Chamberlain *et al.* 2014).

This chapter discusses the role of water reuse and recycling within smart buildings and highlights the need for integrated DSSs which accommodate building water cycles that incorporate these processes. The smart building concept is first introduced in order to affirm the transition from inefficient segregated systems to integrated dynamic systems housed within the building structure. The subsequent discussion regarding the building water cycle identifies opportunities where DSSs may aid building design and operation selections, especially in schemes enhanced by the inclusion of alternative water supplies. Finally, specific functions for future decision support tools are indicated.

19.2 SMART BUILDING

A building generally refers to a single structure and the components that support the structure; however, the term building may also be applied to a group of structures that share the same support network in a campus setting. Buildings that share similar functions and system traits can be categorised by type and include:

- Residential structures (single family homes, multi-family buildings)
- Commercial structures (offices, retail centers, warehouses, distribution centers, data centers)
- Education facilities (schools, universities)
- Healthcare facilities (hospitals, clinics)
- Hospitality facilities (hotels, restaurants)
- Recreational facilities (theaters, fitness centers, aquariums)
- Government facilities (post offices, prisons, courthouses, police stations, firehouses)
- Industrial facilities (factories, laboratories)
- Utilities (water treatment plants, wastewater treatment plants, power stations)

Any of the aforementioned building types has the opportunity to be a smart or intelligent building with the inclusion of prerequisite components that facilitate communication within the building system and integration of building subsystems. Components vary from building to building, but common building subsystems include:

- Structural
- HVAC
- Lighting
- Electrical
- Water
- Sewage

- Security
- Fire suppression

Definitions describing smart buildings vary among sources, but contain shared elements. Table 19.1 outlines a few definitions used by organisations and found in literature. Certain commonalities can be pulled from the definition summary. First, it is evident that technology is a necessary feature of a smart structure. Technology is often synonymous with intelligence regardless of discipline; it is assumed that technology increases the capacity for the collection, organisation, compression, and communication of information. Given the complexity of the building system and associated subsystems, intelligence is desired to accommodate the massive information potential. The implementation of computer technology furnishes a smart building with a synthetic brain that can be programmed to synthesize and share information according to predetermined decision parameters. In this way, the building makes informed decisions regarding daily operations. Engaging this ability supports the second common attribute of smart buildings: efficiency. Operational efficiency is maximized with the aid of technological triggers creating a high-performance structure. Environmental comfort control can be monitored from a central location and immediately altered based on information inputs from remote sensors, or water flow sensors can discover leaks in water features and tag components for repair. By minimising system losses, smart buildings also achieve the goal of cost reduction (Snoonian, 2003). Streamlined operations and maintenance practices help offset the expense of developing a smart building, enhancing the bottom line. Systems integration further reinforces building performance and is the third shared feature of smart building. In particular, integrated computer and communications systems are essential components for smart building as they are responsible for information facilitation to each subsystem (Finley *et al.* 1991). In the case of comfort control, integration of a centralised computerised system allows for efficient command of mechanical ventilation throughout the building. Integration should also exist among other building subsystems, whether directly or through computer and communications components. For example, all subsystems may be wired to a centralised computer control hub where the state of each subsystem is evaluated and altered based on the composite information received.

The fourth important aspect of a smart building is user interaction. Early definitions of smart buildings were solely based on the use of technology and lacked the integrated component of user interaction (Wong *et al.* 2005). Technology is used to increase building performance, but it is the users that benefit from the increased efficiencies; and buildings must be designed to support the occupants. Therefore, how occupants interact with the building and associated subsystems is crucial, and a smart building must allow users to alter the structure's state to their specifications. This leads to the need for flexibility of the building system and subsystems. Smart buildings are networked using technology, and technology is a constantly evolving area. As a result, smart buildings must be able to incorporate technological improvements with limited additional costs and effects to productivity in order to persist (Flax, 1991). The need for flexibility also extends to normal building operations. Modular systems allow smart buildings to quickly and effectively respond to changing environmental conditions. Consequently, this infers that smart buildings should be adaptable dynamic systems in order to meet the changing needs of its users.

Table 19.1 Smart building definitions.

Source	Definition
Smart Buildings Institute (2013)	'Enhances the performance of the building and ease of operation over its life-cycle. The primary goal ... is to minimise the long-term costs of facility ownership to owners, occupants and the environment. In a higher performing building, all components of the building are integrated in order to work together. This improves operational performance, increases occupant comfort and satisfaction and provides the owner with systems, technologies and tools to manage and minimise energy consumption.'
Smart Buildings, LLC (2012)	'A smart building, [also known as an] integrated building, intelligent building, automated building, high performance building or advanced building, is a building that is designed for longevity.'
Flax (1991)	'Creates an environment that maximises the efficiency of the occupants ... while at the same time allowing effective management of resources with minimum life-time costs.'
Finley *et al.* (1991)	'Single building or a complex of buildings which offers a coherent set of facilities to both the building managers and to the occupants, to the building managers, an integrated set of management, control maintenance, and intra- and inter-building communications facilities that allow efficient and cost-effective environmental control, security surveillance, alarm monitoring and communications, both inside the building and out to municipal authorities (police, fire-stations, and hospitals), and to the building occupants in the workplace, an environment ergonomically designed to increase productivity and encourage creativity and in residences and hotels, environments that will foster comfort and a 'humanising' atmosphere as well as provide sophisticated computer and telecommunications services.'
Katz (2012)	'Provides owner, operator and occupant with an environment, which is flexible, effective, comfortable and secure through the use of integrated technological building systems, communications and controls.'

19.2.1 Building automation

Smart buildings require communication between the building system and subsystems, and building automation provides a means to facilitate the transfer of information. Incorporating building automation features allows for increased building efficiency, making automation critical for smart buildings to reduce operations costs (Snoonian, 2003). For example, automated lighting systems ensure that energy is not wasted during building off-hours by shutting down non-emergency lighting systems.

Building automation refers to any technologies applied to building systems that allow for centralised control and communication. However, automated systems often lack integration and operate using separate communication standards and control points (Flax, 1991; Snoonian, 2003). For example, electrical and fire prevention systems may both be automated, yet controlled using two different communication standards thereby preventing the use of a shared centralised control point. In addition, the lack of a shared communications language keeps both systems isolated from each other and disallows an input-response relationship. In the case of a fire, it would be desirable for the fire prevention system to alert the electrical system and shut down building electrical components. In order for such cause and effect relationships to take place, a shared language or central communications 'interpreter' is required to facilitate an integrated systems approach. Completely integrating building systems is challenging due to the wide array of manufacturers involved (Snoonian, 2003). Not only does this expand the number of unique systems and associated controls, but also results in systems, controls, and protocols that are protected property of the manufacturer and cannot be altered. Although a formidable problem, solutions exist that aim to integrate unique building systems.

In the building industry, BACnet and LonWorks represent two common building automation communications standards largely developed in the 1990s that aim to integrate building systems (Snoonian, 2003). BACnet (ASHRAE, 2013) which stands for Building Automation and Control Networks was developed by the American Society of Heating, Refrigerating and Air-Conditioning Engineers (ASHRAE) and is both an American and European standard. LonWorks (Echelon Corporation, 2013), short for Local Operating Network, was developed by Echelon Corp. and is another standard used in the United States and Europe. Both BACnet and LonWorks are standards recognised by the International Organization for Standardization (ISO).

BACnet was originally developed for the mechanical and electrical systems within the building envelope and is a communications-only protocol; however, the generic nature of the protocol allows for the integration of hardware and software associated with other building systems (Snoonian, 2003). The BACnet protocol utilises virtual objects that can be organised and programmed to represent the operations and functionality of the building by describing current operations, desired operating parameters, and resulting commands (Snoonian, 2003). Compatibility with the internet allows BACnet components to be controlled remotely through the web, thus resulting in remote building control from anywhere web-connected; the controller is not tethered to the location of the building systems and has remote and immediate access (Snoonian, 2003). Another benefit of the BACnet protocol is the ability to facilitate communication among diverse building systems; data can be shared and prioritised for system integration and clarity. A command with higher priority, such as shutting down electrical components in the event of a building fire, will be implemented over a command with lower priority, such as running electrical equipment in power-saving mode during building off-hours. Due to BACnet's wide acceptance, it is possible to find devices immediately ready for installation.

LonWorks has been adapted to building applications after being focused on the transportation and utilities industries. Unlike BACnet, the LonWorks standard includes both a communications protocol and a hardware component; BACnet was developed only as a communications protocol. LonWorks uses the Neuron Chip as a link between a device desired to be controlled and a central control system (Snoonian, 2003). Similar to BACnet, LonWorks transmits data using wired connections, as well as web servers. LonWorks utilises network variables in order to create the inputs and outputs of building systems, analogous to the virtual objects comprising the BACnet protocol. However, prioritisation of system commands is not as direct when using LonWorks because it lacks the inherent levels of priority found in the BACnet system. In order to allow for prioritisation in LonWorks, users can define override commands for emergency response or periodic checks to continue normal operations (Snoonian, 2003).

19.2.2 Relationship to green building

Although smart buildings and green buildings are referred to as high-performance buildings, the terms are not interchangeable; a green building does not necessarily need to be smart, and a smart building is not always green (Figure 19.1). The main differences can be attributed to the presence or absence of sustainable aspects. Green buildings aim to limit environmental impacts, and efficient building operations may help achieve this goal; whereas smart buildings focus on efficient operation of the building system, which may result in reduced environmental impacts. A smart building can achieve efficient water use using communications networks; however, if the water source utilised is non-sustainable, such as water from a limited potable supply, then the building is not also green. A green building may be designed with pervious pavement allowing infiltration and a gravity-based rainwater reuse system to offset the potable demand. Both sustainable strategies do not require complex control systems; therefore, this green building is not defined as smart. In addition, while green building requires consideration of the entire building life cycle from design to deconstruction, smart building activities are only applied during the design and operations phases.

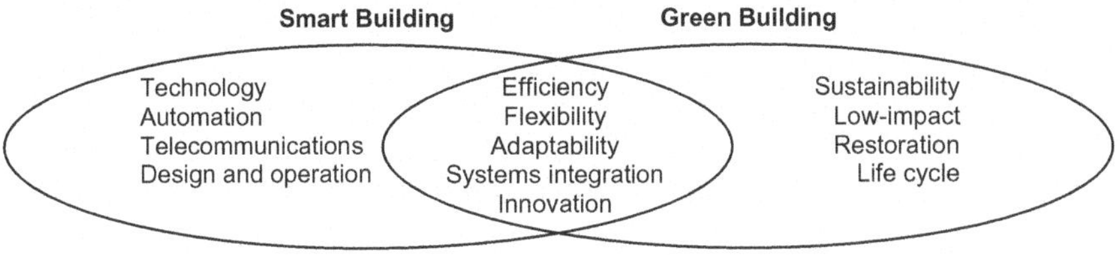

Figure 19.1 Comparison of smart and green building concepts.

Globally, there are more rating systems for green buildings than smart buildings. An industry leader in green building rating systems is the United States Green Building Council (USGBC), which encourages sustainable building practices (USGBC, 2013a–c). Assessment tools for smart buildings are less visible and scarce. However, building intelligence can be evaluated using the Building Intelligence Quotient tool (Katz, 2012). The tool is available online and provides value to smart buildings, integrated design support, and building automation support.

An important similarity between the two building types is the goal of systems integration, and allows for synergy between smart and green strategies. Despite the subtle differences, it is possible for a smart building to also be considered green. Smart building practices can even enhance the sustainable attributes of a green building. Water components controlled and monitored from a centralised computer location allow building operators to verify that alternative water supply systems are functioning properly. Failures or leaks are easily pinpointed resulting in faster repair times. Green building strategies can also make a smart building smarter. Incorporating alternative supply systems within a smart building allows for increased performance through potable water reductions.

19.3 THE BUILDING WATER CYCLE

Each building is a unique system composed of multiple dynamic subsystems, and each subsystem can be separated into multiple smaller components. The building subsystem based upon water utilisation can also be viewed as the building water cycle. Just as the natural, or hydrologic, water cycle maps the flow of water throughout the global system, the building water cycle also contains an inherent map of water flows throughout the building structure. In the former case, the system boundary is global; whereas in the latter cycle, the system boundary is drawn around the building site, which may include the physical building in addition to vegetated spaces, parking areas, and hardscapes. Pathways in the natural water cycle are based on environmental processes in contrast to the physical conveyance and consumption pathways found in the building system; however, the main difference between water cycles is that the global cycle is a closed system while the building cycle is an open system. This fact is illustrated in Figure 19.2. The uninterrupted natural water cycle is represented by a balanced feedback loop (A). In a simplified view of a conventional building, water is fed into the system boundary from the environment, utilised by the building, and discarded back into the environment (B). In this worst-case scenario, the building water cycle is more linear than cyclical and environmental impacts are at their peak. However, smart and green buildings aim to limit disruption to the environmental water cycle through efficient water use measures (C) and recycling practices that mimic the hydrological cycle (D). The ultimate goal is a net-zero water building, a structure with a water cycle that has evolved into a closed system (E).

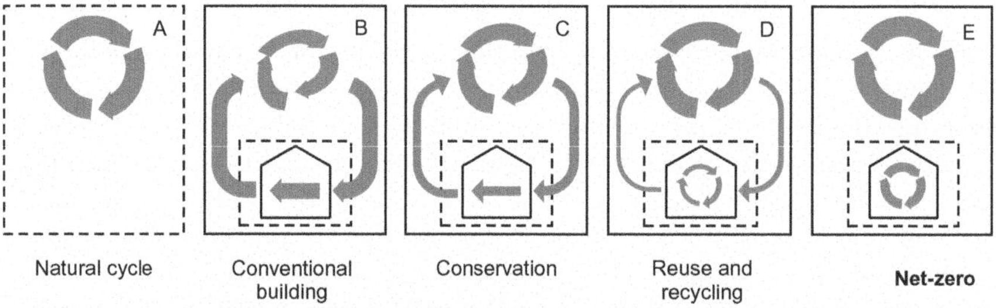

Figure 19.2 Evolution of the building water cycle.

As discussed, the building water cycle ranges from completely open and linear to closed and cyclical. The complexity of the cycle depends on the number of connections and potential routes for water, as well as the magnitude of those flows. Therefore, the building water cycle can be defined based on inherent (1) demands, (2) supply sources, and (3) usage patterns. Water demands, potential supply sources, and allowable interactions among the two are dictated by the building's design. Decisions must be made to first determine the desired water demands of the building system. Then, the connections to available water sources are considered. Finally, based on chosen demands and sources, the connections between each are made. In a conventional building, these decisions are basic; once demands are acknowledged, each is supplied by the potable water source, and remnant water is discarded from the building boundary as wastewater. Incorporating alternative water sources necessitates further decisions regarding where it will be applied, and also creates opportunities for discarded water to be captured and recycled. The increase in choices regarding the building water cycle provides the opportunity for the implementation of DSSs that aid the decision-making process. DSSs are further desired when usage patterns increase the complexity of the cycle. The building design only determines connections, but usage is determined by human behaviour. The changing magnitude of water flows dictated by occupant usage is largely responsible for the dynamic nature of the building water cycle. Estimation of the occupants' effect on water usage further feeds the need for decision support, but understanding of the building water cycle components is a prerequisite for DSS creation.

19.3.1 Building water demands

The number and importance of building water demands is dependent on the type of building. In a restaurant, the demand for water used for cooking is higher than in a retail store; a restaurant may have a water demand associated with an ice machine that is not present in a retail store. Even among buildings of the same type, demands can vary. One residential home may include a swimming pool that creates a water demand due to periodic refilling, whereas a neighbouring home may not. A school containing an on-site garden project would have a water demand for growing crops that would not be included in the water cycles of other schools. Therefore, water demands are site-specific. The quantity of water utilised by a demand depends on the device efficiency used to meet the demand. As a result, the magnitude of the water required for an individual demand can be minimised through conservation strategies.

In regions of limited rainfall, a large portion of a building's water use is directed towards the *irrigation* demand. In the United States, about one third of the water used in the residential sector is for landscaping (USEPA, 2013). Traditionally landscaped building sites incorporate large tracts of water-thirsty turfgrasses, but switching to native landscaping practices by planting water-efficient grasses, groundcover, shrubs, and trees can substantially decrease the demand. Additional irrigation demands may arise in the presence of gardens, whether for aesthetics or food production. In addition to the type of vegetation planted, the density and proximity to the building and other vegetated areas affects the water demand by altering complex evapotranspiration processes. Choices exist regarding existing technologies for irrigation. Sprinkler systems that disperse water through the air are less efficient than drip systems dispensing water underground. The cost to install rainfall or moisture sensing equipment limits inefficiencies and may be offset through water cost savings. Increasing the ratio of vegetated space to hardscapes allows for increased rainwater infiltration rates on-site, thereby reducing the water leaving the building boundary as runoff and moving towards a closed system. However, if the green spaces developed require additional irrigation beyond rainfall, the demand for water sources located outside of the building system boundary may increase, moving away from the closed system goal.

Generally the cultivation of green spaces takes place at the ground level surrounding a building structure, neglecting the remaining hardscape produced by the building itself. However, practices that literally green the building, such as vegetated walls and roofs, soften the effects of the hard building exterior. The green roof is often used as an example of integrated design in sustainable construction because of its effects on the building system. Benefits are seen in water management, energy efficiency, and air quality (Carter & Fowler, 2008; VanWoert *et al.* 2005). Like all vegetated spaces, a green roof mitigates runoff quantities through water retention by plants and substrate. This method also increases the quality of water leaving the green roof, protecting the environment from high pollutant loads. Insulation and evaporation allow a green roof to even out building temperatures over time. In addition to reducing heat outdoors, vegetative roofs may also have positive impacts on the indoor conditions of the building while providing an aesthetically pleasing environment for workers and guests. Here the potential connections among building subsystems is evident. Implementation of a green roof may affect the irrigation demand and alter the building water subsystem. Additional irrigation components will need to be integrated into the existing system. Ensuring the roof system can carry the vegetative roof load and facilitate proper drainage affects the structural system. The evapotranspiration and insulation associated with a green roof alters the HVAC loadings; these effects change the parameters input into the design and operation of the building energy subsystem. Green roofs are especially encouraged in urban areas where green space is limited, such as in the cities of Chicago, Seattle, and New York City. In Toronto, green roofs are required for new construction meeting height and size standards (City of Toronto, 2013). DSSs can easily organise potential vegetation types by water demand and allow users to estimate the total amount of water needed for irrigation based on planted area and placement in order to choose the optimal design. Further effects, such as the potential for shading or insulation of the building structure can be input into energy calculators. However, DSSs may neglect qualitative considerations, such as aesthetics and social acceptance. Overlooking qualitative effects may skew cost-to-benefit results and produce a design choice that is not necessarily the most advantageous. This stresses that DSSs are truly for support, and final decisions require interpretation and assessment by the user.

Within the building structure, most fixtures focus on supplying water for essential human needs, such as *drinking, hygiene, cooking,* and *cleaning.* The fixture type affects water efficiency, and standards determine maximum values allowable by fixture in the form of flow rates or volume per use event; however, green and high-performance buildings aim to install hardware fixtures that exceed the efficiencies set forth by these standards. Low-flow faucets used in kitchens and bathrooms aim to eliminate wasted water by creating a manageable water stream. Most, if not all, buildings include a demand for hygienic practices like hand-washing, but not all will contain a demand for cooking activities, such as washing foods and utensils. Residential buildings will have a higher demand for showering than commercial structures, although this demand is not exempt from all non-residential areas. Schools may provide showering facilities for students, or commercial and industrial structures may include showers for employees. The design choice of how many showering structures to include depends on the expected demand. An urban office building with a large group of employees that commute by bicycle will expect a higher showering demand than an office staffed with all vehicular commuters; in both cases it is unlikely that all occupants will shower, but the demand is steadily expected in residential areas where occupants likely shower daily. In the case of showering, building designers have a choice of fixtures to curb the water demand. Choosing the fixture with lowest flow is assumed to provide the highest water savings, but the initial investment can be higher. A DSS can easily find the optimal balance of water savings and cost for a fixture based on lifetime and payback periods, but personal preference is also a factor.

Sanitation is another essential water demand. Buildings are designed with some form of sewage conveyance for toilet or urinal flushing. Each of these two fixtures is designed with a set water volume utilised to accomplish this goal. Like faucet fixtures, the amount of water needed per event can be reduced using high-efficiency options; ultra low-flow toilets and urinals can use less than half the water per flush as set forth by maximum standards. However, toilet and urinal fixtures exist that eliminate the use of water and still accomplish the sewage conveyance goal, unlike their faucet counterparts. This is possible for two reasons. First, the demands met by faucets are consumptive, and cannot be fulfilled without water; water is consumed for drinking and cooking, and water is a prerequisite for sustainable cleanliness. Second, the delivery of water to faucets requires pressurisation, whereas sewage conveyance can be accomplished using gravity. In the case of waterless urinals, gravity facilitates the movement of liquid waste through a secondary liquid seal. The seal prevents odours from escaping and floats on top of the urine due to a density difference. In this case, the water demand for urinal flushing is eliminated from the building water cycle; however, a limited water stream may still exit the building boundary through the sewer system. A waterless toilet has the ability to eliminate both the water entering and exiting the building system. Also referred to as composting toilets, these fixtures are designed to degrade wastes on-site. User acceptance is crucial for the success of waterless sewage conveyance practices that aid in closing the building water cycle; and additional arrangements for maintenance and nutrient recycling must be considered and integrated. Further impediments to installation include energy and financial costs, which may be significant in vacuum-based drainage systems.

Process water demands vary, but cooling is commonly included. Often mechanical, cooling systems contribute to the comfort of building occupants. Due to fluctuating environmental conditions, the water demand associated with cooling can vary annually, seasonally, and diurnally. For cooling towers, demand is correlated to the makeup water, which depends on multiple losses found within the tower. A portion of water exits through evaporation processes. Water leaving the tower through uptake air flows, rather than through direct evaporation, is referred to as drift. Evaporation and drift cause the concentration of dissolved solids to increase within the cooling tower. In order to reduce the concentration of solids, water is drained periodically in a process referred to as bleed-off and replaced by clean water. Bleed-off is an intermittent process, whereas evaporation and drift constantly occur. Although challenging, limiting evaporation and bleed-off will decrease the cooling demand. Chemical additives can inhibit scaling within the tower; thereby prolonging residence time of the recycled water and reducing the frequency of bleed-off events. Decisions regarding cooling tower design and operation require optimisation of chemical use, costs, and water savings.

Additional process water demands include thermal cooling, boilers, steamers, industrial dishwashers, ice machines, and pre-rinse spray valves.

Buildings require a degree of *safety* in order to protect the structure, interior elements, and human occupants. Fire suppression systems dispense water when activated under emergency circumstances; and therefore, the demand associated with firefighting is rarely incurred. However, if the water demand is activated, the volume required to meet the demand is appreciable and causes this demand to be notable in the building water cycle. Unlike the other water demands discussed, conservation measures cannot be applied to the fire suppression system.

Often overlooked are water demands regarding *recreation and aesthetics.* Examples such as sports fields or flower gardens are better listed under irrigation demands; rather this category focuses on aspects such as swimming pools, fountains, and other water features. After supplying the initial water volume needed in order to enact each feature, evaporation, infiltration, and usage losses consequently fuel a consistent operational water demand. This category best demonstrates the trade-offs between function and form. Aesthetic demands focus on leisure and beautification of the building site over practical and essential functions, but this does not mean they are without value. Increasing the building appeal can add financial value to the property and increase occupant productivity through heightened morale. Some of these benefits can be quantified in economic terms, while qualitative benefits based on psychological benefits are difficult to assess. DSSs can weigh the costs of implementing aesthetic water features, but the final decision cannot be made without consideration of immeasurable qualities. In this case, the information organised and presented by a DSS then becomes an informational input and assists the decision maker.

19.3.2 Building water sources

Once demands have been established, available sources must be investigated and chosen to meet the demands. The accessibility of water sources depends on the location of the building site, available infrastructure connections, and the demands outlined as part of the building's water cycle. The meteorology and hydrology of the region encompassing the building dictates whether certain conventional or alternative sources exist. Structures atop natural water reserves may be able to bore through the surface and construct on-site wells for groundwater recovery. Areas with substantial precipitation events provide buildings with potential rainwater and stormwater sources. Rainwater is assumed to be the water captured before interacting at the ground level, and is therefore assumed to have a higher water quality than stormwater and with appropriate treatment can be used for potable or non-potable applications (further discussed in Chapters 1–8). The state of the infrastructure supplied to the building determines potential municipally supplied sources, such as potable or reclaimed water. Reclaimed water is a high-quality water source produced after intensive treatment of municipal wastewater at (de) centralised treatment facilities (examples discussed in Chapters 14–17 and Chapter 20). Additional alternative water sources are produced within the building boundary by the fixtures associated with demands. Wastewaters can be separated into two streams, greywater and blackwater, depending on the discharge quality. Greywater exits from sinks, showers, and other low-strength sources (further details in Chapters 10 to 13), whereas blackwater contains higher amounts of organic material and includes water flushed from toilets and urinals. Buildings with a cooling demand also contain a potential condensate water source (as discussed in Chapter 9). The quality of condensate collected from air handling equipment is comparable to distilled water, requiring little to no treatment for non-potable applications (Licina & Sekhar, 2012). Conventionally, easily-attainable high quality sources are pursued for all building water demands; however, focus has shifted to alternative water sources to meet the needs of green and smart buildings.

Traditionally buildings are designed to shed rainwater from the building site as stormwater runoff and lose this volume as an alternative water source. Regulations regarding treatment and mitigation of runoff volumes that mimic the predevelopment hydrologic cycle also form the basis for augmentation within the building water cycle by capturing the water in cisterns, rain barrels, detention and retention ponds, or other natural water bodies. One barrier to rainwater use within buildings, especially in the United States, is the lack of regulation regarding application of this source. As a result, codes and statutes often limit uses to irrigation. However, in many island nations, such as the United States Virgin Islands, water is an especially scarce resource, and rainwater is the primary and sometimes only available source for potable applications (Solomon & Smith, 2007).

Sources dependent upon wastewater streams have the benefit of being continuous, as opposed to natural sources dependent upon meteorological and hydraulic conditions. Limitations are often imposed on these sources due to associated human and ecological risks in order to ensure public safety (Anderson *et al.* 2001). Reclaimed water is generally considered a safe and sustainable option within water-critical regions (Wintgens *et al.* 2005). The majority of reclaimed water in the United States is applied to landscaping, both in residential and commercial structures. However, dual-plumbing systems can serve other non-potable fixtures with this source. Nearly all reclaimed water produced at the city's wastewater treatment facility in Dunedin, Florida makes it to lawns within the city limits. During dry months, demand can even surpass available supply. Wastewater treatment and reuse can also be accomplished on-site by compact packaged systems. Membrane bioreactors (MBR) accomplish wastewater treatment within a small footprint by replacing secondary and tertiary treatment trains found in municipal facilities with membranes. High quality MBR effluents produced from either greywater or blackwater influents have the potential for recycling within the building water cycle (Atasoy *et al.* 2007; Boehler *et al.* 2007; Ghisi & Ferreira, 2007; Sorgini, 2004). The Helena Building in New York City includes an MBR system that recycles wastewater for cooling, toilets flushing, and irrigation applications, reducing the demand for municipal potable water (Clerico, 2007). Cooling and dehumidification of buildings in warm climates produces a high-quality condensate source usually considered a waste stream. Condensate flows can be directed to existing storage components, such as a rainwater cistern, or collected and distributed separately. In San Antonio, condensate capture systems have become standard; the shopping mall produces 950 litres per day, and the central library produces about 163,000 litres per month (Guz, 2005). Common recycling applications include cooling tower makeup water, irrigation, and aesthetic water features, although the high quality of the source allows for varied applications. Further details on condensate recovery and reuse projects can be found in Chapter 9.

Having on-site alternative sources implies a need for storage since the time of source production does not necessarily coincide with the time of demand; and any treatment following collection delays the delivery of the source to the demand. Storage builds flexibility into the building water cycle by allowing it to respond to changes in the magnitude of water demands using alternative sources. Even storage of conventional sources, such as municipal potable water, provides flexibility and security. Elevated water towers also ensure delivery of a water source by creating pressure within the building water system when the pressure within municipal pipelines is intermittent.

19.3.3 Usage patterns

The building design component partially contributes to the expected interaction of the building water cycle. The remaining element largely affecting the movement of water is human behaviour. The same individual has different interactions with unique building types, and even among buildings of the same type depending on the role of the individual within that system. An individual in their residence will create a higher overall demand for water than in a commercial building. Unique demands within each cycle are also affected differently. In a residence, the occupant is assumed to use more water for kitchen and bathroom uses, as well as discharge more water from these applications than in the commercial structure. The effect that the role taken by the individual has on the building water cycle is evident in a retail structure. As an employee, the individual would spend most of the day within the building system and exert a higher stress on the water subsystem than a visitor. It is possible for the visitor to have no effect on the building water cycle within their short stay, whereas the employee is likely to interact with drinking, cooking, sanitation, and hygiene demands. The application of certain water demands also depends on the individual's preference. One employee may prefer to take a premade meal to work from home, whereas another employee may prefer to make lunch at the office, thereby shifting the associated demands from the residential to the commercial building water cycle.

Human behaviour further affects the performance of individual water fixtures. Fixtures are rated based on the amount of water they are designed to use per application, and building owners install fixtures under the pretence that each application will fulfil the design standard. However, human interaction can override expected water demand operations. It is assumed that low-flow faucets reduce overall water consumption, and this is true if the time required to fulfil a demand is the same for the low-flow feature as it would be for a conventional faucet with higher flow. In reality, the low-flow faucet may be active for a longer time period to accomplish a similar task due to the lower magnitude flowrate. Even the installation of automatic features does not guarantee design performance. For example, a sensor-activated toilet flushes with a predetermined volume after activation by the sensor. False sensor readings can result in multiple flushes per use event. A delayed flush response can cause a human user to override the flushing mechanism causing an additional volume to be lost during the application. Further human interaction affecting automated flush volumes was verified in a school study (Joustra, 2010). Automated toilets installed as part of a rainwater collection and reuse system at a green-certified school were rated to use 4.8 litres per flush (lpf). However, data collection based on individual flush events found multiple instances of flushes that exceeded the rating. An investigation found that the pressure exerted on the manual flush *override button* changed the volume consumed, and holding the button down caused a continuous flow of water. In addition, students were urged to utilise the *override button* to eliminate all waste as a social courtesy. This example shows how human interference alters the design state. It is important to acknowledge that the magnitude of water use according to demand can be estimated based on the building type, role of the occupant, and water fixture design, but precise usage patterns would best be evaluated using sensors tied to smart building networks. Collecting the usage patterns unique to the building system would allow for better decisions regarding water efficiency.

19.3.4 Integrated Building Water Management (IBWM)

Integrated building water management acknowledges the interrelationships among water sources and demands and aims to operate the building water sector on a systems level. This requires that the water demands and potential sources for an individual building are first inventoried. Then, decisions regarding proper allocation of water sources to meet specific demands can be accomplished. In a conventional building, potable water is often the sole source used to meet all demands. However, water meeting potable standards is not necessary to accomplish non-potable applications, such as flushing of toilets or urinals. Utilising an integrated systems approach, IBWM first observes the potential for alternative water sources to meet demands as part of a fit-for-purpose approach, before drawing from on-site or municipally supplied potable water. Efficient and integrated source allocation manages the inflow of water into the building system, the recycling of water throughout the building system, and outflow of water from the building site. Measures taken as part of an IBWM aim to decrease the inflow of water, particularly potable water, as well as decrease the outflow of water using efficient wastewater and infiltration processes (Lazarova *et al.* 2001). The efficiency pursued as part of an

IBWM approach is shared with both green and smart building concepts. IBWM implementation strongly aligns with green building goals by promoting sustainable management through water reuse and recycling practices. Due to the increased complexity of the green building water cycle, total water use is reduced by reusing water for non-potable demands and recycling wastewater streams after treatment. Closing these flows transitions the building water cycle toward a net-zero water system leading to the maintenance of the natural hydrologic cycle and lowered environmental impact.

As discussed, the building water cycle is comprised of a complex web connecting water sources with demands. Deciding how to match sources to each demand creates the need for prioritisation based on preference. When the same source is available for multiple demands, prioritisation by demand is necessary. Drivers affecting demand prioritisation are based around public acceptance and include perception of alternative water sources, knowledge about the source, previous experience with the water source, and interaction or influence from friends, family, and colleagues (Dolnicar *et al.* 2011). Public acceptance is also driven by the perceived cleanliness of the water source; for recycled water allocation, the aesthetic quality is an important consideration factor (Jefferson *et al.* 2004). In decreasing order of preference, potential demands met by alternative water sources include irrigation, cooling, industrial processes, recreational water use, non-potable public water uses, and potable public water uses (Howell, 2004; Asano, 2002). Surveys conducted regarding alternative water use are in general agreement; the highest support again focuses on irrigation followed by toilet flushing, laundry, cooking, and drinking, respectively (Browning-Aiken *et al.* 2011; Campbell & Scott, 2011). However, the aesthetics of a particular water source may alter the demand preference. Jefferson *et al.* (2004) observed that recycled water with a poor appearance caused the allocation preference to change from irrigation to toilet flushing. The highest priority demand for greywater alternatives between irrigation and flushing of fixtures. Ludwig (2006) prefers applying greywater for landscaping due to treatment processes that occur within the soil, whereas Jamrah *et al.* (2006) argue the best use is for flushing toilets. However, both agree that demands with higher human interaction, such as clothes laundering, have a lower priority. According to Hauber-Davidson (2007), acceptable uses for rainwater include irrigation, cooling, bathroom uses, laundry, and refilling swimming pools; less acceptable demands include kitchen use and food preparation. Condensate is a high quality alternative water source, and due to its proximity to the cooling system, Licina and Sekhar (2012) propose cooling make up water as the top priority for allocation. The preferences discussed demonstrate that the preference of utilising alternative water sources for reuse and recycling is highest for water demands with the least amount of direct human contact, although additional social factors can alter the desired prioritisation.

Additional prioritisation based on source is required when multiple sources can meet one demand. The logic employed by green building and IBWM assumes a higher preference for alternative water sources over potable sources. Often a demand served by alternative water sources also contains a potable water backup supply. In this case, the potable supply is given the lowest use priority. The prioritisation by source should be defined given the number of potential alternative water sources, diverse water quality parameters, and public perception. For example, the priority given to greywater use may be elevated because its treated quality quickly degrades over time (Al-Jayyousi, 2003). Rainwater may be assigned a lower priority than greywater due to its longer storage potential with proper collection. Although public views of water sources tend to drive prioritisation, green buildings often challenge this perception by pioneering new technologies. Building designers have the opportunity to change the building water cycle prioritisation framework based on their own preferences and decision-making aids. IBWM forms the foundation for a DSS capable of taking an integrated systems approach towards this goal.

19.4 DECISION SUPPORT SYSTEMS

Decision support systems (DSSs) come in various forms and complexities utilising multiple programs and platforms. Models can be qualitative, quantitative, or a combination of both. Qualitative DSS models include decision-making trees and or diagrams outlining strengths, weaknesses, opportunities, and threats (SWOT); information is based on observable data. Quantitative DSS models utilise mathematical inputs in order to produce numerical outputs used for decision-making. Both qualitative and quantitative attributes can be incorporated into water management models. The volume and rate of water delivery is quantifiable; evaluation of water quality depends on quantifiable parameters that can be assessed using analytical methods and qualitative parameters such as colour, taste, and odour. The quality of water can also be assessed qualitatively based on treatment (primary standards, secondary standards, tertiary standards) or regulated and accepted end uses.

19.4.1 Advantages and disadvantages

The inherent advantages of DSSs result in widespread application (Power, 2002). Time savings are accomplished by quick decision-making accomplished by using DSS models. Creation of a user-controlled model can be quicker than waiting for and recording real-time observations. For example, the decision to enact water conservation measures prior to a drought can be made earlier and faster using prediction models rather than waiting for deteriorating conditions to reach a critical point. In addition, the use of models can be cost-effective due to lower infrastructure, technology, and labour costs. Building a computer model for decision support is less intensive than constructing a pilot-scale system; modelling the impact of various alternatives can eliminate poor solutions from being considered for further studies or final implementation, saving time, labour, and cost. Further savings are accomplished with flexible DSS models that allow input parameters to be easily changed for running multiple scenarios. Increased effectiveness of decision making and improved communication are two additional advantages. A DSS model organises information and presents a scenario as one complete picture that is shared with all users; everyone is given the same results from which to form a decision.

Potential disadvantages can decrease the value of DSS outputs (Power, 2002). It is important to remember that DSSs should be used as a support tool, and not as the sole source for decision-making. Generally DSSs do not incorporate social and political impacts of a potential decision; and therefore, consequences related to these areas must be taken into account when using purely technical forms of DSSs. Decision authority may be applied to DSS tools, but final decisions should be made by humans using input from the DSS outcomes. Users must also acknowledge the boundary wherein information input and output by a DSS is applicable as decisions made outside of these bounds lose validity. It is possible for systems to be overloaded with information, or provide excess information that interferes with coherent decision-making. However, properly formed support systems organise vast information inputs for simplicity. Information outputs depend on the information inputs; bad inputs result in bad outputs. Therefore, care should be taken to reduce poor information from entering the support system and producing bad results. Users of DSSs must also prevent over-reliance on support systems, which can reduce the effectiveness of decision-making. If reliance is high, it is also possible that users may overlook low quality results or place high importance on complex results. In both instances, decision-making effectiveness is reduced. This increases the potential for false objectivity. The ultimate responsibility regarding decisions lies with people and not computers. DSSs are assumed to be rational and objective, but the same assumption cannot be made of people. The manipulative nature of DSSs can allow users to come to subjective decisions rationalised using the support systems outcomes. When implementing a DSS model, all advantages and disadvantages should be considered and addressed.

19.4.2 Role of DSSs in smart building water reuse and recycling

The building water cycle consists of a labyrinth of connections among demands and sources, which increases in complexity in smart building systems that incorporate water reuse and recycling strategies, elevating the appeal of support systems to aid in the decision-making process regarding design and operation. Aiming for efficient resource consumption and utilising green building practices to meet high-performance standards, such as alternative water allocation, creates a number of variant building water cycle combinations. DSSs provide the opportunity for water cycle optimisation based on user-defined parameters of interest including water savings, energy use, cost reduction, and social acceptance. Individual decisions to be made include:

- Water demands served by the building
- Potential water sources available to meet demands, including alternative supplies
- Connections between demands and sources
- Priority of demands met by same source
- Priority of sources meeting same demand
- Design components
- Alternative water management strategies
- Operation parameters
- Estimated water usage

Certain decisions regarding water demands and sources within the building water cycle are made implicitly and are historically expected. Building codes and statutes outline required fixtures based on building type and number of occupants. Residential structures are expected to have fixtures for bathing, cooking, cleaning, hygiene, and sanitation; whereas small commercial structures may only be mandated to include bathroom facilities. Required inclusion of water fixtures based on regulations introduces the associated demands into the building facility and often includes minimum performance standards for each fixture. Although installation of specific fixtures cannot be eliminated, the opportunity exists to choose devices that limit water consumption, and this is where DSSs can aid users in choosing appropriate hardware. The addition of other demands remains at the discretion of the building design team. These largely include demands associated with building aesthetics, such as water features and decorative landscaping. Decisions regarding these features balance water consumption with measurable quantitative benefits including worker productivity and financial value of the building site (Montalto *et al.* 2007). DSSs that compare expected benefits to buildings with similar aesthetic features may aid the design of these additional non-essential demands, although immeasurable social benefits require a human component to synthesize all benefits before weighting against potential costs.

Control regarding connections between water sources and demands is also largely regulated. In most cases it is expected that a potable water supply exists to meet all demands. At a minimum, all demands are supplied by the potable water source and discharge to a sanitation system. However, the inclusion of alternative water sources resulting from water reuse and recycling schemes presents designers with a myriad of water cycle arrangements based on choices that direct water throughout the subsystem. This also leads to decisions associated with the prioritisation of unique demands and sources. Designers must decide which demand or demands should be met by each alternative source, or whether more than one alternative source should be grouped to meet a demand. Variables affecting these decisions include the magnitude of the alternative source, quality of the source, costs of implementing the alternative water supply system, and public acceptance of the source. These variables not only dictate which sources will be viable in general, but also which sources are viable for each water demand. DSSs that compile information and present the best potential alternative water strategies still require a final weighting based on human interpretation in order to rank the best connection scenarios.

In addition to flow connections and design components, water reuse and recycling strategies modify the movement and quality of water within the building water cycle. Certain sources may be established as acceptable for a set group of demands, but the wisest allocation method can depend on the volume of the source attainable, cost, and energy use based on the technology or strategy considered. A packaged wastewater treatment and recycling system may provide enough water to offset half of all sewage conveyance needs, but implementation of low-flow and waterless fixtures may accomplish the same goal at a lower initial and annual cost. Based on the efficiency standards pursued by the building, a compromise involving both strategies may help achieve higher performance goals. The opportunity to define and alter design components helps determine the best methods to achieve target goals and identifies specific design parameters required to ensure proper building operation. The computer and communication network within a smart building will require boundary conditions for individual systems to run, as well as triggers based on shared information. For example, a water equalisation tank installed as part of a rainwater collection system may need to have the pressure monitored to ensure the alternative water can be supplied to interior fixtures. In addition, the lowest water level allowed should be determined and programmed into the intelligent system to allow for the inclusion of makeup water when the supply is low. Emphasis on integration within smart buildings further encourages support that identifies water connection relationships and the relationships among the water subsystem and other building subsystems. For example, the operation parameters set for the cooling system will dictate the water bled from and added to the system in order to meet desired environmental conditions.

Even the best designed water system is still susceptible to fluctuations and environmental changes. Occupation by building inhabitants and visitors will create a dynamic and sometimes unexpected demand profile, thereby creating a need to estimate human behaviour effects on the system. Using DSSs, the water subsystem can be tested against a range of potential demand arrangements and magnitudes in order to verify flexibility and strength. The establishment of maximum and minimum loadings determines the limits of the designed building water cycle and can be re-evaluated under different design conditions. An estimation of water patterns is a prerequisite to accomplish these goals. Current use cycles can be described through the use of meters which can be implemented within the smart building framework. Using current information, projection scenarios developed using DSSs can prepare building owners for potential changes or upgrades to the system to meet future demands, building the adaptive capacity expected of a smart building.

19.4.3 Tools for building water management

The development and use of support tools aimed at the building water cycle are limited (Table 19.2). Although still scarce, research on DSSs focusing on sustainable water management at larger scales has produced more detailed and integrated frameworks. For example, Chamberlain *et al.* (2014) presents a DSS prototype capable of evaluating the environmental, economic, and social effects for sustainable wastewater strategies at the community level. The inclusion of impacts beyond measurable water use is an important component often lacking when the scope is narrowed to building structures and further limited to the water subsystem.

The trend for building-specific water support tools consists of calculators that track estimated water consumption, and thereby view the building water subsystem as a series of divided inflows. These water use calculators are prevalent online, with many published by organisations linked to water awareness and conservation. Homeowners are the main audience for simple calculators; current design patterns or human habits are exposed by informing water users of their water consumption habits. The most basic calculators use estimated volumes and flows for water demand applications and allow users to fill in the number of times each application occurs within a given time frame. For example, a user may be asked how often laundry is done or how often a bath is taken during a week. The input parameters provided by the user are fed into equations that produce the amount of water used by the individual either by water sector, all household activities, or both. The time frame may also be changed to reflect daily, weekly, monthly or annual usage. These tools focus on water consumption by demand and are generally not concerned with alternative water sources.

Support tools that incorporate water reuse and recycling or relationships to energy and costs are generally separated from software addressing the entire building. In the case of rainwater, some calculators consider annual precipitation that meets a portion of the irrigation demand, while other programs provide the option for rainwater collection, storage, and use for landscaping or interior building water demands. However, it is easier to find calculators specifically programmed around the design of a rainwater storage and collection system. Some allow users to input specific parameters regarding their building footprint and potential collection area, resulting in the maximum possible volume of rainwater that could be collected. Other tools incorporate storage and cost components to provide better information to users. Calculators developed around a specific water component, such as cooling or irrigation, tend to include a higher level of detail used to model water use for that demand. The Leadership in Energy and Environmental Design (LEED) series of rating systems produced by the USGBC includes calculations outlined to determine water reductions for landscaping and interior building fixtures (USGBC, 2013a–c; USGBC, 2012). As a green building rating system, alternative water supplies are incorporated as strategies to offset potable water demands. However, the current system still relies on a budget approach where water volumes for demands are

Table 19.2 Building water support tools.

Name	Scale, software	Alternative sources	Description
Water Footprint Calculator (National Geographic, 2013)	Single residential, web-based	Natural rainwater for landscaping	Daily water use calculated based on household water consumption, personal diet considerations, energy consumption, and consumer spending
WECalc (Pacific Institute, 2010)	Single residential, web-based	Greywater for landscaping	Extensive questionnaire that calculates total water demand, hot water demand, energy demand, and carbon footprint by end use; also includes costs associated with energy use. Suggestions that reduce energy and water are presented with benefits, costs, and payback periods.
HouseWater Expert (CSIRO, 2004)	Single residential, web-based	Rainwater storage, greywater diversion, on-site treated wastewater	Water consumption, wastewater generation, and runoff amounts are calculated for Australian regions based on a graphical platform that allows users to choose water demands and sources found both inside and outside of the building structure. The tool includes options for alternative water sources. Wastewater applications are limited to landscaping and toilet flushing.
Assessment tool (Fidar *et al.* 2010)	Single residential, basic tool	Greywater reuse, rainwater harvesting	Water consumption, energy use, and greenhouse gas emissions are calculated and compared for 8000 scenarios for an average residence in England. Interior micro-components (water demands) are varied.
WaterSmart Scenario Builder (POLIS, 2010)	Community, spreadsheet	Input for undisclosed non-potable source (rainwater, greywater, wastewater recycling)	Impacts associated with future water use scenarios for a community are estimated. The community is broken down into residential, commercial and institutions, industrial, agricultural, and non-revenue sectors. Users view impacts of chosen water efficiency scenarios in terms of water, energy, and greenhouse gas emissions reductions.
IBWM Model for Green Building (Joustra, 2010)	Generic building, STELLA model (iSee systems)	Rainwater, stormwater, recycled wastewaters, reclaimed water	The model provides users with the ability to analyse the effects that water management options have on a building's water cycle. Various building types are evaluated by changing demand portfolios. Alternative water supplies are incorporated. All demands and sources are networked.

tallied and compared to available water sources; alternative sources are subtracted from the total demand to determine the total potable water needed by water sector and the percent reduction.

The thoroughness and amount of information both received from and presented to the user dictates the amount of options the user perceives. Calculators that present water usage by sector allow the user to view areas of highest consumption and decide whether design or habitual changes can alter the usage patterns. The user is increasingly exposed to parameters affecting the water cycle when DSS tools require more information from the user. Exposure to alternative water sources and demands that can be met by those sources can open the design possibilities available to the user. The fragmented nature of tools that address alternative water supply systems and links to energy and cost hinders the potential for decision-making based on integration. A systems approach allowing for the complete interaction among water sources and demands while identifying the affects to other subsystems will result in DSS tools that are robust and flexible.

19.4.4 Incorporating IBWM into smart building DSSs

Existing DSS tools specifically addressing water within buildings contain deficiencies that limit their implementation potential. Easily accessible programs addressing building water use are often directed at the residential level, although all building types exert a water demand. The models assume a limited variety of building systems and lack the ability to accommodate buildings with different occupant loadings. Models also tend to separate building demands and focus only on water consumption when wastewater generation is an integral part of the water subsystem. Inclusion of alternative water sources is extremely limited; even when sources are available, demand applications are controlled. Smart building and IBWM share concepts related to systems integration, and the intelligence of buildings can be enhanced with DSS models that combine IBWM practices. Smart buildings recognise relationships among building subsystems and aim to manage the building as a coordinated system, whereas IBWM accomplishes the same goal at the building water subsystem level.

The perception of the building water subsystems should be similar to that of the hydrologic cycle, where all outputs are potential inputs for other components. In this view, wastes become potential resources. DSSs utilising an IBWM approach should monitor all inflows and outflows from each water demand and note the change in water quality that occurs. Water quality parameters affect how sources will be allocated by the user, and whether decisions regarding treatment or disposal will be made. All potential water demands and sources should be allowed to interact in order to fully incorporate all potential water cycle arrangements and easily alter connections to create new configurations. More options built into the DSS result in more possibilities for the user to investigate and allow for models that cover conventional water cycles to potential compositions that result in a net-zero water structure.

Inputs fed into IBWM DSSs should allow for flexibility. Tools capable of modelling different building types with various water demands, sources, and flow magnitudes decrease the development of repetitive models which can be costly and time-consuming. Options presented to users allow for comparisons between different building types or variations of the same building type to be made. This flexibility also allows modelling of future scenarios, such as company growth, space utilisation changes, or building additions. Additional scenarios can be created that evaluate the adaptive capacity of the building water cycle to short-term or long-term changes. An example of a short-term stressor is the loss of a water supply source due to a pipe break, whereas decreased precipitation events due to drought conditions is an example of a long-term event.

IBWM and smart building operations both benefit from monitoring equipment and sensors. Incorporating submetering practices provides information about water usage and operation parameters that can be fed into DSSs. With respect to IBWM, submetering assesses whether water cycle design goals are being met by logging information about the amount of water directed towards specific fixtures and applications (Tamaki *et al.* 2001). This data accounts actual water use within the subsystem which can be compared to the expected amounts estimated from support models. The resolution resulting from submetered water systems aids building operators in tagging inefficiencies in the system. Usage patterns captured by the monitoring system can also be used to improve modelling of the building in DSSs and produce better results when evaluating future scenario projections.

19.5 CONCLUSION

Decision support systems are powerful tools that organise and present information to users for improving the quality and effectiveness of decision-making; however, the development of DSSs addressing the intricacy of the building water cycle is limited. Building-level DSSs regarding the building water cycle should follow the concepts of IBWM and:

- recognise potential water demands, sources, and the connections between them,
- incorporate the use of alternative water supply systems,
- simulate building water cycles for multiple building types and buildings of different magnitude,
- be dynamic,
- project outputs based on input scenarios,
- consider effects on related subsystems, and
- enhance building automation procedures.

Smart buildings encourage increased efficiency and adaptability of building systems, thereby creating a demand for buildings that are flexible and dynamic. Incorporating water reuse and recycling systems within the building water cycle assists in achieving these goals. Inclusion of alternative water supplies to meet non-potable water demands increases the efficiency of potable water use and protects potable sources. The increased water use efficiency also allows the building water cycle to better adapt to changes in potable water availability, whether due to varying natural or regulatory conditions, and to changes within the building, such as fluctuating occupancy and behaviour. The complexity and dynamic nature of the building water cycle means frequent decisions are required regarding (re) design and operation. DSSs should be used to efficiently determine optimum design parameters and to adeptly direct building automation operations. Operating parameters (e.g., irrigation schedules, storage volumes, overflow triggers, treatment specifications, cooling tower cycles) can be determined based on outcomes from decision support tools, and information collected from smart building computer monitoring should form the basis for DSS inputs. Finally, DSSs should increase the intelligence of smart buildings, and smart buildings should be flexible enough to support the adaptation of integrated alternative water systems.

REFERENCES

Al-Jayyousi O. R. (2003). Greywater reuse: towards sustainable water management. *Desalination*, **156**(1–3), 181–192.

Anderson J., Adin A., Crook J., Davis C., Hultquist R., Jimenez-Cisneros B., Kennedy W., Sheikh B. and Van der Merwe B. (2001). Climbing the ladder: A step by step approach to international guidelines for water recycling. *Water Science Technology*, **43**(10), 1–8.

Asano T. (2002). Water from (waste)water – the dependable water resource. *Water Science Technology*, **45**(8), 23–33.

ASHRAE (2013). BACnet. http://www.bacnet.org/ (accessed 31 May 2013).

Atasoy E., Murat S., Baban A. and Tiris M. (2007). Membrane bioreactor (MBR) treatment of segregated household wastewater for reuse. *Clean – Soil, Air, Water*, **35**(5), 465–472.

Boehler M., Joss A., Buetzer S., Holzapel M., Mooser H. and Siengrist H. (2007). Treatment of toilet wastewater for reuse in a membrane bioreactor. *Water Science and Technology*, **56**(5), 63–70.

Browning-Aiken A., Ormerod K. J. and Scott C. A. (2011). Testing the climate for non-potable water reuse: opportunities and challenges in water-scarce urban growth corridors. *Journal of Environmental Policy and Planning*, **13**(3), 253–275.

Campbell A. C. and Scott C. A. (2011). Water reuse: policy implications of a decade of residential reclaimed water use in Tucson, Arizona. *Water International*, **36**(7), 908–923.

Carter T. and Fowler L. (2008). Establishing green roof infrastructure through environmental policy instruments. *Environment Management*, **42**(1), 151–164.

Chamberlain B. C., Carenini G., Öberg G., Poole D. and Taheri H. (2014). A decision support system for the design and evaluation of sustainable wastewater solutions. *IEEE Transactions on Computers*, **63**(1), 1–14.

City of Toronto (2013). Green Roof Bylaw. http://www.toronto.ca/greenroofs/overview.htm (accessed 31 May 2013).

Clerico E. (2007). The future of water reuse in NYC. http://allianceenvironmentalllc.com/pdfs/AE_NYC_Incentives_White_Paper_final.pdf (accessed 16 July 2014).

CSIRO (2007). House Water Expert Software. http://www.csiro.au/products/House-Water-Expert.html (accessed 29 May 2013).

Dolnicar S., Hurlimann A. and Grün B. (2011). What affects public acceptance of recycled and desalinated water? *Water Research*, **45**(2), 933–943.

Echelon Corporation (2013). LonWorks. http://www.echelon.com/technology/lonworks/ (accessed 31 May 2013).

Fidar A., Memon F. A. and Butler D. (2010). Environmental implications of water efficient microcomponents in residential buildings. *Science of the Total Environment*, **408**(23), 5828–5835.

Finley M. R., Karakura A. and Nbogni R. (1991). Survey of intelligent building concepts. *IEEE Communications Magazine*, **29**(4), 18–23.

Flax B. M. (1991). Intelligent buildings. *IEEE Communications Magazine*, **29**(4), 24–27.

Ghisi E. and Ferreira D. F. (2007). Potential for potable water savings by using rainwater and greywater in a multi-storey residential building in southern Brazil. *Building and Environment*, **42**(7), 2512–2522.

Guz K. (2005). Condensate water recovery. *ASHRAE Journal*, **47**(6), 54–56.

Hauber-Davidson G. (2007). Large commercial rainwater harvesting projects. In *Proc. of the 13th IRCSA Conf. on Rainwater and Urban Design*, Sydney, 20–23 August 2007.

Howell J. A. (2004). Future of membranes and membrane reactors in green technologies and for water reuse. *Desalination*, **162**(10), 1–11.

Jamrah A., Al-Omari A., Al-Qasem L. and Ghani N. A. (2006). Assessment of availability and characteristics of greywater in Amman. *Water International*, **31**(2), 210–220.

Jefferson B., Palmer A., Jeffrey P., Stuetz R. and Judd S. (2004). Grey water characterisation and its impact on the selection and operation of technologies for urban reuse. *Water Science and Technology*, **50**(2), 157–164.

Joustra C. (2010). An Integrated Building Water Management Model for Green Building. M.Sc. thesis, Dept Civil & Environ. Eng., Univ. of South Florida, Tampa, United States.

Katz D. (2012). Building intelligence quotient 2.0 development update. Presented at *High-Performance Buildings Conf.*, Falls Church, 11–12 September.

Lazarova V., Levine B., Sack J., Cirelli G., Jeffrey P., Muntau H., Salgot M. and Brisaud F. (2001). Role of water reuse for enhancing integrated water management in Europe and Mediterranean countries. *Water Science and Technology*, **43**(10), 25–33.

Licina D. and Sekhar C. (2012). Energy and water conservation from air handling unit condensate in hot and humid climates. *Energy and Buildings*, **45**, 257–263.

Ludwig A. (2006). Create an Oasis with Greywater, 5th edn, Oasis Design, Santa Barbara.

Montalto F., Behr C., Alfredo K., Wolf M., Arye M. and Walsh M. (2007). Rapid assessment of the cost-effectiveness of low impact development for CSO control. *Landscape Urban Planning*, **82**(3), 117–131.

National Geographic Society (2013). Change the Course: Reducing Water Use. http://environment.nationalgeographic.com/environment/freshwater/change-the-course/water-footprint-calculator/ (accessed 29 May 2013).

Pacific Institute (2010). WECalc: Your Home Water-Energy-Climate Calculator. http://www.wecalc.org/ (accessed 31 May 2013).

POLIS Project on Ecological Governance. (2010). WaterSmart Scenario Builder. http://poliswaterproject.org/publication/362 (accessed 29 May 2013).

Power D. J. (2002). Decision Support Systems: Concepts and Resources for Managers. Quorum Books, Westport.

Smart Buildings Institute (2013). Overview. http://www.smartbuildingsinstitute.org/ (accessed 31 May 2013).

SmartBuildings, LLC (2012). Smart Building FAQs. http://www.smart-buildings.com/faqs.html (accessed 31 May 2013).

Snoonian D. (2003). Smart Buildings. *IEEE Spectrum*, **40**(8), 18–23.

Solomon H. and Smith H. H. (2007). Effectiveness of mandatory law of cistern construction for rainwater harvesting on supply and demand of public water in U.S. Virgin Islands. In: *Proceedings Of the 7th Caribbean Island Water Resources Congress*, St. Croix, 25–26 October.

Sorgini L. (2004). Packaged membrane systems offer cost-effective solutions. *Journal American Water Works Association*, **96**(11), 48–50.

Tamaki H., Silva G. and Goncalves O. (2001). Submetering as an element of water demand management in water conservation programs. In *Proceeding of the 27th International Symposium on Water Supply and Drainage for Buildings*, Portoroz, 17–20 September.

USEPA (2013). Water Sense: Outdoor Water use in the United States. http://www.epa.gov/WaterSense/pubs/outdoor.html (accessed 31 May 2013).

USEPA (2009). Buildings and Their Impact on the Environment: A Statistical Summary. http://www.epa.gov/greenbuilding/pubs/gbstats.pdf (accessed 31 May 2013).

USGBC (2013a). LEED 2009 for Existing Buildings Operations and Maintenance Rating System http://www.usgbc.org/sites/default/files/LEED%202009%20Rating_EBOM-GLOBAL_07-2012_8d_0.pdf (accessed 31 May 2013).

USGBC (2013b). LEED 2009 for New Construction and Major Renovations Rating System. http://www.usgbc.org/sites/default/files/LEED%202009%20RS_NC_04.01.13_current.pdf (accessed 31 May 2013)

USGBC (2013c). LEED for Homes Rating System. http://www.usgbc.org/sites/default/files/LEED%20for%20Homes%20Rating%20System_updated%20ApriA%202013.pdf (accessed 31 May 2013).

USGBC (2012). Water Use Reduction Additional Guidance. http://www.usgbc.org/Docs/Archive/General/Docs6493.pdf (accessed 31 May 2013).

Van Woert N. D., Rower D. B., Andersen J. A., Rught C. L., Fernandez R. T. and Xiao L. (2005). Green roof stormwater retention: effects of roof, surface, slope, and depth. *Journal of Environment Quality*, **34**(3), 1036–1044.

Wintgens T., Melin T., Schäfer A., Khan S., Muston M., Bixio D. and Thoeye C. (2005). The role of membrane processes in municipal wastewater reclamation and reuse. *Desalination*, **178**(1–3), 1–11.

Wong J. K. W., Li H. and Wang S. W. (2005). Intelligent building research: a review. *Automation in Construction*, **14**(1), 143–159.

Chapter 20

A blueprint for moving from building-scale to district-scale – San Francisco's non-potable water programme

Paula Kehoe, Sarah Rhodes and John Scarpulla

20.1 INTRODUCTION

The San Francisco Public Utilities Commission (SFPUC) is a department of the City and County of San Francisco, California, that provides retail drinking water and wastewater services to San Francisco, green hydroelectric and solar power to San Francisco's municipal departments, and wholesale water to 27 cities, water districts, and private utilities within three neighbouring counties. The SFPUC operates the Regional Water System, which delivers water from the Hetch Hetchy watershed in Yosemite National Park west to San Francisco, serving over 2.6 million customers along the way. In 2002, the SFPUC launched a $4.6 billion Water System Improvement Programme to repair, replace, and seismically upgrade the system's deteriorating pipelines, tunnels, reservoirs, pump stations, storage tanks, and dams. In 2008, the SFPUC adopted a goal of developing an additional 38,000 m^3 per day of locally available water resources in lieu of importing additional drinking water from the Regional Water System. This includes 'active' conservation and the development of local groundwater and recycled water sources.

Concurrently, developers and designers have shown increasing interest in incorporating innovative on-site non-potable water use systems, such as treating greywater for toilet flushing or using rainwater for spray irrigation, into their projects. The main hurdle they identified was the lack of direction for the on-site use of alternative water sources, including which agencies have authority over such systems and what are the regulations for implementing on-site reuse. Before 2013, the state of California only had regulations for municipally-supplied recycled water from publically owned treatment works (POTWs) and for on-site untreated greywater used for subsurface irrigation on a residential property. Recycled water is regulated at the state level through the California Department of Public

Health (CDPH) and the Regional Water Quality Control Boards (RWQCB), while greywater was included in the California Plumbing Code and regulated by local county or city health and building officials. Therefore, developers trying to navigate the regulatory pathway were sent to multiple municipal and State departments and were left confused on the next steps. What permits are required, who issues the permits (state, local), and what water quality is required to reuse on-site sources to help new buildings dramatically lower their potable water consumption? The SFPUC saw this as an opportunity and joined together with the City's Departments of Public Health (SFDPH) and Building Inspection (SFDBI) to:

- Develop a streamlined local regulatory pathway to permit on-site non-potable water systems and provide for on-going monitoring and reporting to protect both public health and the public water system; and
- Permit and install a blackwater recycling system at the new SFPUC headquarters.

20.2 ALTERNATIVE WATER SOURCES AND END USES AVAILABLE ON-SITE

20.2.1 Alternative water sources

There are a number of on-site alternative water sources available for treatment and reuse. The alternative water sources discussed here are defined as follows:

- *Rainwater* – precipitation collected from roof surfaces or other manmade, aboveground collection surfaces;
- *Stormwater* – Precipitation collected from grade or below grade surfaces;
- *Foundation Drainage* – nuisance groundwater that is extracted to maintain a building's or facility's structural integrity and would otherwise be discharged to a sewer system. Foundation drainage does not include non-potable groundwater extracted for a beneficial use that is subject to groundwater well regulations;
- *Greywater* – untreated wastewater that has not been contaminated by any toilet discharge, has not been affected by infectious, contaminated, or unhealthy bodily wastes, and does not present a threat from contamination by un-healthful processing, manufacturing, or operating wastes. 'Greywater' includes, but is not limited to, wastewater from bathtubs, showers, bathroom sinks, clothes washing machines, and laundry tubs, but does not include wastewater from kitchen sinks or dishwashers;
- *Blackwater* – wastewater containing bodily or other biological wastes, as from toilets, dishwashers, kitchen sinks and utility sinks.

20.2.2 Non-potable end uses

As stated previously, the focus of the SFPUC's programme is on non-potable water end-use applications. The major non-potable applications include: toilet and/or urinal flushing, irrigation, cooling/heating applications, decorative fountains and water features, dust control and soil compaction, and process water. In California, rainwater is allowed for use in commercial and residential clothes washing while only municipal recycled water is allowed for commercial clothes washing. Specific source and end use application regulations or alternatives such as this add to the confusion around allowed uses and regulations.

20.3 WATER USE REDUCTION

To help calculate a building's total water use and the potential on-site alternative water sources available by volume, the SFPUC developed a Water Use Calculator (San Francisco Public Utilities Commission, 2014a) that allows developers to input basic information about their building and generate estimates of on-site supply availability and non-potable water demand. Using the calculator, the SFPUC has estimated that the combined reuse of rainwater and greywater in a typical office building for toilet flushing can offset approximately 65% of the indoor potable water use in the building, and potentially more if more abundant foundation drainage or blackwater supplies are available and utilized. In a typical multi-family residential building, greywater reuse for toilet flushing can offset approximately 22% of the indoor potable water use in the building. Note that the percentages assume minimal irrigation and no cooling demand for San Francisco buildings. A larger percentage reduction in potable water use is observed for office buildings because toilet flushing makes up the large majority of all water demands. The percentage reduction is lower in residential buildings; however, the volume of potable water offset can be much greater since residential buildings use much more water overall. Table 20.1 below provides examples of the range of potable water offset potential based on building size and alternative water source.

Table 20.1 Potential potable water offset by building type.

Building type	Alternative water supply used	Potable water offset (m^3 per year) (% reduction in total water use)			
		Building size (m^2)			
		3715	9290	18,580	46,450
Office	Rainwater & greywater	413 (77%)	760 (74%)	980 (61%)	1336 (46%)
	Foundation drainage	450 (78%)	1079 (78%)	2127 (78%)	5270 (77%)
Mixed use development	Greywater or foundation drainage	662 (22%)	1605 (22%)	3183 (21%)	7911 (21%)

Implementation of on-site water use would result in a considerable decrease in a building's water and sewage bill. On average, implementing on-site non-potable reuse for toilet flushing and irrigation demands can potentially reduce potable water consumption anywhere from 20% to 75% when compared to a building of similar size and use.

20.4 GREEN BUILDING MOVEMENT AS A DRIVER FOR ON-SITE NON-POTABLE WATER USE

Drivers such as regulations requiring sustainable site and building development and the growth in public awareness of green technologies are elevating the appeal of green building to owners and tenants. Increasingly, developers are seeing the value in pursuing the United States Green Building Council's (USGBC) Leadership in Energy and Environmental Design (LEED) credit programmes to validate and quantify each project's environmental attributes (United States Green Building Council, 2005). The USGBC is working with the commercial real estate market to promote the economic advantages of green building and its implementation in new and existing buildings and tenant spaces. There is a section in the USGBC LEED Reference Guide for Building Design and Construction (2009 Edition) that describes the benefits associated with non-potable reuse, specifically in Water Efficiency Credit 2 (WEc2): Innovative Wastewater Technologies.

In 2008, San Francisco implemented its ground-breaking Green Building Ordinance (SFGBO) for newly constructed residential and commercial buildings, and major renovations to existing buildings. All new construction in San Francisco must meet California's Green Building Standards Code (Cal Green), exceed California's energy code requirements by at least 15% and provide on-site facilities for recycling and composting. New high-rise residential and many common types of new non-residential buildings (such as office, retail, assembly, and institutional buildings) over 2320 m^2, as well as certain major alterations and first time tenant improvements, must also be built to LEED Silver and municipal buildings must be certified to LEED Gold (which has more requirements than LEED Silver). Also, the San Francisco Stormwater Design Guidelines, codified in 2010 through a City ordinance, requires all new development and redevelopment projects disturbing 465 m^2 or more of the ground surface to meet the requirements of either LEED Sustainable Sites credits 6.1 or 6.2 (SSc6.1 or SSc6.2).

The implementation of non-potable on-site reuse augments the environmental objectives of a green building by reducing potable water consumption and the amount of stormwater and wastewater runoff to the City's Water Pollution Control Plants (WPCPs) for treatment. Non-potable reuse is listed as one of the three Water Efficiency strategies in the LEED rating system and projects that can either demonstrate 100% reduction of potable water consumption or 100% on-site reuse or infiltration of generated wastewater can be awarded an Innovation credit for exemplary performance that would add to the LEED credits and increase the green marketing profile of the building. Rainwater harvesting systems can also be implemented to meet the credit requirements for LEED SSc6.1 and LEED SSc6.2. A summary of the LEED Credit potential is shown in Table 20.2.

Table 20.2 LEED Credit potential through on-site reuse.

LEED Credit	Description	Points available
Prerequisite, Water Use Reduction	• Employ strategies that in aggregate use 20% less water than the water use baseline calculated for the building (not including irrigation)	Required
WEc1, Water Efficient Landscaping	• Reduce potable water consumption for irrigation by 50% from baseline • Use only non-potable water for irrigation	4
WEc2, Innovative Wastewater Technologies	• Reduce potable water use for building sewage conveyance by 50% through the use of water-conserving fixtures or non-potable water • Treat 50% of wastewater on-site to tertiary standards. Treated water must be infiltrated or used on-site	2
WEc3, Water Use Reduction	• Employ strategies that in aggregate use at least 30% less water than the baseline	4
SSc6.1, Stormwater Design – Quantity	• For sites with existing imperviousness greater than 50%, implement measures that will decrease the volume of stormwater runoff from a 2-year 24-hour design storm by 25%	1
SSc6.2, Stormwater Design – Quality	• Capture and treat the rainfall from a design storm of 19.05 mm	1
Innovation in Design	• Demonstrate 100% on-site reuse or infiltration of generated wastewater to receive credit for exemplary performance.	1
Regional Priority Credits	• Incentivize achievements of credits that address geographically specific environmental priorities. In San Francisco, bonus points are awarded for achieving full points under WEc2 and WEc3	2
Total potential points that can be achieved by implementing on-site non-potable reuse		*15*

Source: United States Green Building Council (2005).

While LEED is the most prominent green building guidance and accreditation process in the United States, there are other programmes such as The Living Building Challenge™ in which projects must meet a series of ambitious performance requirements, including net zero energy, waste, and water over a minimum of 12 months of continuous occupancy. Programmes like this are increasing pressure for greater regulatory action as they are including on-site alternative water sources for potable water uses to facilitate achievement of zero net potable use.

20.5 CURRENT REGULATION OF ALTERNATIVE WATER SOURCES

In the United States, there are no overarching national standards for water quality and wastewater treatment. There are a number of guidance documents and policies that exist, including but not limited to:

- *National laws* – such as the Clean Water Act which established the National Pollutant Discharge Elimination System (NPDES) for permitting wastewater treatment plants,
- *National guidelines* – such as the Guidelines for Water Reuse developed by Environmental Protection Agency (EPA), and
- *International codes* – such as those developed by the International Code Council (ICC) or the International Association of Plumbing and Mechanical Officials (IAPMO).

However, each state establishes their own interpretation of the laws, guidelines, and codes to develop state regulations which can then be further refined when setting water quality numerical limits in individual project permits.

California, like most states within the United States, uses IAPMO's Uniform Plumbing Code (UPC) as the model for the state's plumbing code. While many states adopt the UPC directly, California adopts the UPC with amendments to create the California Plumbing Code (CPC) which is typically more restrictive based on California's laws and specific geographic, environmental, or other regulatory concerns.

The 2012 UPC expanded code language by adding Chapter 16: Alternative Water Sources for Non-potable Applications, incorporating a number of alternative water sources, including some of those defined in Section 20.2.1, and others like cooling tower blow down and boiler condensate. Following the release of the 2012 UPC, California began its tri-annual code revision cycle to incorporate UPC updates into the 2013 CPC. While California has yet to adopt the full range of alternative water sources, the 2013 CPC includes and/or expands regulations for greywater and rainwater for multiple end-use applications in both residential and commercial occupancies. Prior to 2013, the state of California regulations only covered two types of non-potable water: 1) municipally-supplied recycled water that is collected, treated, and distributed by an NPDES permit holder;

2) residential greywater that is collected and distributed on-site for subsurface irrigation.

The 2013 CPC is moving in the right direction, however a gap in regulation still exists; namely, the ongoing operation and maintenance of alternative water source systems to ensure the protection of public health and the public water system post-construction. Building codes, including the plumbing code, are generally enforced at the time of construction and are not intended to mandate on-going operation and maintenance. Ensuring proper monitoring and compliance during continued operation was a major concern raised by municipal and state agencies during the code update process.

20.6 WORKING TOGETHER – A THREE-PRONGED APPROACH TO COLLABORATION

The SFPUC spearheaded an effort to create a local programme for regulating on-site water use, the Non-potable Water Programme, which was codified in September 2012 through a City ordinance. When developing a regulatory framework for onsite water reuse, it is critical to have the participation of all applicable entities so that all parties feel ownership for the programme. In San Francisco, this task involved coordination between three City agencies to provide a streamlined permitting, review, and approval process for on-site system installation and operation. SFPUC staff worked with the City's Departments of Building Inspection and Public Health to develop a regulatory pathway to approve alternative water source projects. Below are general descriptions of each agency's role; additional information is available in the ordinance (Chiu & Mar, 2013).

San Francisco Department of Building Inspection (SFDBI): SFDBI oversees construction and inspection within San Francisco. As part of the programme, developers installing on-site non-potable water systems need to submit plans to SFDBI for plan review and the issuance of plumbing permits for construction. In addition, SFDBI provides the final inspection to ensure the system was constructed in accordance with the submitted plans and all applicable federal, state, and local buildings codes.

San Francisco Department of Public Health (SFDPH): SFDPH oversees the protection of public health. As part of the programme, SFDPH reviews a Non-potable Water Engineering Report prepared by the system applicant that describes the on-site non-potable water system, including collection, distribution, and appropriate treatment. SFDPH prescribes water quality criteria in accordance with state code and has developed requirements for alternative water sources that are not addressed at the state level. SFDPH permits the systems through a 'Permit to Operate' and an annual licensing fee. This includes provisions for on-going monitoring of water quality and reporting requirements for the system operator. The Permit to Operate involves 3 phases post-construction: start-up, temporary use, and final use.

San Francisco Public Utilities Commission (SFPUC): The SFPUC serves in two primary capacities: programme administration and cross connection control. Programme administration involves providing outreach and technical assistance to developers as well as tracking on-site non-potable water use projects and quantifying potable water offset potential. The SFPUC administers non-potable water audits to dialogue with system operators on non-potable water use and ways to further improve outreach and assistance efforts. Cross connection control involves protecting the public water supply and includes backflow prevention and cross-connection control, testing, certification, and tracking.

20.7 WATER QUALITY REQUIREMENTS FOR ON-SITE NON-POTABLE SYSTEMS

Water quality criteria and monitoring and reporting requirements are the topics that generate the most questions and concern as projects move forward not only in San Francisco, but across the country. Through San Francisco's Non-potable Water Programme, SFDPH developed the *Director's Rules and Regulations Regarding the Operation of Alternate Water Source Systems* which established water quality criteria and permit requirements. The SFDPH rules and regulations are consistent with available state regulations and provide additional water quality criteria for alternative water sources or non-potable applications that are not addressed in the state code. For rainwater and greywater reuse, the local regulations are consistent with the 2013 CPC. For blackwater reuse, SFDPH is consistent with the bacteriological limits in California Code of Regulations [CCR] Title 22. For stormwater and foundation drainage, SFDPH has utilized existing regulations and staff expertise to bridge the gap and develop appropriate water quality criteria.

SFDPH also developed the monitoring and reporting regime to ensure proper operation of the on-site systems after construction. This includes a graduated monitoring regime based on source water and includes three phases: start-up, temporary operation, and final operation. The start-up phase includes treating the non-potable water and bypassing to the sewer while demonstrating the system's ability to meet the specified water quality criteria. The temporary phase has the system in full use, supplying non-potable water and verifying continued system performance. Lastly, the final use permit or annual license will be issued for continued operation. This allows for more extensive oversight initially as the system is first put into operation and the building staff are learning the ins and outs of the operation.

Most systems will be required to monitor system water quality through parameters such as turbidity, total coliform or *E.coli*, and chlorine residual – this would not apply to most subsurface irrigation systems. Stormwater and foundation drainage systems may be required to test for Volatile Organic Compounds (VOCs), determined on a case-by-case basis (San Francisco Department of Health, 2014). Annual reports are required for all systems, describing system operation and maintenance over the past year; providing the monitoring reports at a frequency prescribed by SFDPH is a condition of the permit to operate.

20.8 THE SFPUC AS A RESOURCE

The SFPUC provides outreach on the programme and both technical and financial assistance. The SFPUC developed a Water Use Calculator that helps developers estimate the volume of on-site non-potable supplies and demands available for their project based on general size, occupancy type and fixture rate, rainfall and evapotranspiration assumptions. This tool is publically available online (San Francisco Public Utilities Commission, 2014a) and allows developers to input basic information about their building and generate estimates of supply availability and non-potable water demand. In addition, the SFPUC created a developer's guidebook that provides an overview on alternative water sources, non-potable applications, and how to navigate the City regulatory landscape to successfully implement a project (San Francisco Public Utilities Commission, 2014b).

Lastly, the SFPUC Commission approved the Grant Assistance for Large Alternate Water Source Projects in June of 2012. The programme will provide up to $250,000 for projects implementing on-site non-potable water use. The grant programme has limited funding and is on a first-come, first-serve basis. To be eligible, a project must:

- Be 9290 m^2 or more of residential or commercial occupancy;
- Complete the SFPUC's Water Use Calculator;
- Replace at least 1,000,000 gallons of potable water per year of the project's potable water use.

The programme will allow the SFPUC to gather data on potable water offset, capital costs, operation and monitoring costs, and other important data to track the effectiveness of on-site non-potable water use. With the addition of the financial grant programme, the SFPUC hopes to gather valuable data to expand the programme and serve as a model for other jurisdictions grappling with the same tough topics – who should regulate/permit these systems and how should it be done?

20.9 ON-SITE NON-POTABLE REUSE AT THE SFPUC HEADQUARTERS

In 2009, the SFPUC began planning for the construction of a new, LEED Platinum headquarters to be located in San Francisco's Civic Center district – a dense, National Historic Landmark District, composed primarily of a collection of historic buildings and Beaux Arts civic architecture. During the early planning stages, implementing a water reuse system within the building became a primary goal for the SFPUC. However, at the time, the permitting process for a blackwater reuse system at the building-scale was not clearly defined, and there were uncertainties about which government agency would be charged with regulating the system.

20.9.1 Permitting the system

As stated previously, jurisdiction over the use of recycled water in California is shared by the State Water Resources Control Board (SWRCB), the Regional Water Quality Control Boards (RWQCBs), and the California Department of Public Health (CADPH). Under California Water Code, the SWRCB establishes general policies governing the permitting of recycled water projects consistent with its role of protecting water quality and sustaining water supplies. The SWRCB exercises general oversight over recycled water projects, including review of RWQCBs permitting practices. The CADPH is charged with protection of public health and drinking water supplies, and is statutorily required to establish uniform state-wide regulations for recycled water (Water Code Section 13521) (State Water Resources Control Board). The CADPH's regulations (in California Code of Regulations [CCR] Title 22, Division 4, Chapter 3, Section 60301 et seq.) provide specified approved uses of recycled water, numerical limitations and requirements, treatment method requirements, and performance standards. The CADPH also establishes permit conditions needed to protect human health (California Department of Public Health, 2009). The RWQCBs protect surface and groundwater resources by issuing permits that implement CADPH recommendations and applicable laws.

In 1996, the San Francisco RWQCB adopted Order No. 96-011, which authorizes 'Producers' and 'Distributors' to deliver recycled water for reuse by 'Users,' the expectation being that municipal agencies are responsible for producing and distributing recycled water (California Regional Water Quality Control Board San Francisco Bay Region, 1996). This order both implements the CADPH's public health protection criteria and imposes requirements designed to protect surface water and groundwater. By its terms, the order does not apply to individual, closed-looped treatment systems that produce recycled water for indoor uses, where the 'Producer,' Distributor,' and 'User' are one and the same – such as the SFPUC headquarters. Similarly, on-site treatment systems that do not discharge to land or water bodies are not addressed by current RWQCB policies or permits.

The authority of the CADPH to regulate internal uses of recycled water is established by current laws and regulations, as noted above. Furthermore, the CADPH may delegate all or part of the duties that it performs regarding the uses of recycled water within a county to a local health agency authorized by the board of supervisors to assume these duties, if, in the judgment of the CADPH, the local health agency can perform these duties (Water Code 13554.2) (State Water Resources Control Board).

Based on the above, the SFPUC and SFDPH proposed to the RWQCB and CADPH that the authority for regulating the water recycling system at the

SFPUC headquarters be the SFDPH, and not the RWQCB as is typical with most recycled water systems. The SFDPH is authorized to perform duties associated with regulating the internal uses of recycled water. Under Section 4.110 of the San Francisco Charter, the Health Commission and the Health Department have authority to provide for the preservation, promotion, and protection of the health of the inhabitants of the City and County (City and County of San Francisco, 1996).

Additionally, Articles 11 and 12A of the City's Health Code authorize the 'county health officer' to investigate and abate any nuisance, activity, or condition that the county health officer deems to be a threat to public health and safety, and to investigate and abate any cross-connection risks between potable and non-potable water and sanitation systems in both public and private facilities (City and County of San Francisco, 2008). The Health Code provides authority to order the vacating of property, the cessation of prohibited activities, the abatement of unsafe or unsanitary conditions, and the assessment and collection of penalties.

The Charter and the Health Code provide sufficient authority to the SFDPH to regulate on-site wastewater treatment and recycling systems and the use of non-potable water for appropriate purposes within buildings and structures located in the City, provided that these waters are not applied to land or water or have the potential to runoff to land or water.

Typically, state water quality criteria would apply to the on-site production and use of treated sewage; authority for such activities is not addressed by existing state permits and regulations. Therefore, the authorizing agency would default to the SFDPH. Delegating oversight to the City's county health officer fills that gap by providing meaningful local regulation and control.

The authorization approach for the on-site wastewater system at the site requires SFDPH to develop a regulatory programme and monitoring protocol to manage and control the on site production and on-site, internal use of reclaimed water. The SFDPH did so – creating a 'Permit to Operate' for SFPUC headquarters which detailed:

- Required effluent water quality values;
- Water quality monitoring and sampling regime;
- Reporting requirements;
- Operation and maintenance manual requirements; and
- Public notification requirements.

With the regulatory framework and permitting process established, the SFPUC now had the vehicles needed to install a blackwater recycling system at its new headquarters.

20.9.2 The treatment system at SFPUC headquarters

The energy and public education goals for the SFPUC's new headquarters made the selection of a low-energy, high visiblity water reuse system critical if it was to be included in the building's final design. After researching several types of systems, SFPUC staff recommended installing an ecological wastewater treatment and reuse technology. Specifically constructed wetland treatment systems that could be installed in the right-of-way surrounding the building. The challenge was fitting a treatment and reuse system serving 930 employees and approximately 150 visitors per day could be sited in a dense urban area and meet SFDPH's water quality requirements.

The SFPUC selected this technology because of its ability to blend function and aesthetics (Figure 20.1) – the system treats the building's wastewater to SFDPH reuse standards while providing aesthetically-pleasing wetlands installed in the right-of-way and building's lobby.

Figure 20.1 The ecological wastewater treatment and reuse system at SFPUC headquarters contains constructed wetlands that treat wastewater in the sidewalks surrounding the building.

20.9.2.1 Treatment process

The treatment system at SFPUC headquarters treats all of the building's wastewater, up to 19 m^3 per day, and then distributes the treated water throughout the building for toilet flushing purposes. On average, the system provides approximately 15 m^3 of water per workday for toilet flushing.

The treatment system utilizes a series of diverse ecologically engineered environments (Figure 20.2) for treatment. The treatment process begins with the building's wastewater being directed to a dual-chambered, 10,000 gallon primary treatment tank, where the solids are separated and the influent is clarified and screened. The first chamber, called the trash chamber, receives sewage flow and screens coarse solids from the waste stream. From there, the water flows to the settling chamber. This chamber separates settleable and floatable solids from the liquid that would cause problems in downstream processes. After passing through a

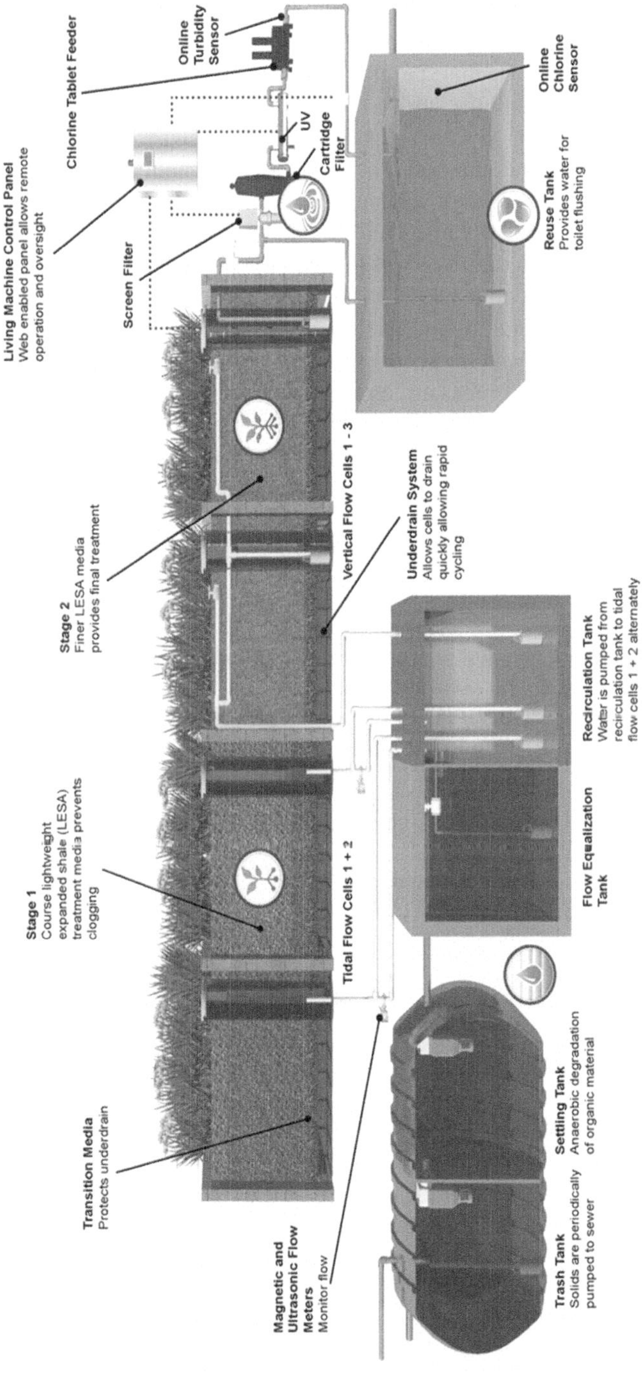

Figure 20.2 The process schematic for the treatment system at SFPUC headquarters (Living Machine Systems, L3C).

filtration strainer in the settling tank, the liquid stream enters an equalization tank. This tank allows for influent flows to be stored during periods when they exceed treatment capacity and to be dosed to the system when it is ready for more water.

The treatment process of the system operates best on a relatively steady influent flow pattern. Wastewater flows are equalized by pumping from the equalization tank into the recirculation tank over a period of 12 to 20 hours. The goal is to have maximum equalization volume available so that daytime flows, which typically are the majority of the influent volume, can be metered into the system.

From the recirculation tank, water is pumped to tidal flow wetland cells, located in the right-of-way, which contain engineered media, vegetation, and micro-organisms. These wetlands remove pollutants from wastewater using microbial biofilms attached to the wetland media (gravel). Through programmemed cycling of the water levels, the microbes are repeatedly introduced to the wastewater food supply, which cleans the water. Cells are filled from the bottom until the water level reaches a predetermined level, then drained by gravity back to the recirculation tank. The fill and drain process provides all of the oxygen necessary for high-treatment performance in a small footprint, without additional mechanical aeration. Cycle rates are adjusted based on influent loading, wastewater characteristics, and environmental conditions.

Table 20.3 Living Machine water quality – requirements and performance to date.

Parameter	SFDPH requirement	Average to date	Maximum to date
BOD_5 (influent to wetland)	N/A	220.4 mg/l	367 mg/l
BOD_5 (effluent)	<45 mg/l	4.9 mg/l	5.0 mg/l
Suspended Solids (influent to wetland)	N/A	68.2 mg/l	129 mg/l
Suspended Solids (effluent)	<45 mg/l	7.4 mg/l	12.5 mg/l[1]
pH	6.0–9.0	7.6	8.13
Turbidity	10 NTU	.53 NTU	2.90 NTU
Chlorine Residual	0.5–4.0 mg/l	1.2 mg/l	3.9 mg/l
Total Coliform (since reuse began)	7-day median: <2.2 MPN/100 ml 30-day average: <23 MPN/100 ml	1 MPN/100 ml	2 MPN/100 ml

[1]The lowest Suspended Solids value SFPUC labs can detect is 7 mg/l – the majority of drawn samples (93%) have a value of <7 mg/l; MPN – most probable number.

Water flows from the tidal flow wetland cells to the vertical polishing flow wetland cells, which are located in the right-of-way and inside the building lobby. These wetland cells remove any remaining organic material, ammonia nitrogen, and suspended solids. The polishing wetlands contain subsurface piping arrangement that supply the water for vertical flow through the treatment media. Water trickles down through the media and plant roots, and collects in the underdrain system. Periodically, the polishing recirculation pump pumps water from the underdrain back to the surface dosing manifold. The water is recirculated through the polishing wetland media to ensure that the effluent leaving it is of high quality which will require minimal disinfection for reuse.

From the polishing vertical flow wetlands, the water is pumped to the disinfection room where it goes through two filters – first a 100 micron screen filter, then a 5 micron cartridge filter. Filtering of the effluent is necessary to remove fine particles that will inhibit ultraviolet (UV) disinfection. After filtration, the water flows through UV disinfection, and then through a tablet (trichlor) Chlorinator. From there it is placed in a 19 m^3 reclaimed water storage tank, and then when needed, it circulates through the building for toilet flushing purposes.

20.9.2.2 Water quality results

SFPUC staff began water quality monitoring and sampling of the system's influent and effluent in September 2012, when the new headquarters became fully occupied. From September 4, 2012, to December 3, 2012, the system was operated in 'start-up' mode. This meant that the influent received primary and secondary treatment, but was then discharged to the municipal sewer system. This enabled the microorganism population in the wetlands to increase in size, while also allowing SFPUC operators to become comfortable with system operation. On December 4, SFPUC staff began disinfecting the water; however, the effluent was still discharged to the municipal sewer system. It was not until December 10, 2012, that water was first supplied to the building's toilets for reuse.

As required by SFDPH, SFPUC staff sampled the influent for BOD_5 and suspended solids, and the effluent for BOD_5, suspended solids, pH, turbidity, chlorine residual, and total coliform. Since the water started being used in the toilets, the system's effluent has never failed to meet the water quality requirements of the SFDPH. A summary of results is shown in Table 20.3.

The graphs below (Figures 20.3 to 20.5) provide more detailed water quality data for turbidity, Total Coliform and BOD_5 from September 4, 2012 to April 4, 2013. SFPUC staff will continue to monitor and sample influent and effluent throughout the life of the system, as required by SFDPH.

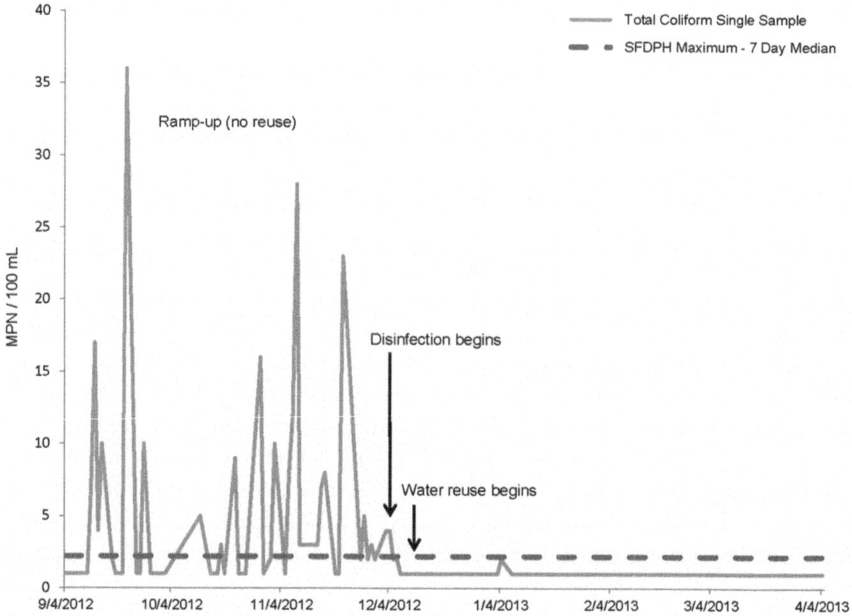

Figure 20.3 Total coliform sampling results – September 4, 2012–April 4, 2013.

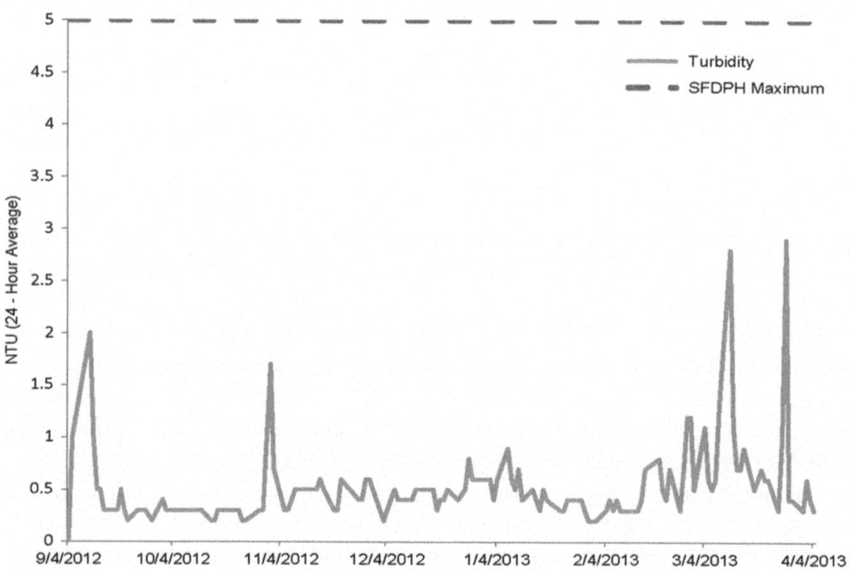

Figure 20.4 Effluent turbidity sampling results – September 4, 2012–April 4, 2013.

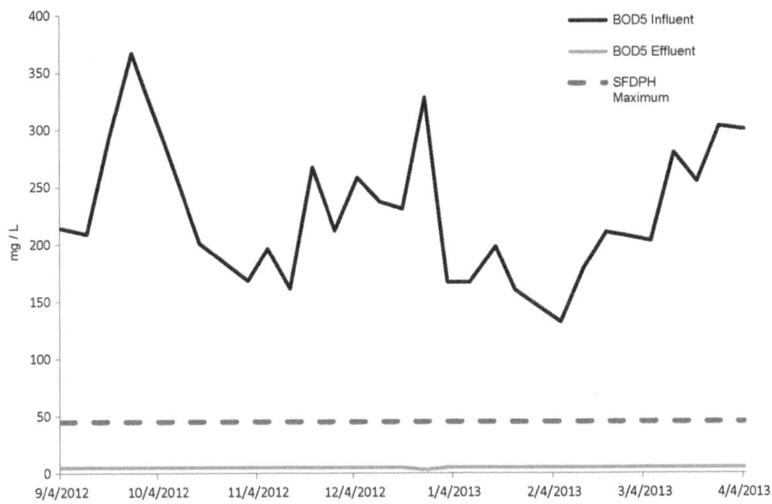

Figure 20.5 BOD_5 sampling results – September 4, 2012–April 4, 2013.

20.10 MOVING TOWARDS DISTRICT-SCALE WATER SHARING IN SAN FRANCISCO

Moving forward, the SFPUC is looking to expand on-site water use and researching district-scale water use, where 'district' is simply defined as more than one property pooling or sharing their non-potable water resources. The SFPUC is continuously meeting with developers that are proposing larger water reuse schemes that encompass multiple adjacent parcels within the City. Furthermore, the City's Planning Department is developing certain neighbourhoods of the City as 'Eco-districts', defined as a neighbourhood or district where neighbours, community institutions, and businesses join with city leaders and utility providers to meet ambitious sustainability goals and co-develop innovative projects. This also promotes the ideas of neighbourhood scale water resource solutions.

In 2013, the SFPUC completed a study that included a review of California law surrounding district-scale water sharing and detailed case studies from district-scale water reuse systems around the world. The research showed that private developers or parties may partake in district-scale water reuse provided that appropriate legal arrangements were established between the participating entities. Three of the legal issues around district-scale water sharing are discussed further below.

20.10.1 Crossing property lines

Code restrictions related to moving water across property lines have previously been cited as a major challenge to district-scale applications. However, any current code specifically limiting the location of water lines includes subsequent exception clauses that allow water to cross property lines if appropriate legal agreements or necessary easements are in place. This sets a precedent for collaborative water management across parcels. Some of the relevant Codes include:

- The California Plumbing Code (CPC), which does not address district-scale water use, focusing mainly on work within the property line (building-scale);
- Section 609.6 of the 2010 CPC, which states the potable supply from the meter to the building should not cross property lines, but allows exceptions with legal easements in place. This does not apply to non-potable water; and
- Section 1602.4 of the 2013 CPC, which requires greywater systems to stay on-site but includes an exception when there is a legal agreement between property owners.

20.10.2 Selling water and public utilities

Generally, the CA Public Utility Code does not define private entities or a group of private entities who collect and reuse water for their own purposes as a public utility and therefore does not regulate their practices. The law allows neighbours and adjoining land, as well as corporations or associations organized for to serve their members to operate without becoming public utilities. However, there is not a clear tipping point between 'a group of neighbours' and the 'public', therefore case-by-case review of district-scale proposals is required.

The two relevant public utility code sections are as follows:

- *Section 2704* – On-site water can be sold by owners to neighbours/adjoining lands without becoming a public utility.
- *Section 2705* – A corporation or association that is organized for the purposes of delivering water to its members or others at cost is not a public utility.

20.10.3 Water rights

Some states in the U.S. have laws prohibiting developing such sources due to potential impacts to downstream water rights holders. Specific to the City and County of San Francisco, there are no impacts on downstream water rights holders from water reuse, as most alternative water sources are collected into the City's combined sewer system and discharged into the San Francisco Bay or Pacific Ocean.

20.10.4 Next steps

In October 2013, the Mayor of San Francisco signed into law an amendment to the ordinance, which expanded the Non-potable Water Programme to include projects that go beyond 'on-site' and include multiple parcels. The SFPUC will work with developers to ensure their projects will operate in concert with on-going work being completed by the SFPUC, including low impact development (LID) stormwater management and recycled water planning.

The SFPUC is excited about the future potential of on-site reuse at the building and district scale. There are many developers interested in lowering water use and developing innovative systems for reuse. With a system in place to properly permit and oversee the safe construction and operation, the SFPUC hopes the interest among the development community continues to grow.

20.11 CONCLUSIONS

As populations increase, diversifying water portfolios with onsite water reuse systems will become critical for meeting future water demands. Creating a regulatory framework and streamlined permitting process for the installation and ongoing operation of these systems is critical if their implementation is going to be widespread and to fully ensure the continued protection of public health. San Francisco has provided a blueprint for jurisdictions throughout the world to follow to successfully implement an onsite water reuse programme. While the blueprint will need to be adapted by jurisdictions to fit their local needs, a key message is the importance of forming partnerships with all applicable stakeholders, including local health departments and private developers.

REFERENCES

California Department of Public Health. (2009). Regulations Related to Recycled Water. http://www.cdph.ca.gov/certlic/drinkingwater/Documents/Lawbook/RWregulations-01-2009.pdf (accessed 02 June 2014).

California Regional Water Quality Control Board San Francisco Bay Region (1996). Order 96-011 General Water Reuse Requirements for: Municipal Wastewater and Water Agencies. http://www.waterboards.ca.gov/sanfranciscobay/publications_forms/documents/orderno96-011.doc (accessed 13 May 2013).

Chiu D. and Mar E. (2013). Health, Business and Tax Regulations Codes – On-Site Water Reuse for Commercial, Multi Family, and Mixed-Use Developments. http://www.sfwater.org/modules/showdocument.aspx?documentid=4779 (accessed 13 May 2013).

City and County of San Francisco (1996). City and County of San Francisco Municipal Code – 1996 Charter. http://archive.org/details/gov.ca.sf.charter (accessed 15 May 2013).

City and County of San Francisco (2008). San Francisco Health Code. http://www.archive.org/stream/gov.ca.sf.health/ca_sf_health_djvu.txt (accessed 12 May 2013).

Living Machine Systems, L3C (2013). Living Machine Process Schematic.

San Francisco Department of Public Health (2014). Director's Rules and Regulations Regarding the Operation of Alternate Water Source Systems. http://www.sfdph.org/dph/files/EHSdocs/ehsWaterdocs/NonPotable/Rules-RegsJan2014.pdf (accessed 6 January 2014).

State Water Resources Control Board (2013). California Water Code. http://codes.lp.findlaw.com/cacode/WAT/1/d7/7/7/s13554.2 (accessed 14 May 2013).

San Francisco Public Utilities Commission (2014a). Non-potable Water Calculator. www.sfwater.org/np (accessed 24 August 2014).

San Francisco Public Utilities Commission (2014b). Non-potable Water Programme Guidebook. www.sfwater.org/np (accessed 24 August 2014).

United States Green Building Council (2005). LEED for New Construction & Major Renovations. http://www.usgbc.org/Docs/Archive/General/Docs1095.pdf (accessed 14 May 2013).

Chapter 21

The socio-technology of alternative water systems

Sarah Bell

21.1 INTRODUCTION

Water reuse and rainwater harvesting systems provide water system managers and water users with technical alternatives for supplying water under conditions of scarcity and uncertainty. In changing the scale, source and use of water, alternative water systems also challenge long standing social, institutional, economic and environmental relationships that support conventional modes of provision of water infrastructure services. Engineers and decision makers commonly consider infrastructure provision to be a technical problem to be solved to support economic development. A broader analysis of infrastructure systems shows that they have co-evolved with society, politics, local environments and other factors (van Vliet *et al.* 2005). Infrastructure systems might therefore be considered to be socio-technical systems (Marvin & Graham, 2001; Hughes, 1989). The growth of new modes of water provision, such as reuse and rainwater harvesting, is likewise contributing to the co-evolution of alternative patterns of relationships between people, water, technology and their environment. The extent to which these new patterns of provision are more or less sustainable than current infrastructure systems will be the outcome of system design, governance and social change.

Everyday use of water by people in homes, businesses and public places in most developed countries has evolved in the context of continuous supply of clean water, irrespective of the weather, state of the environment or availability of water resources in the local catchment. As water use and populations have increased, water infrastructure has expanded to meet demand. Limits to water resources and the cost of implementing new supplies have prompted renewed interest in alternative water systems, including decentralised rainwater harvesting and non-potable reuse. However, these systems are not necessarily inherently more sustainable than existing water infrastructure. Moreover, in urban areas in developed countries these systems are usually used to supplement, rather than replace, existing supplies. Consequently, it is important to understand how alternative water systems are positioned in relation to conventional socio-technical infrastructure systems.

Sustainability assessment of infrastructure is usually based on indicators that address social, economic and environmental impacts. A socio-technical approach can help to deepen sustainability assessment by addressing interactions between different elements of sustainability indicators, and widening the analysis to include governance, regulation and other contextual factors. Analysis of infrastructure as socio-technical systems by scholars from a range of fields, including geography, history, sociology and science and technology studies, contributes to the development of a theoretical framework that can serve the basis for a more critical evaluation of sustainability than is usually achieved through indicator based approaches. Whilst this analysis is largely qualitative, it serves to inform the development of alternative water systems by showing how they both challenge and reinforce conventional approaches to water and infrastructure.

This chapter presents a framework for assessing the sustainability of water systems, based on critical perspectives on infrastructure and the relationship between society and the environment. General trends in infrastructure provision are summarised before reviewing recent theoretical developments in understanding the socio-technical nature of infrastructure. This forms the basis of a framework for assessing and comparing the sustainability of different forms of water provision. The framework is applied to conventional water systems, potable reuse, district scale non-potable reuse and rainwater harvesting. For each system the key socio-technical elements are analysed in general terms and then applied to specific case studies from South East England, UK and South East Queensland, Australia. The analysis highlights opportunities and challenges to sustainability with alternative reuse systems, as they both potentially reconfigure and/or stabilise conventional relationships between water, technology, society and the environment.

21.2 INFRASTRUCTURE, SOCIETY AND THE ENVIRONMENT

Infrastructure systems, such as those for water, transport, energy and communications, underpin modern societies and economies. These systems are essentially technical, but their existence and functioning depends upon political, economic, social and natural environments. Infrastructures also fundamentally change the environments and societies in which they operate. Thus the relationships between infrastructure, technology, society and the environment can be characterised as co-evolutionary (Shove, 2004; van Vliet *et al.* 2005).

For much of the twentieth century infrastructure provision was considered a function of the state, with major utilities owned and operated by government authorities (Marvin & Graham, 2001). The provision of these services and the need to establish institutions to finance, build and operate infrastructure had considerable influence in shaping the nature of municipal governance, balancing democratic oversight with technical expertise (Ben Joseph, 2011; Melosi, 2008; Halliday, 1999). The public provision of infrastructure reflected its importance in underpinning economic development and growth, as well as the modern social imperative to connect all members of society to essential services to improve public health and standards of living (Marvin & Graham, 2001).

Infrastructure systems emerged on a model of centralised, universal provision. Centralised provision of services through large technical systems provided for levels of control and standardisation that were absent from earlier efforts to implement water and energy systems (Hughes, 1985). Public financing of large systems of universal provision also recognised the social and economic benefits of water, electricity, transport and other services (Marvin & Graham, 2001). Management and provision of infrastructure services grew on a predict-and-provide basis, with demographers and economists forecasting demand for services and engineers and utility managers expanding systems accordingly.

The expansion of infrastructure to meet ever increasing demand has had considerable impacts on the environment. Conventional models of infrastructure effectively assume that natural resources will be available to meet growing demand and that the environment is capable of absorbing waste and pollution. The continued expansion of infrastructure to abstract water, fossil fuels and other resources from the environment, and the impacts of pollution from burning fuels and disposing of wastewater and municipal waste, have had considerable impacts on local and global environments.

In the last two decades of the twentieth century ownership, management and financing infrastructure changed significantly. During the 1980s and 1990s privatisation of infrastructure was seen as the means to increase efficiency of operation and provide access to private capital to upgrade and expand networks. This also reflected wider political changes, which emphasised the role of markets and the individual preferences in development, rather than the role of the state and universal provision (Marvin & Graham, 2001; Swyngedouw, 1999). For individuals, access to infrastructure services, such as transport or communication, and to some extent water and energy, became more dependent on ability to pay. Thus changes in infrastructure systems reflect wider changes in society and politics at the end of the twentieth century.

The last decades of the twentieth century also revealed environmental limits to the continued expansion of infrastructure systems. Volatility in energy prices and their impacts on transport systems reflected constraints on fossil fuel supplies. Growing population and changing rainfall patterns contributed to water scarcity in cities including Sydney, Las Vegas, London and Athens (Kaika, 2006; Sofoulis, 2013). Climate change targets for reducing carbon emissions also provide constraints on continued expansion of infrastructure. These trends have contributed to increased focus on demand management as an alternative to continued expansion of infrastructure. Demand management programmes aim to reduce per capita consumption of resources by reducing distribution losses, improving efficiency of appliances and changing user behaviour, to enable existing systems to meet the needs of a growing population without increasing resource use, and to continue to meet the needs of current populations under conditions of resource scarcity (Butler & Memon, 2006).

Provision of infrastructure services, including energy, water and communications, has led to dramatic, unanticipated transformation of everyday life and social norms. Elizabeth Shove has shown the interaction between systems, technologies and social norms, in her 'co-evolutionary triangle' used to explain the 'racheting-up' of consumption of energy and water resources in homes (Shove, 2004). While provision of clean water and electricity provide unquestionable benefits to public health, these infrastructures have enabled the development of new domestic technologies, such as washing machines, that in turn contributed to changing social norms, such as wearing freshly washed clothes every day, or wearing a fresh change of clothes for different activities within the same day. Water and energy infrastructure were not built with constantly changing, clean clothes in mind, but social expectations have shifted as laundry has become more convenient.

Environmental and resource constraints have also prompted increased attention on decentralised technologies, in contrast to centralised infrastructure systems (van Vliet *et al.* 2005). Decentralised systems are often assumed by environmentalists to be more efficient than centralised systems by avoiding conveyance losses from large scale distribution networks. Local systems have also been promoted as being inherently more sustainable, encouraging people to live within their locally available, renewable resources. Such systems for self-sufficiency or local management of resources have been associated with the alternative technology movement, which promotes technologies that are able to be operated and maintained by local communities, with reduced requirement for centralised, expert led design and management (Schumacher, 1973).

21.3 SUSTAINABILITY, TECHNOLOGY AND WATER

A socio-technical perspective on infrastructure helps to identify the broader conditions and assumptions required in order for systems to exist and operate effectively. Modes of infrastructure provision reflect and stabilise assumptions about water, the environment, technology, society, governance and economics (van Vliet *et al.* 2005). Different technical options for water supply may require different economic and governance arrangements, and they might reflect different understandings about how people use water and relate to their local environment. Achieving sustainable water systems requires consideration of these wider socio-technical aspects of water supply and use. New technologies can be used in different ways to either reinforce unsustainable patterns of water supply and use, or to support the transition to more sustainable systems and lifestyles.

Sustainability assessment usually focuses on the impacts of developments or technologies on the environment, economy and society. Pressure-state-response indicators expand this perspective to address the wider systemic interactions between different elements. A socio-technical analysis of sustainability provides more contextual, cultural and political knowledge about proposed systems. It reveals the deeper assumptions underpinning the development and operation of the systems, including assumptions about society and behaviour, values, the environment and the nature of water itself. For example, beyond the quantitative environmental impacts, a water reuse system is fundamentally based on an assumption that water is a limited resource, while conventional dam construction assumes that water can be captured and stored to meet social demands. Similarly, a tap connected to a conventional pipe network embodies a message that water is limitless, as the water keeps flowing unless user turns off the tap, while a water butt for garden watering presents water as a limited resource dependent on rainfall. Alternative water systems may present short term or small scale improvements in environmental performance or water resource conservation. However, if they reinforce behaviours based on an understanding of water as limitless then these improvements may be ultimately undermined.

The sustainability of infrastructure systems is also dependent on appropriate governance and financing arrangements. Regulation and ownership for alternative water systems can reinforce the role of centralised utility providers, or allow for wider participation in the water sector by different actors, including building owners and water technology and service providers. Different scales of technology and a diversity of service providers is a clear challenge for regulation and governance. In some cases, this may lead to greater public participation and deliberation in decision-making, in line with sustainability principles, whilst in others models of expert-led decision making are enhanced. The provision of water and sanitation services has shifted as political ideologies have changed. The introduction of alternative water systems provides opportunities for reform of infrastructure governance, but it may also re-enforce wider trends towards market and individualistic governance.

These themes are explored in the following sections, which analyse the socio-technology of conventional water systems, potable reuse, district scale reuse and rainwater harvesting. Each system is analysed in terms of its assumptions and requirements regarding water, the environment, technology, society, governance and economics. Comparison of different socio-technical arrangements for water highlights the challenges and opportunities for alternative water systems to contribute to sustainability.

21.4 CONVENTIONAL SUPPLY

Conventional water supply infrastructure is based on an assumption that supply will always be able to expand to meet demand. Thus, water is assumed to be a limitless resource. Consumers have developed uses for water accordingly, under the expectation that water will always flow from the tap.

Water supply and sanitation infrastructure developed largely to address significant public health risks (Halliday, 2001; Melosi, 2008). Clean water is produced and supplied under centralised management and dirty water is drained from homes as quickly as possible, before centralised treatment and discharge back to the environment. Water is thus either clean or dirty within conventional infrastructure, with no scope for water of multiple qualities for different uses. Centralised control of water infrastructure is essential to minimise risks to public health.

Conventional infrastructure provision assumes private control over water demand. Following the predict-and-provide model of provision, infrastructure managers traditionally anticipate demand but make no interventions in how people use water in the privacy of their own homes. In the exceptional circumstance of drought, water utilities may restrict outdoor water use, but indoor water use is usually assumed to be private and difficult to change (Allon & Sofoulis, 2006).

Provision of water infrastructure is capital intensive. Most water systems were initially constructed by the state, but in recent years the private sector has become involved in operating, maintaining and owning water infrastructure. Arrangements for funding water infrastructure and supply vary globally. Some jurisdictions fund water from centralised taxation revenues, but it is more common for revenue to be raised from users in the form or water rates or charges for water use. Water users are customers of water utilities, paying for the service of uninterrupted supply.

Governance arrangements for water infrastructure vary around the world. Large water utilities, whether privately or publically owned, are usually subject to regulation of water quality, abstraction from and discharge to the environment and the prices charged to customers. Regulation and governance of water utilities balances the needs for environmental and public health controls, with the economic impact on customers and investors. As the private sector has become more involved in provision of water infrastructure, the governance arrangements have become more complex, as water has moved from a public service to maintain good public health and economic development, to a source of profit for shareholders.

21.4.1 Case study: London, England

London and the Thames Valley are located in the water scarce region of South East England. Average rainfall is around 600 mm per year, with high population growth rates. Water provision in London has shifted between public and private ownership since the initial construction of the system in the nineteenth century. Before the 1890s water was supplied by private companies, between the 1890s and 1980s water was supplied by municipal authorities, and since 1987 water and sewerage services have been supplied by the privately owned Thames Water Utilities Limited. England and Wales are unique in the world in having a fully privatised water sector, regulated by three key regulators dealing with economics, environment and drinking water quality.

A number of important rivers and streams in the region are over-abstracted with the environmental regulator aiming to address over-licensing in vulnerable catchments. Addressing future water security for a growing population is a key concern for Thames Water and others. Key options include constructing a new reservoir to maintain environmental flows in the Thames during dry periods, expanding desalination capacity and potable reuse. The UK government has also set a target of reducing per capita water consumption from an average of 150 litres per person per day to 120 litres by 2030, which is reflected in a target for water companies to reduce daily customer demand by 1 litre each year. Most customers currently pay for water through a flat rate based on property values, with a programme underway to install water meters for all customers in coming decades.

21.5 POTABLE REUSE

Potable reuse involves treating wastewater to a very high standard, usually using membrane filtration and reverse osmosis, and returning it to the drinking water system rather than discharging to the environment. The treated water can be directly re-introduced to the drinking water system at the water treatment works, or indirectly introduced by aquifer recharge, discharge into raw water reservoirs or into rivers immediately upstream from abstraction points.

Technically, potable reuse involves a relatively minor adaptation of water supply infrastructure. Treated wastewater becomes another resource for conventional supply systems, with no changes required to water treatment and distribution systems or to how consumers use water. Potable water reuse maintains the water utility as a centralised owner and operator of the system, subject to the same water quality, economic and environmental regulations. Membrane technologies require much higher energy consumption to produce the raw water than abstraction of conventional water resources from the environment (Cooley & Wilkinson, 2012). Potable water reuse maintains centralised control of water quality. It presents the technical possibility for endless supply of water, as infinitely reusable, although this may be limited in practice to manage risks of recirculating micro-contaminants.

Socially, potable reuse has proved to be highly contentious (Hartley, 2006). Whilst potable reuse appears to present minimal changes to the overall structure of water supply networks, public acceptability of potable reuse has been a significant hurdle to implementation (Dolnicar & Schäfer, 2009). Public concerns with potable reuse include emotional ‘yuck factor’ responses, concerns about health risks associated with recirculating micro-contaminants, wider concerns about unknown risks associated with new technologies and the high energy consumption of water treatment (Dolnicar *et al.* 2011). Public backlash against potable reuse has been responsible for the failure of proposed systems, such as in Toowoomba,

Australia and has delayed implementation in other cases, such as in San Diego, USA (Hurlimann & Dolnicar, 2010).

Public controversy about potable reuse highlights fundamental changes in the role of water utilities in society, as well as the relationship between consumers, infrastructure and water (Bell & Aitken, 2008; Colebatch, 2006). Whilst potable reuse represents minimal technical and institutional change to conventional water infrastructure, the impact of public opposition on proposed schemes and the strength of controversy shows that under conditions of water scarcity the public are no longer willing to accept expert decisions about water supply.

Potable reuse cannot succeed as a technical proposition, without taking account of social factors (Chilvers *et al.* 2011). This requires engineers and water managers to consider social factors in the design of systems and in decision making about water resource options. Deliberative decision-making processes have been proposed as a means of achieving a higher quality of decision about potable reuse and other water management options. This moves beyond public relations or education campaigns that aim to convince the public of the benefits and safety of potable reuse, to stronger engagement and involvement of the public in decision making. Involving the public at early stages of proposals and designs for potable reuse may lead to higher acceptability, but more importantly can help water utilities and regulators identify at an early stage if potable reuse is not a viable option for water supply (Bell & Aitken, 2008; Russell & Lux, 2009).

21.5.1 Case study: South-East Queensland, Australia

A prolonged drought in the 2000s and continued population growth in the South East of Queensland resulted in reduced water storage in dams and the need to evaluate options for alternative water supplies. The Western Corridor Recycled Water Project (WCRWP) was implemented between 2007 and 2009 to provide reclaimed water to power stations and other industrial users and to allow for potable reuse during drought conditions. The WCRWP is wholly owned and operated by the government of Queensland, through independent entities. The project was funded by the Australian Federal Government through the National Water Commission. Funding and ownership of the project reflect conventional public interest and benefit from provision of water infrastructure.

The role of potable reuse in this region has been highly controversial. In 2007 an indirect potable reuse scheme proposed for the town of Toowoomba was rejected in a referendum of residents, after a highly adversarial campaign. Under worsening drought conditions the Premier of the State of Queensland announced that future potable reuse, through the WCRWP, would go ahead without further referenda, as a result of the seriousness of water shortages. A change in leadership of the government and a break in the drought led to a further change in direction, to the current arrangement that recycled water is used for industrial uses, except under conditions of extreme water shortage.

The changing role of recycled water in South-East Queensland and the changes in government decisions highlight the challenges that this new source of water poses for conventional institutional arrangements for delivering water infrastructure. The complex issues associated with the technical and social elements of this source of water are not amenable to conventional expert-led decision making that until recently has been largely free from public scrutiny or controversy. Efforts to involve the public through referenda failed to deliver robust decisions about water recycling, leading to changing positions for government decision making in order to achieve a socially acceptable outcome for water reuse.

21.6 DISTRICT NON-POTABLE WATER REUSE

Non-potable reuse at a district scale involves distributing treated municipal wastewater for landscape irrigation, toilet flushing and other non-potable uses. Early implementation of district scale reuse involved irrigation of sports fields and parks with primary or secondary treated effluent. More recently developments have involved dual reticulation of housing developments and public buildings to supply water treated to a high quality using membrane bioreactors or other advanced technologies. In such cases, non-potable water supply becomes a new infrastructure service, delivered through its own network, with separate systems for treatment and management.

As a new infrastructure service, non-potable water supply largely conforms to the conventional institutional arrangement for water supply. In most cases to date, the supplier of recycled water has been the incumbent water utility. However, in some jurisdictions it may be possible for new suppliers to enter the market providing non-potable water in competition with potable supply. Non-potable water is usually supplied to customers at a lower price than potable water, however this does not yet reflect the relative costs of supply, requiring economic subsidy.

The two sets of pipes for potable and non-potable water signify the multiple qualities of water and its scarcity in the environment. However, potable backup for non-potable supply can undermine recycling efforts and continue to support an understanding of water supply as limitless. Control of risk in non-potable reuse schemes shifts beyond the centralised authority, as customers, plumbers and others must take account of the different supply systems and manage risks of cross-connection of potable and non-potable systems, or misuse of non-potable water.

The energy balance of non-potable reuse schemes is comparable to conventional systems. Treatment of municipal wastewater and pumping through the local distribution network is usually comparable to conventional treatment and pumping for drinking and wastewater. The energy requirement for treatment is significant compared to conventional water treatment, but can be comparable with the combined energy required for both water and wastewater treatment, which are displaced by reuse (Hills & James, 2014).

Non-potable reuse has been shown to be more acceptable to the public than potable reuse. Use of water for landscape irrigation, fire suppression, agricultural irrigation and toilet flushing are more acceptable than for cleaning, bathing and drinking. Non-potable systems at the district scale are more likely to be publically acceptable than potable reuse. Thus entirely new infrastructure systems may be needed to enable water recycling within the current social arrangement, rather than simply incorporating recycled water into the existing potable supply.

21.6.1 Case study: Old Ford water recycling plant, London

The Old Ford water recycling plant was built to supply non-potable water to the Queen Elizabeth II Olympic Park for the 2012 Olympic and Paralympic Games, as well as for the legacy period during which the site is to be redeveloped for housing, community and sports facilities. This case study is explained in further detail in Chapter 15 in this volume (Hills & James, 2014). The plant abstracts water from a main sewer running close to the site and treats it using a membrane bioreactor to non-potable standards. The water is also chemically dosed to remove phosphorous, filtered through activated carbon to remove colour and disinfected using sodium hypochlorite before distribution to the site. Reclaimed water is used for landscape irrigation and toilet flushing in a number of venues on the park, with the intention of expanding use to additional venues and new developments.

The plant is owned and operated by Thames Water, the privately owned water utility supplying water and sewerage services to London and surrounding regions. The plant is also financed by Thames Water, whose investment plans and customer charges are regulated by the Office for Water (Ofwat, The Water Services Regulation Authority). The recycled water is charged at a lower cost than potable water, but the operating costs of the plant are higher than conventional water and wastewater treatment and distribution. The UK does not have regulations for non-potable water quality and the plant is designed and managed according to the US EPA standard for use of reclaimed water for landscape irrigation. The system is backed up by the potable mains, so that supply of water through the non-potable water network is not disrupted when demand is high or the plant is out of operation. Outside the Olympic period, demand has been driven largely by requirements for landscape irrigation, with very low demand during winter months. The overall energy intensity for treatment of the non-potable water is comparable to the combined energy intensity of potable and wastewater treatment through the conventional infrastructure system. Water is supplied mostly to public buildings and used for public landscaping, with one housing development currently supplied, and future housing developments being targeted for supply. The water is treated to a high standard, including disinfection, to reduce health risks from cross connection to the potable supply on customers' premises.

21.7 RAINWATER HARVESTING

Rainwater harvesting is a decentralised form of non-potable water supply, ranging from simple water butts for garden watering to building scale systems with dedicated pipe networks and automated control systems (Hassell & Thornton, 2014). As a non-potable source of water, rainwater harvesting is mostly used for landscape irrigation, toilet flushing and fire suppression. Harvested rainwater is relatively clean, allowing for the development and implementation of decentralised treatment systems to produce potable water in remote locations (Adler *et al.* 2014; Thayil-Blanchard & Mihelcic, 2014).

Rainwater harvesting is a significant departure from conventional water supply systems operated by water utilities. The systems are usually owned and operated by building owners, with suppliers providing management and maintenance support in some cases. Thus the ownership and operation of water supply systems are decentralised, as well as the technology and the water source.

Rainwater harvesting systems recognise that water resources are limited. They present the opportunity for users to maintain current patterns of water use by providing an alternative source, rather than directly driving changes in consumption behaviour. Where rainwater harvesting systems are backed up by potable supply, this can undermine water savings potential, particularly during dry weather. This can be addressed by applying restrictions during drought events on outdoor use from all water sources, including rainwater harvesting, in order to avoid individualist perceptions that rainwater harvesting allows users complete control over their water supply and use.

Regulation of rainwater harvesting challenges conventional institutional arrangements for water supply. Standards for water quality and technology have developed to manage public health risks, which have contributed to increasing complexity of technology and increasing energy consumption. Requirements for pumping for supply and recirculation of water through distribution systems contribute to high energy demands. In the UK, some types of rainwater harvesting system have been shown to be more energy intensive than mains supply, due to the relative efficiencies of pumping. Rainwater harvesting has been driven by policy interventions in several jurisdictions, as described by Ward *et al.* (2014).

21.7.1 Case study: Pimpama Coomera, Australia

The Pimpama Coomera development in Australia incorporates rainwater harvesting as well as dual reticulated district scale reuse. Rainwater is harvested from individual houses into tanks owned and managed by home owners. Rainwater tanks are above ground and external to the houses, and are built according to two mandatory minimum sizes (5 m^3 for detached homes, 3 m^3 for semi-detached homes and townhouses). Pumping requirements are minimised by above ground storage and due to most houses being single storey bungalows.

Rainwater is used for outdoor irrigation and for cold water supply to washing machines. The rainwater system is backed up by potable supply and is subject to the water restrictions for outdoor use during drought events. Rainwater harvesting is promoted as part of the stormwater management for the development, which also includes swales and other elements of water sensitive urban design. Rainwater harvesting at Pimpama Coomera is also integrated with non-potable supply and stormwater management and is compulsory for all homes, with standard requirements for tank size. The overall strategy for water management is delivered by the municipal water utility, Gold Coast Water and the individual water user has minimal involvement in technology choice, management or other decisions. Water supply is maintained, but with restrictions during drought. Individual consumers who use more than the rainwater supply during normal (non-drought) periods are not restricted, due to the potable backup. Rainwater harvesting is effectively a buffer for the non-potable and potable supply networks, providing an additional source of water with relatively low energy requirements, but with limited impact on user experience or behaviour.

21.8 DISCUSSION

A socio-technical analysis of alternative water systems enables comparison with existing infrastructure to assess the extent to which they reinforce or challenge conventional arrangements for relationships between technology, society and water. A framework for sustainability analysis is presented in Table 21.1: categories in the column on the left and criteria in the columns to the right. Whilst alternative systems present opportunities for improving the sustainability of urban water infrastructure, this is not inevitable. Indicator-based comparison provides useful data on environmental impacts and economic costs and benefits, whilst more qualitative socio-technical analysis, such as that presented here, reveals underlying assumptions and values that are embodied in different infrastructure arrangements.

Conventional water supply systems are adapting to resource constraints and population growth, and ownership and regulation arrangements vary around the world. Despite demand management efforts, the essential assumption of water as an endless resource to be provided by expert-led decision making is maintained in most efforts to adapt to changing social and environmental conditions. Potable reuse of water is effectively a supply side solution for conventional water infrastructure. However, controversy surrounding public acceptability demonstrates that governance arrangements for conventional supplies must adapt to changing public expectations and concerns about risks associated with new technologies and contaminants. Expert-led decision making has moved tentatively towards more democratic forms of decision making about infrastructure, but the structures and governance arrangements are still being confirmed and in most jurisdictions remain to be stabilised.

Table 21.1 Comparison of conventional and alternative water infrastructure systems.

	Conventional supply	Potable reuse	District non-potable reuse	Rainwater harvesting
Water	Limitless Pure or contaminated Dangerous	Endlessly recyclable Industrial product Water cycle can be short circuited	Scarce	Scarce
Environment	Resource Sink for waste Must be controlled	For human use Energy is limitless Outside city	Humans part of water cycles	Humans part of water cycles
Technology	Universal standards Big systems Requires expert knowledge	Complex Engineer control Centralised	Simple to intermediate complexity Lay to engineer controlled Simple to complex Decentralised and intermediate scale	Simple Lay expertise Designer and community led Decentralised and intermediate scale
Society	Consumption is private Infrastructure serves society	Consumption is private Accepts technology	Consumption is private	Consumption is private
Governance	Expert led decision making Large utilities (public and private) Independent regulation	Regulated utility Municipal to national governance Technocratic, with some recognition of public concerns	Expert led decisions municipal governance Regulated utility	Minimal regulation or governance beyond installation Municipal scale
Economics	Capital intensive	Centralised Public and private Capital intensive	Capital intensive, ownership dominated by utilities, with potential for diverse ownership and management	Decentralised Minimal capital requirement Initial investment by household, municipality or other funded programme

District scale non-potable reuse and rainwater harvesting shift conventional assumptions about water to recognise multiple qualities for multiple uses. However, to date most significant cases of non-potable reuse at district scale have been owned and operated by conventional water utilities. Governance arrangements for non-potable reuse are still being formulated to allow the entrance of a wider range of providers for non-potable water. Although it is recognised that water for non-potable use can be of a lower quality than potable water, standards have not yet been confirmed and countries such as the UK have relied on the US EPA standards for landscape irrigation. Managing risks of cross connection mean that non-potable reuse water is treated to a much higher standard than required for its intended end uses, increasing energy and chemical requirements and undermining its sustainability potential. The embodied energy and resources in the distribution network must also be accounted for in assessing the overall sustainability of reuse compared to other water resource options.

Rainwater harvesting most clearly shifts responsibility for water provision to householders and building owners, as owner and operators of non-potable water supply systems. Whilst this provides an additional distributed source of water, where rainwater tanks are backed up by piped supply, they maintain and even amplify the expectation that water is a constant resource. Regulation of rainwater systems also presents challenges to public health and local government authorities, with the need to balance health risks with technical complexity.

The case studies presented in this chapter demonstrate the extent to which alternative systems re-enforce conventional arrangements for water infrastructure and the degree to which they limit their potential contribution to more sustainable water systems (Table 21.2). Widening participation in decision making in relation to the provision of services and infrastructure has the potential to improve the overall sustainability of alternative systems, but requires more complex and adaptable governance arrangements and risk management. Whilst integration of urban water systems is desirable, backup of non-potable water systems with potable water systems undermines their potential to transform social norms of water use. Reuse and rainwater harvesting systems can reinforce the idea of water as a limitless resource to be delivered by technical systems and managed by technical experts. Encouraging a shift in behaviour to live within local water resources and environmental conditions may require rethinking the integration of potable and non-potable systems.

21.9 CONCLUSION

Alternative water systems present fundamental challenges to conventional modes of infrastructure provision. Successful implementation of alternative water systems requires development of new economic, social and governance arrangements, as well as the design and commercialisation of technologies. In order for these systems to be sustainable, it is important to consider their environmental and social

Table 21.2 Case studies of conventional and alternative supply.

	London, UK	**Western Corridor, Australia**	**Old Ford, UK**	**Pimpama Coomera, Australia**
Water	Resource for human benefit produced and delivered by large infrastructure system Water is either pure of dirty	Resource for human benefit produced and delivered by large infrastructure system Water is scarce but recyclable	Resource for human benefit produced and delivered by large infrastructure system Water is scarce Water quality is fit-for-purpose	Resource for human benefit. Fresh water is scarce. Local supplies supplement centralised provision. Water quality is fit-for-purpose
Environment	Constraint on options for future development	Resource for human benefit. Environment is unpredictable, requiring infrastructure systems to buffer extreme events	Constraint on development, providing limits to energy and water	Resource for human use. Multiple sources of water provide resilience against extreme events
Technology	Complex, centralised control of technology, requiring high levels of expertise	Complex, centralised control of technology, requiring high levels of expertise	Complex, centralised control of technology, requiring high levels of expertise	Domestic scale systems backed up by centrally controlled supply. Higher risk water recycling managed by experts

(*Continued*)

Table 21.2 Case studies of conventional and alternative supply (*Continued*).

	London, UK	**Western Corridor, Australia**	**Old Ford, UK**	**Pimpama Coomera, Australia**
Society	Limited capacity to chance social expectations and demand for water	Public acceptability of technology is a risk to development Technology can be reconfigured to address public concerns, within environmental constraints	Limited capacity to chance social expectations and demand for water Environmental credential of Olympic Games were important to local and international community	Shared responsibility for ownership and management of water supply technology Higher risk managed by expert authorities. Social demand for alternative water sources
Governance	Private water company regulated for price, environment and quality. Limited public engagement.	Public ownership of independent company	Private water company regulated for price, environment and quality. Limited public engagement.	Planning and urban development processes used to drive uptake of alternative water systems
Economics	Capital intensive, monopoly provision, price and investment plans subject to regulation	Water infrastructure is of national importance. Government investment and ownership	Price of water subsidised Demonstration site not economically feasible under normal conditions	Individual ownership of alternative water systems

implications, as well as conventional concerns about supply and demand and costs and benefits of investment. A socio-technical approach to water infrastructure and alternative technologies helps to highlight the potential for different supply options to contribute to sustainability, or to re-enforce unsustainable relationships between people, technology and the environment.

Infrastructure systems stabilise relationships between people, technology, institutions and the environment. Alternative water systems such as rainwater harvesting and non-potable reuse introduce new technical elements into urban water systems, which renegotiate these relationships. The extent to which these system are incorporated into existing institutional arrangements will influence their sustainability in the long term. Alternative systems backed up by a mains potable supply maintain the user expectation of unlimited continuous supply of water, independent of weather and hydrological conditions, which is ultimately unsustainable.

Water reuse and rainwater harvesting systems have the potential to contribute to a fundamental restructuring of the relationships between people and water, to support the transition to sustainability. However, these technologies may not be inherently sustainable. The technical configuration of alternative water systems and their integration with existing systems is central in determining whether they contribute to a transformation of urban water systems or further stabilise conventional infrastructure systems and social norms.

REFERENCES

Adler I., Campos L. and Bell S. (2014). Community participation in decentralised rainwater systems: a Mexican case study. In: Alternative Water Supply Systems, F. A. Memon and S. Ward (eds), IWA Publishing, London, Chapter 6, pp. 117–130.

Allon F. and Sofoulis Z. (2006). Everyday water: cultures in transition. *Australian Geographer*, **37**, 45–55.

Bell S. and Aitken V. (2008). The socio-technology of indirect potable water reuse. *Water Science & Technology: Water Supply*, **8**, 441.

Ben-Joseph E. (2005). The Code of the City. MIT Press, Cambridge.

Butler D. and Memon F. A. (2006). Water Demand Management. IWA Publishing, London.

Chilvers A., Bell S. and Hillier J. (2011). The socio-technology of engineering sustainability. *Proceedings of the Institution of Civil Engineers – Engineering Sustainability*, **164**, 177–184.

Colebatch H. K. (2006). Governing the use of water: the institutional context. *Desalination*, **187**, 17–27.

Cooley H. and Wilkinson R. (2012). Implications of Future Water Supply Sources for Energy Demands. Water Reuse Association, Alexandria.

Dolnicar S., Hurlimann A. and Grün B. (2011). What affects public acceptance of recycled and desalinated water? *Water Research*, **45**, 933–943.

Dolnicar S. and Schäfer A. (2009). Desalinated versus recycled water: Public perceptions and profiles of the accepters. *Journal of Environmental Management*, **90**, 888–900.

Halliday S (2001). The Great Stink of London. The History Press, Abingdon.

Hartley T. (2006). Public perception and participation in water reuse. *Desalination*, **187**, 115–126.

Hassell C. and Thornton J. (2014). Rainwater harvesting for toilet flushing in UK schools; opportunities for combining with water efficiency education. In: Alternative Water Supply Systems, F. A. Memon and S. Ward (eds), IWA Publishing, London, Chapter 5, pp. 85–115.

Hills S. and James C. (2014). The queen elizabeth olympic park water recycling system, London. In: Alternative Water Supply Systems, F. A. Memon and S. Ward (eds), IWA Publishing, London, Chapter 15.

Hughes T. (1985). Networks of Power. Johns Hopkins University Press, Baltimore.

Hughes T. (1989). The evolution of large technical systems. In: The Social Construction of Technological Systems, W. Bijker, T. Hughes and T. Pinch (eds), MIT Press, London, pp. 45–76.

Hurlimann A. and Dolnicar S. (2010). When public opposition defeats alternative water projects – The case of Toowoomba Australia. *Water Research*, **44**, 287–297.

Kaika M. (2006). The political ecology of water scarcity. In: In the Nature of Cities, N. Heynan, M. Kaika and E. Swyndgedow (eds), Taylor & Francis, London, pp. 157–172.

Marvin S. and Graham S. (2001). Splintering Urbanism. Routledge, London.

Melosi M. (2008). The Sanitary City. University of Pittsburgh Press, Pittsburgh.

Russell S. and Lux C. (2009). Getting over yuck: moving from psychological to cultural and sociotechnical analyses of responses to water recycling. *Water Policy*, **11**, 21.

Schumacher E. F. (1973). Small is Beautiful. ABACUS, London.

Shove E. (2004). Comfort, Cleanliness and Convenience. Berg Publishers, Oxford.

Sofoulis Z. (2013). Water systems adaptation: an australian cultural researcher's perspective. *Water Resources Management*, **27**, 949–951.

Swyngedouw E. (1999). Modernity and hybridity: nature, regeneracionismo, and the production of the Spanish waterscape, 1890–1930. *Annals of the Association of American Geographers*, **89**, 443–465.

Thayil-Blanchard J. and Mihelcic J. (2014). Assessing domestic rainwater harvesting storage cost and geographic availability in Uganda's Rakai district. In: Alternative Water Supply Systems, F. A. Memon and S. Ward (eds), IWA Publishing, London, Chapter 7, pp. 131–152.

van Vliet B., Chappells H. and Shove E. (2005). Infrastructures of Consumption. Earthscan, London.

Ward S., Dornelles F., Giacomoni M. and Memon F. A. (2014). Incentivising and charging for rainwater harvesting – three international perspectives. In: Alternative Water Supply Systems, F. A. Memon and S. Ward (eds), IWA Publishing, London, Chapter 8, pp. 153–168.

Index

IWA Publishing's authorised EU representative for General Product Safety Regulations is Diane D'Arras, 15 rue Duret, 75116 Paris, France, e-mail: safety@iwap.co.uk.

Printed and bound by CPI Group (UK) Ltd, Croydon, CR0 4YY
16/06/2026
02140363-0004